Differential Geometry and Topology

With a View to Dynamical Systems

Studies in Advanced Mathematics

Titles Included in the Series

John P. D'Angelo, Several Complex Variables and the Geometry of Real Hypersurfaces

Steven R. Bell, The Cauchy Transform, Potential Theory, and Conformal Mapping

John J. Benedetto, Harmonic Analysis and Applications

John J. Benedetto and Michael W. Frazier, Wavelets: Mathematics and Applications

Albert Boggess, CR Manifolds and the Tangential Cauchy–Riemann Complex

Keith Burns and Marian Gidea, Differential Geometry and Topology: With a View to Dynamical Systems

Goong Chen and Jianxin Zhou, Vibration and Damping in Distributed Systems
    Vol. 1: Analysis, Estimation, Attenuation, and Design
    Vol. 2: WKB and Wave Methods, Visualization, and Experimentation

Carl C. Cowen and Barbara D. MacCluer, Composition Operators on Spaces of Analytic Functions

Jewgeni H. Dshalalow, Real Analysis: An Introduction to the Theory of Real Functions and Integration

Dean G. Duffy, Advanced Engineering Mathematics with MATLAB®, 2nd Edition

Dean G. Duffy, Green's Functions with Applications

Lawrence C. Evans and Ronald F. Gariepy, Measure Theory and Fine Properties of Functions

Gerald B. Folland, A Course in Abstract Harmonic Analysis

José García-Cuerva, Eugenio Hernández, Fernando Soria, and José-Luis Torrea,
    Fourier Analysis and Partial Differential Equations

Peter B. Gilkey, Invariance Theory, the Heat Equation, and the Atiyah-Singer Index Theorem, 2nd Edition

Peter B. Gilkey, John V. Leahy, and Jeonghueong Park, Spectral Geometry, Riemannian Submersions, and the Gromov-Lawson Conjecture

Alfred Gray, Modern Differential Geometry of Curves and Surfaces with Mathematica, 2nd Edition

Eugenio Hernández and Guido Weiss, A First Course on Wavelets

Kenneth B. Howell, Principles of Fourier Analysis

Steven G. Krantz, The Elements of Advanced Mathematics, Second Edition

Steven G. Krantz, Partial Differential Equations and Complex Analysis

Steven G. Krantz, Real Analysis and Foundations, Second Edition

Kenneth L. Kuttler, Modern Analysis

Michael Pedersen, Functional Analysis in Applied Mathematics and Engineering

Clark Robinson, Dynamical Systems: Stability, Symbolic Dynamics, and Chaos, 2nd Edition

John Ryan, Clifford Algebras in Analysis and Related Topics

John Scherk, Algebra: A Computational Introduction

Pavel Šolín, Karel Segeth, and Ivo Doležel, High-Order Finite Element Method

André Unterberger and Harald Upmeier, Pseudodifferential Analysis on Symmetric Cones

James S. Walker, Fast Fourier Transforms, 2nd Edition

James S. Walker, A Primer on Wavelets and Their Scientific Applications

Gilbert G. Walter and Xiaoping Shen, Wavelets and Other Orthogonal Systems, Second Edition

Nik Weaver, Mathematical Quantization

Kehe Zhu, An Introduction to Operator Algebras

Differential Geometry and Topology
With a View to Dynamical Systems

Keith Burns
Northwestern University
Evanston, Illinois, USA

Marian Gidea
Northeastern Illinois Univeristy,
Chicago, USA

Chapman & Hall/CRC
Taylor & Francis Group
Boca Raton   London   New York   Singapore

Published in 2005 by
Chapman & Hall/CRC
Taylor & Francis Group
6000 Broken Sound Parkway NW, Suite 300
Boca Raton, FL 33487-2742

© 2005 by Taylor & Francis Group, LLC
Chapman & Hall/CRC is an imprint of Taylor & Francis Group

No claim to original U.S. Government works
Printed in the United States of America on acid-free paper
10 9 8 7 6 5 4 3 2 1

International Standard Book Number-10: 1-58488-253-0 (Hardcover)
International Standard Book Number-13: 978-1-58488-253-4 (Hardcover)

This book contains information obtained from authentic and highly regarded sources. Reprinted material is quoted with permission, and sources are indicated. A wide variety of references are listed. Reasonable efforts have been made to publish reliable data and information, but the author and the publisher cannot assume responsibility for the validity of all materials or for the consequences of their use.

No part of this book may be reprinted, reproduced, transmitted, or utilized in any form by any electronic, mechanical, or other means, now known or hereafter invented, including photocopying, microfilming, and recording, or in any information storage or retrieval system, without written permission from the publishers.

For permission to photocopy or use material electronically from this work, please access www.copyright.com (http://www.copyright.com/) or contact the Copyright Clearance Center, Inc. (CCC) 222 Rosewood Drive, Danvers, MA 01923, 978-750-8400. CCC is a not-for-profit organization that provides licenses and registration for a variety of users. For organizations that have been granted a photocopy license by the CCC, a separate system of payment has been arranged.

Trademark Notice: Product or corporate names may be trademarks or registered trademarks, and are used only for identification and explanation without intent to infringe.

Library of Congress Cataloging-in-Publication Data

Catalog record is available from the Library of Congress

Taylor & Francis Group
is the Academic Division of T&F Informa plc.

Visit the Taylor & Francis Web site at
http://www.taylorandfrancis.com

and the CRC Press Web site at
http://www.crcpress.com

To Peter, Sonya and Imke – K.B.

To Claudia – M.G.

Preface

This book grew out of notes from a differential geometry course taught by the second author at Northwestern University. It aims to provide an introduction, at the level of a beginning graduate student, to differential topology and Riemannian geometry. The theory of differentiable dynamics has close relations to these subjects. We introduce basic concepts from dynamical systems and try to emphasize interactions of dynamics, geometry and topology.

We have attempted to introduce important concepts by intuitive discussions or suggestive examples and to follow them by significant applications, especially those related to dynamics. Where this is beyond the scope of the book, we have tried to provide references to the literature.

We have not attempted to give a comprehensive introduction to dynamical systems as this would have required a much longer book. The reader who wishes to learn more about dynamical systems should turn to one of the textbooks in that area. Three excellent recent books, with different emphases, are the texts by Brin and Stuck (2002), by Katok and Hasselblatt (1995), and by Robinson (1998).

The illustrations in this book were produced with Adobe Illustrator, DPGraph, Dynamics Solver, Maple, and Sierpinski Curve Generator. We thank Victor Donnay, Josep Masdemont, and John M. Sullivan for permission to reproduce some of the illustrations.

Contents

1 Manifolds — **1**
  1.1 Introduction 1
  1.2 Review of topological concepts 4
  1.3 Smooth manifolds 9
  1.4 Smooth maps 16
  1.5 Tangent vectors and the tangent bundle 19
  1.6 Tangent vectors as derivations 27
  1.7 The derivative of a smooth map 30
  1.8 Orientation 33
  1.9 Immersions, embeddings and submersions 36
  1.10 Regular and critical points and values 45
  1.11 Manifolds with boundary 48
  1.12 Sard's theorem 53
  1.13 Transversality 59
  1.14 Stability 62
  1.15 Exercises 66

2 Vector Fields and Dynamical Systems — **71**
  2.1 Introduction 71
  2.2 Vector fields 74
  2.3 Smooth dynamical systems 80
  2.4 Lie derivative, Lie bracket 86
  2.5 Discrete dynamical systems 94
  2.6 Hyperbolic fixed points and periodic orbits 97
  2.7 Exercises 106

3 Riemannian Metrics — 109
- 3.1 Introduction 109
- 3.2 Riemannian metrics 112
- 3.3 Standard geometries on surfaces 121
- 3.4 Exercises 125

4 Riemannian Connections and Geodesics — 127
- 4.1 Introduction 127
- 4.2 Affine connections 131
- 4.3 Riemannian connections 136
- 4.4 Geodesics 142
- 4.5 The exponential map 149
- 4.6 Minimizing properties of geodesics 155
- 4.7 The Riemannian distance 162
- 4.8 Exercises 167

5 Curvature — 171
- 5.1 Introduction 171
- 5.2 The curvature tensor 176
- 5.3 The second fundamental form 184
- 5.4 Sectional and Ricci curvatures 195
- 5.5 Jacobi fields 201
- 5.6 Manifolds of constant curvature 208
- 5.7 Conjugate points 210
- 5.8 Horizontal and vertical sub-bundles 213
- 5.9 The geodesic flow 217
- 5.10 Exercises 222

6 Tensors and Differential Forms — 225
- 6.1 Introduction 225
- 6.2 Vector bundles 227
- 6.3 The tubular neighborhood theorem 231
- 6.4 Tensor bundles 233
- 6.5 Differential forms 238
- 6.6 Integration of differential forms 247
- 6.7 Stokes' theorem 251
- 6.8 De Rham cohomology 257
- 6.9 Singular homology 263
- 6.10 The de Rham theorem 271
- 6.11 Exercises 276

7 Fixed Points and Intersection Numbers — 279
- 7.1 Introduction 279
- 7.2 The Brouwer degree 282
- 7.3 The oriented intersection number 291
- 7.4 The fixed point index 293
- 7.5 The Lefschetz number 303
- 7.6 The Euler characteristic 306
- 7.7 The Gauss-Bonnet theorem 313
- 7.8 Exercises 324

8 Morse Theory — 327
- 8.1 Introduction 327
- 8.2 Nondegenerate critical points 329
- 8.3 The gradient flow 337
- 8.4 The topology of level sets 340
- 8.5 Manifolds represented as CW complexes 348
- 8.6 Morse inequalities 351
- 8.7 Exercises 356

9 Hyperbolic Systems — 357
- 9.1 Introduction 357
- 9.2 Hyperbolic sets 359
- 9.3 Hyperbolicity criteria 368
- 9.4 Geodesic flows 373
- 9.5 Exercises 376

References — 379

Index — 385

Chapter 1

Manifolds

1.1 Introduction

A manifold is usually described by a collection of 'patches' sewed together in some 'smooth' way. Each patch is represented by some parametric equation, and the smoothness of the sewing means that there are no cusps, corners or self-crossings.

As an example, we consider a hyperboloid of one sheet $x^2 + y^2 - z^2 = 1$ (see Figure 1.1.1 (a)). The hyperboloid is a surface of revolution, obtained by rotating the hyperbola $x^2 - z^2 = 1$, lying in the (x, z)-plane, about the z-axis. The hyperbola can be parametrized by $t \to (\cosh t, 0, \sinh t)$, so the hyperboloid of revolution is given by the differentiable parametrization

$$\phi(t, \theta) = (\cosh t \cos \theta, \cosh t \sin \theta, \sinh t), \quad -\infty < t < \infty, \; -\infty < \theta < \infty.$$

We would like to have each point (x, y, z) of the hyperboloid uniquely determined by its coordinates (t, θ) and, conversely, each pair of coordinates (t, θ) uniquely assigned to a point. This does not work for the above parametrization, since the points of the hyperbola $x^2 - z^2 = 1$, $y = 0$, correspond to all (t, θ) with θ an integer multiple of 2π. We can get parametrizations that are one-to-one by restricting the mapping ϕ to certain open subsets of $\mathbb{R}^2$:

$$\phi_1(t, \theta) = (\cosh t \cos \theta, \cosh t \sin \theta, \sinh t), \quad -\infty < t < \infty, \; 0 < \theta < 3\pi/2,$$

$$\phi_2(t, \theta) = (\cosh t \cos \theta, \cosh t \sin \theta, \sinh t), \quad -\infty < t < \infty, \; \pi < \theta < 5\pi/2.$$

Note that the image of each ϕ_i is the intersection of the hyperboloid with some open set in $\mathbb{R}^3$. In cylindrical coordinates (r, θ, z) on $\mathbb{R}^3$, the

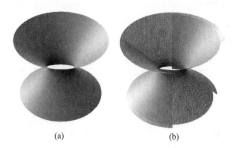

(a) (b)

FIGURE 1.1.1
Hyperboloid of one sheet.

image of ϕ_1 represents the portion of the hyperboloid inside the open region $0 < \theta < 3\pi/2$, and the image of ϕ_2 represents the portion of the hyperboloid inside the open region $\pi < \theta < 5\pi/2$.

Since the mappings ϕ_1 and ϕ_2 are differentiable, the images of ϕ_1 and ϕ_2 are smooth patches of surface.

The following properties are at the core of the general definition of a manifold:

- Each ϕ_i is an injective map, and ϕ_i^{-1} is continuous, that is, ϕ_i^{-1} is the restriction to the hyperboloid of a continuous map defined on an open set in $\mathbb{R}^3$. This condition ensures that the surface does not self-intersect.

- For each ϕ_i, the vectors $\partial \phi_i/\partial t$, $\partial \phi_i/\partial \theta$ are linearly independent. This condition ensures that there is a well defined tangent plane to the surface, spanned by these two vectors, at each point.

A subset S of $\mathbb{R}^3$ together with a collection of smooth parametrizations whose images cover S and which satisfy the above properties is called a regular surface.

The images of ϕ_1 and ϕ_2 are sewed together along two regions corresponding to $0 < \theta < \pi/2$ and to $\pi < \theta < 3\pi/2$, in the following sense:

- In the regions where the images of ϕ_1 and ϕ_2 overlap, the mapping ϕ_1 can be obtained from the mapping ϕ_2 by a smooth change of coordinates, and ϕ_2 can be obtained from ϕ_1 by a smooth change of coordinates. This means that there exist mappings θ_{12} and θ_{21},

1.1. INTRODUCTION

defined on appropriate open domains in $\mathbb{R}^2$, such that $\phi_2 = \phi_1 \circ \theta_{12}$ and $\phi_1 = \phi_2 \circ \theta_{21}$. Moreover, θ_{12} and θ_{21} are each the inverse mapping of the other.

Indeed, $\phi_2(t, \theta) = \phi_1(t, \theta)$ for all (t, θ) with $t \in \mathbb{R}$ and $\pi < \theta < \pi/2$, and $\phi_2(t, \theta) = \phi_1(t, \theta - 2\pi)$ for all (t, θ) with $t \in \mathbb{R}$ and $2\pi < \theta < 5\pi/2$. The corresponding smooth change of coordinates

$$\theta_{12} : \mathbb{R} \times [(\pi, 3\pi/2) \cup (2\pi, 5\pi/2)] \to \mathbb{R} \times [(\pi, 3\pi/2) \cup (0, \pi/2)]$$

is given by

$$\theta_{12}(t, \theta) = \begin{cases} (t, \theta), & \text{for } t \in \mathbb{R} \text{ and } \pi < \theta < 3\pi/2, \\ (t, \theta - 2\pi), & \text{for } t \in \mathbb{R} \text{ and } 2\pi < \theta < 5\pi/2. \end{cases}$$

Similarly, the change of coordinates

$$\theta_{21} : \mathbb{R} \times [(\pi, 3\pi/2) \cup (0, \pi/2)] \to \mathbb{R} \times [(\pi, 3\pi/2) \cup (2\pi, 5\pi/2)]$$

is given by

$$\theta_{21}(t, \theta) = \begin{cases} (t, \theta), & \text{for } t \in \mathbb{R} \text{ and } \pi < \theta < 3\pi/2, \\ (t, \theta + 2\pi), & \text{for } t \in \mathbb{R} \text{ and } 0 < \theta < \pi/2. \end{cases}$$

These coordinate changes provide essential information about the surface. As we will see in the next section, the above property is a cornerstone of the definition of a manifold.

Finally, we notice that same surface can be described through different collections of parametrizations. For example, consider a third local parametrization of the hyperboloid, which agrees with the previous ones:

$$\phi_3(t, \theta) = (\cosh t \cos \theta, \cosh t \sin \theta, \sinh t), \quad -\infty < t < \infty, \ \pi/2 < \theta < 2\pi.$$

This patch lies on top of the other two, and it does not supply any new information about the surface. Indeed, $\phi_3(t, \theta) = \phi_1(t, \theta)$ for all (t, θ) with $t \in \mathbb{R}$ and $\pi/2 < \theta < 3\pi/2$, and $\phi_3(t, \theta) = \phi_2(t, \theta)$ for all (t, θ) with $t \in \mathbb{R}$ and $\pi < \theta < 2\pi$. Thinking of the hyperboloid as a collection of smooth parametrizations, we have two equivalent representations: one consisting of $\{\phi_1, \phi_2\}$, and a second one consisting of $\{\phi_1, \phi_2, \phi_3\}$. There are, in fact, infinitely many collections of equivalent parametrizations describing the same surface. We can always choose one collection of parametrizations as a representative.

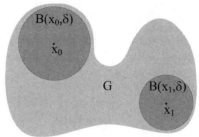

FIGURE 1.2.1
Open set in the plane.

1.2 Review of topological concepts

In the next section we will define manifolds. Unlike surfaces in Section 1.1, manifolds are not necessarily embedded in Euclidean spaces. Therefore, the ideas of nearness and continuity on a manifold need to be expressed in some intrinsic way.

Recall that a mapping $f : U \subseteq \mathbb{R}^m \to \mathbb{R}^n$ is continuous at a point $x_0 \in U$ if for every $\epsilon > 0$ there exists $\delta > 0$ such that, for each $x \in U$,

$$\|f(x) - f(x_0)\| < \epsilon \quad \text{provided} \quad \|x - x_0\| < \delta.$$

A mapping is said to be a continuous if it is continuous at every point of its domain. A sufficient (but not necessary) condition for a mapping $f : U \subseteq \mathbb{R}^m \to \mathbb{R}^n$ defined on an open set U in $\mathbb{R}^m$ to be continuous is that f is differentiable at every point $x_0 \in U$.

Continuity can be expressed in terms of open sets. A set $G \subseteq \mathbb{R}^m$ is open provided that for every $x_0 \in G$ one can find an open ball

$$B(x_0, \delta) = \{x \in \mathbb{R}^m \mid \|x - x_0\| < \delta\},$$

that is contained in G, for some $\delta > 0$. See Figure 1.2.1. If X is a subset of $\mathbb{R}^m$, a set $G \subseteq X$ is said to be (relatively) open in X if there exists an open set $H \subseteq \mathbb{R}^m$ such that $G = H \cap X$. A set is said to be closed if its complement is an open set. One can easily verify that a mapping is continuous on its domain if and only if, for every open set $V \subseteq \mathbb{R}^n$, the set $f^{-1}(V)$ is an open set in U. Equivalently, a mapping is continuous on its domain if and only if for every closed set $F \subseteq \mathbb{R}^n$, the set $f^{-1}(F)$ is a closed set in U.

1.2. REVIEW OF TOPOLOGICAL CONCEPTS

Open sets are therefore essential in studying continuity. It turns out that all familiar properties of continuous mappings can be proved directly from only a few properties of open sets, with no reference to the $\epsilon - \delta$ definition. Those properties are at the core of the concept of a topological space:

DEFINITION 1.2.1

A topological space is a set X together with a collection $\mathcal{G}$ of subsets of X satisfying the following properties:

(i) The empty set $\emptyset$ and the 'total space' X are in $\mathcal{G}$;

(ii) The union of any collection of sets in $\mathcal{G}$ is a set in $\mathcal{G}$;

(iii) The intersection of any finite collection of sets in $\mathcal{G}$ is a set in $\mathcal{G}$.

The sets in $\mathcal{G}$ are called the open sets of the topological space. The collection $\mathcal{G}$ of all open sets is referred to as the topology on X.

We will often omit specific mention of $\mathcal{G}$ and refer to a topological space only by the total space X. Given a set $A \subseteq X$, the union of all open sets contained in A is called the interior of A and is denoted by $\text{int}(A)$. The interior of a set is always an open set, possibly empty.

Example 1.2.2

(i) The Euclidean space $\mathbb{R}^m$ with the open sets defined as above is a topological space.

(ii) If X is any set, the collection of all subsets of X is a topology on X; it is called the discrete topology.

(iii) If X is a topological space and S is a subset of X, then the set S together with the collection of all sets of the type $\{S \cap G \,|\, G \in \mathcal{G}\}$ is a topological space. This topology is referred to as the relative topology induced by X on S.

(iv) If X and Y are topological spaces, then the collection of all unions of sets of the form $G \times H$, with G an open set in X and H an open set in Y, is a topology on the product space $X \times Y$. This is called the product topology. This definition extends naturally to the case of finitely many topological spaces.

(v) Assume that X is a topological space and $\sim$ is an equivalence relation on X. We define the quotient set $X/\sim$ as the set of all equivalence classes on X, and the canonical projection $\pi : X \to X/\sim$ that sends every element $x \in X$ into its equivalence class $[x] \in X/\sim$. The set of all $U \subseteq X/\sim$ for which $\pi^{-1}(U)$ is an open set in X defines a topology on $X/\sim$, called the quotient topology.

(vi) A distance function (also called a metric) on a set X is a function $d : X \times X \to \mathbb{R}$ satisfying the following properties for all $p, q, r \in X$:

(1) positive definiteness and non-degeneracy: $d(p,q) \geq 0$ and $d(p,q) = 0$ if and only if $p = q$;

(2) symmetry: $d(p,q) = d(q,p)$;

(3) triangle inequality: $d(p,q) \leq d(p,r) + d(r,q)$.

A set X together with a distance function on it is called a metric space. A metric space has a natural topology: a set G in X is defined to be open provided that for every $x_0 \in G$ there exists $\delta > 0$ such that the open ball
$$B(x_0, \delta) = \{x \in X \,|\, d(x, x_0) < \delta\}$$
is contained in G. A metric space is said to be complete if every Cauchy sequence is convergent.

The natural topology on a Euclidean space is the topology induced by the Euclidean distance. □

In many instances, it is easier to describe the topology of a space by specifying a certain sub-collection of open sets. A basis for a topology $\mathcal{G}$ of X is a collection $\mathcal{B} \subseteq \mathcal{G}$ of open sets with the property that every set in $\mathcal{G}$ can be obtained as a union of sets from $\mathcal{B}$. As an example, in a metric space, the collection of all open balls is a basis for the metric space topology. A sub-basis for a topology $\mathcal{G}$ of X is a collection $\mathcal{S} \subset \mathcal{G}$ with the property that the collection of all finite intersections of sets in $\mathcal{S}$ is a basis for $\mathcal{G}$.

Example 1.2.3
Let X_i be an infinite collection on topological spaces, whose topologies are denoted by $\mathcal{G}_i$, respectively. On the cartesian product $\Pi_i X_i$ we define a topological basis as the collection of all sets of the type $\Pi_i U_i$, where U_i is an open set in X_i for each i, and only finitely many of the sets U_i are different from X_i. The product topology of $\Pi_i X_i$ is defined as consisting of unions of sets of the above type. In the finite case, this topology is the same as the one described in Example 1.2.2 (iv). □

The complement of an open set is said to be a closed set. Given a set $A \subseteq X$, the intersection of all closed sets containing A is called the closure of A, and is denoted by $\text{cl}(A)$. The closure of a set is always a closed set. For a set A, the boundary set is defined by $\text{bd}(A) = \text{cl}(A) \setminus \text{int}(A)$. The boundary of a set is always a closed set. If the closure of a set is the total space, that set is said to be a dense set. A closed set with an empty interior is called nowhere dense. For example, if $X = \mathbb{R}$ with the natural topology, and $A = \mathbb{Q}$, the set of all rational numbers, then $\text{cl}(A) = X$, so A is dense. The set $B = \mathbb{Z}$ of all integers is a nowhere dense set in X.

If $f : X \to \mathbb{R}$ is continuous, the support of f is

$$\text{supp}(f) = \text{cl}\left(\{x \mid f(x) \neq 0\}\right).$$

A set $N \subseteq X$ is said to be a neighborhood of a point $x \in X$ provided that there exists an open set G with $x \in G \subseteq N$. A basis of neighborhoods of a point $x \in X$ is a collection $\mathcal{V}_x$ of neighborhoods of x with the property that every neighborhood of x contains some set from $\mathcal{V}_x$. As an example, the balls of the type $B(x, 1/n)$ in a metric space form a basis of neighborhoods of x. One can completely describe a topology by specifying a basis of neighborhoods for each point of the space.

The idea of nearness in a topological space can be expressed through convergent sequences. A sequence $(x_n)_{n \geq 0}$ in X is said to be convergent to a point $z \in X$ provided that for every neighborhood V of z, there exists an integer n_V such that all terms of the sequence x_n with $n \geq n_V$ are contained in V. Unlike in $\mathbb{R}$, the limit of a convergent sequence in a topological space may not be unique. In order for the limit to be unique, it is sufficient that for every pair of points $x \neq y$ there exits a pair of disjoint neighborhoods V_x of x and V_y of y. A topology satisfying this condition is said to be Hausdorff.

There is a natural definition of continuity in the context of topological spaces.

DEFINITION 1.2.4
Let X and Y be topological spaces. A map $f : X \to Y$ is continuous at a point x_0 in X provided that $f^{-1}(V)$ is a neighborhood of x_0 for every neighborhood V of x_0. A map $f : X \to Y$ is said to be continuous if for each open set V in Y, the set $f^{-1}(V)$ is an open set in X.

From calculus, we know that every continuous function on a closed bounded interval is bounded and attains its minimum and maximum values. We also know that continuous functions satisfy the intermediate value property. These properties have natural generalizations for continuous maps on topological spaces.

Compact sets in a topological space generalize closed bounded subsets of $\mathbb{R}^m$. It is well known that every sequence in a closed bounded subset of $\mathbb{R}^m$ has a convergent subsequence. This can be proved by 'dichotomy' as follows: enclose the set in an m-cube, and divide this cube into 2^m cubes of half size. At least one of the sub-cubes will contain infinitely many terms of the sequence. Divide such a sub-cube into 2^m cubes of half size. Again, at least one of the resulting sub-cubes contains infinitely many terms of the subsequence from the previous step. Repeating this process indefinitely, we obtain a nested sequence of cubes that shrink down to a point. That point is a limit of some subsequence. The key idea of this proof is that any nested sequence of closed sets has a non-empty intersection. To define compactness, one needs to ensure a stronger property: for every collection of closed sets with the property that every finite sub-collection has a non-empty intersection, the intersection of all the elements in the family has to be non-empty. An equivalent definition can be given in terms of open sets.

DEFINITION 1.2.5
A set $K \subseteq X$ is compact if every collection of open sets whose union contains K has a finite sub-collection whose union still contains K. A topological space X is said to be compact if the set X is a compact set.

In a compact set, every sequence contains a convergent subsequence. This property is equivalent to compactness for metric spaces.

Connected sets in a topological space generalize intervals in $\mathbb{R}$ (finite or infinite). A set is connected if it cannot be separated into two 'globs' (see Figure 1.2.2).

DEFINITION 1.2.6
A set $C \subseteq X$ is connected if there is no pair G_1, G_2 of disjoint open sets in X with $C \cap G_1 \neq \emptyset$ and $C \cap G_2 \neq \emptyset$, such that $C \subseteq G_1 \cup G_2$. A topological space X is said to be connected if the set X is a connected set.

A continuous map always takes compact sets into compact sets, and connected sets into connected sets. A continuous map is called a homeo-

1.3. SMOOTH MANIFOLDS

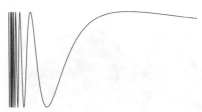

FIGURE 1.2.2
Connected set in $\mathbb{R}^2$: $(\{0\} \times [-1, 1]) \cup \{(x, \sin(1/x)) \,|\, x > 0\}$.

morphism if it is bijective and its inverse is also continuous.

We will discuss some other topological facts throughout the text. As a general reference, we recommend Munkres (2000).

1.3 Smooth manifolds

A more general concept than a regular surface is a smooth manifold.

DEFINITION 1.3.1
A smooth manifold of dimension m is a set M together with a collection of maps $\phi_\alpha : U_\alpha \to M$, with each U_α an open subset of $\mathbb{R}^m$, satisfying the following conditions:

(i) each map $\phi_\alpha : U_\alpha \to V_\alpha = \phi(U_\alpha)$ is injective;

(ii) if $V_\alpha \cap V_\beta \neq \emptyset$, then there exists a smooth mapping

$$\theta_{\alpha\beta} : \phi_\beta^{-1}(V_\alpha \cap V_\beta) \to \phi_\alpha^{-1}(V_\alpha \cap V_\beta)$$

such that $\phi_\beta = \phi_\alpha \circ \theta_{\alpha\beta}$;

(iii) $M = \bigcup_\alpha \phi(U_\alpha)$.

Manifolds with dimension 2 are called surfaces.

The maps $\phi_\alpha : U_\alpha \to V_\alpha$ above are called local parametrizations, and $\mathbb{R}^m$ is referred to as the parameter space. The coordinates $(x_1, \ldots, x_m)$ in $\mathbb{R}^m$ corresponding through ϕ_α^{-1} to a point $p \in V_\alpha$ are called the local coordinates of p, and the inverse map $\phi_\alpha^{-1} : V_\alpha \to U_\alpha$ is called a local

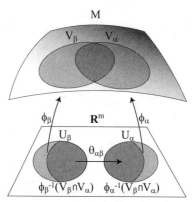

FIGURE 1.3.1
A smooth re-parametrization.

coordinate system or chart. We will often refer to the set V_α itself as a chart.

The set M inherits a natural topology. A basis of neighborhoods of a point $p \in M$ is defined by taking the image of a basis of neighborhoods in $U_\alpha \subseteq \mathbb{R}^m$ through some ϕ_α with $p \in V_\alpha$. Different mappings ϕ_α and ϕ_β may yield different bases of neighborhoods of the same point p. These bases, however, generate the same topology due to condition (ii).

The first condition implies that M is locally homeomorphic to the m-dimensional Euclidean space. The second condition says that if the images of ϕ_α and ϕ_β overlap, then ϕ_β results from ϕ_α by a smooth coordinate change. Whenever two local coordinate systems satisfy this condition, we say that they are compatible. Note $\theta_{\alpha\beta} = \phi_\alpha^{-1} \circ \phi_\beta$ on $\phi_\beta^{-1}(V_\alpha \cap V_\beta) \subseteq \mathbb{R}^m$, $\theta_{\beta\alpha} = \phi_\beta^{-1} \circ \phi_\alpha \subseteq \mathbb{R}^m$ on $\phi_\alpha^{-1}(V_\alpha \cap V_\beta)$, and thus $\theta_{\beta\alpha} = \theta_{\alpha\beta}^{-1}$. See Figure 1.3.1. When all coordinate changes $\theta_{\alpha\beta}$ are C^k-differentiable, we say that M is a C^k manifold. For simplicity, we will consider only C^∞ manifolds, unless we specify otherwise.

Different collections of local parametrizations may be used to define the same manifold. We define two collections of local parametrizations to be equivalent when every local parametrization from the first is compatible with every local parametrization from the second.

- A smooth structure on M is a choice of an equivalence class of local parametrizations. A smooth structure can be specified by any single representative from that class.

1.3. SMOOTH MANIFOLDS

Without any further assumptions, it is possible that the topology of M is rather pathological. See Exercises 1.15.6 and 1.15.7. In differential geometry, we want to study manifolds satisfying the following two reasonable properties:

- The topology of the manifold M is Hausdorff.
- The topology of the manifold M has a countable basis of open sets.

We will always assume that M has these properties. The first property means that the limit of any convergent sequence in M is unique. The second property implies that one can always find a smooth structure on M consisting of a countable collection of local parametrizations. Another important consequence is the existence of a differentiable partition of unity subordinate to an open covering. This means that for every open covering $\{V_\alpha\}_\alpha$ of M (where V_α does not have to be the domain of a chart), there exists a family of smooth real valued functions $\{f_i\}_i$, with $0 \leq f_i \leq 1$ such that:

(1) for each point $p \in M$, there are only finitely many i with $f_i(p) > 0$;

(2) for each i, there is an α with $\mathrm{supp}(f_i) = \mathrm{cl}\left(\{x \mid f(x) \neq 0\}\right) \subseteq V_\alpha$;

(3) $\sum_i f_i(p) = 1$ at each $p \in M$.

See Fathi (1997) for a proof.

REMARK 1.3.2 One can alternatively define a manifold M as a subset of a Euclidean space $\mathbb{R}^N$ of certain dimension N, endowed with the topology induced by $\mathbb{R}^N$, and locally diffeomorphic to $\mathbb{R}^m$. A smooth map is called a diffeomorphism if it is bijective, and its inverse is also smooth. By saying that M is locally diffeomorphic to $\mathbb{R}^m$ we mean that each point $p \in M$ has a neighborhood in M diffeomorphic to some open set in $\mathbb{R}^m$.

This type of manifold will be referred to as an 'embedded' manifold, as opposed to one satisfying the conditions in Definition 1.3.1, that we will refer to as an 'abstract' manifold. There is a theorem of Whitney saying that the two notions are equivalent. See Section 1.8.

The notion of 'embedded manifold' is easier to visualize, since it describes an object placed in an environment (a Euclidean space) endowed with a natural topology. Therefore, the technical requirements on the topology are no longer needed. Nevertheless, there are a couple of reasons why we should prefer the notion of an 'abstract manifold'. First,

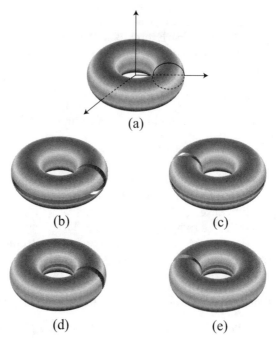

FIGURE 1.3.2
The torus with a smooth structure.

there are examples for which the conditions in Definition 1.3.1 are easier to check. Second, the notion of an abstract manifold is intrinsic, that is, its properties can be understood through observations made entirely within the manifold, with no reference to an ambient space. ∎

Example 1.3.3 *(The torus)*
(i) The following parametric equation

$$\phi(x_1, x_2) = ((R + r\cos x_1)\cos x_2, (R + r\cos x_1)\sin x_2, r\sin x_1),$$

with $(x_1, x_2) \in \mathbb{R}^2$, represents a 2-dimensional torus $\mathbb{T}^2$ in $\mathbb{R}^3$, obtained by revolving about the z-axis a circle of radius r in the yz-plane, with its center at a distance of $R > r$ from the z-axis. See Figure 1.3.2 (a).

The map ϕ is not injective. We have to consider restrictions of ϕ to open subsets of $\mathbb{R}^2$ in order to abide by injectivity. Let ϕ_1 be the restriction of ϕ to the open set

$$U_1 = \{(x_1, x_2) \in \mathbb{R}^2 \mid 0 < x_1 < 2\pi \text{ and } 0 < x_2 < 2\pi\}.$$

1.3. SMOOTH MANIFOLDS

The image V_1 of this open set through ϕ_1 is the portion of the torus shown in Figure 1.3.2 (b). In order to cover the whole torus we produce three more local parametrizations: ϕ_2, ϕ_3 and ϕ_4, defined as the restrictions of ϕ to the open sets

$$U_2 = \{(x_1, x_2) \in \mathbb{R}^2 \,|\, 0 < x_1 < 2\pi \text{ and } -\pi < x_2 < \pi\},$$
$$U_3 = \{(x_1, x_2) \in \mathbb{R}^2 \,|\, -\pi < x_1 < \pi \text{ and } 0 < x_2 < 2\pi\},$$
$$U_4 = \{(x_1, x_2) \in \mathbb{R}^2 \,|\, -\pi < x_1 < \pi \text{ and } -\pi < x_2 < \pi\},$$

respectively, whose images are shown in Figure 1.3.2 (c), (d), (e).

We need to show that the appropriate coordinate changes are diffeomorphisms. This is easy to see since they are all 'piecewise translations'; for example

$$\theta_{12}(x_1, x_2) = \begin{cases} (x_1, x_2 + 2\pi) & \text{for } 0 < x_1 < 2\pi \text{ and } -\pi < x_2 < 0, \\ (x_1, x_2) & \text{for } 0 < x_1 < 2\pi \text{ and } 0 < x_2 < \pi. \end{cases}$$

Therefore, this family of local parametrizations provides $\mathbb{T}^2$ with a smooth structure.

(ii) We can represent a 2-dimensional torus as a smooth manifold differently. Consider the equivalence relation on $\mathbb{R}^2$ defined by $(x_1, x_2) \sim (x_1', x_2')$ provided that $(x_1' - x_1, x_2' - x_2) \in \mathbb{Z}^2$. We denote the quotient set $\mathbb{R}^2/\mathbb{Z}^2$ by M. Each equivalence class of $\mathbb{R}^2/\mathbb{Z}^2$ has at least one representative in the unit square $[0, 1] \times [0, 1]$. Consider the open disks D_1, D_2 D_3, D_4 of radius r with $1/\sqrt{2} < r < 1$, centered at the corners of the unit square. See Figure 1.3.3. Denote by $[x_1, x_2]$ the equivalence class of a point (x_1, x_2) in $\mathbb{R}^2$, and by ϕ_i the mapping $(x_1, x_2) \to [x_1, x_2]$ restricted to D_i.

The mappings ϕ_i are all injective. The re-parametrizations are $\theta_{12}(x_1, x_2) = (x_1 - 1, x_2)$, $\theta_{13}(x_1, x_2) = (x_1 - 1, x_2 - 1)$, $\theta_{14}(x_1, x_2) = (x_1, x_2 - 1)$, and so on. They are all diffeomorphisms. In this way we put a smooth structure on $\mathbb{T}^2$. The topology induced on $\mathbb{R}^2/\mathbb{Z}^2$ is the quotient topology, hence $\mathbb{R}^2/\mathbb{Z}^2$ is homeomorphic to the torus described in (i). See Figure 1.3.3. The image of each vertical line segment under the above local parametrizations is a 'meridian' circle and the image of each horizontal line segment is a 'parallel' circle on the torus. This construction does not involve any embedding of $\mathbb{T}^2$ in an ambient space, everything is done 'abstractly'. □

Here is another example of 'abstract' manifold.

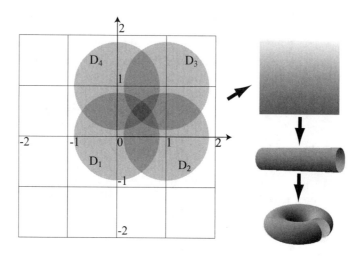

FIGURE 1.3.3
The lattice $\mathbb{R}^2$ with $\mathbb{Z}^2$ nodes.

Example 1.3.4 *(The real projective space)*
Let $\mathbb{RP}^n$ be the quotient set of $\mathbb{R}^{n+1} \setminus \{0\}$ by the equivalence relation:

$$(y_1, \ldots, y_{n+1}) \sim (\lambda y_1, \ldots, \lambda y_{n+1}), \text{ for any non-zero real number } \lambda.$$

This equivalence relation collapses each line through the origin in $\mathbb{R}^{n+1}$ to a single point of $\mathbb{RP}^n$, which can be though of as the direction of the line. Alternatively, $\mathbb{RP}^n$ can be obtained by identifying each point of the unit sphere S^n in $\mathbb{R}^{n+1}$ with its antipodal point.

Let $[y_1, \ldots, y_{n+1}]$ denote the equivalence class of $(y_1, \ldots, y_{n+1})$, and consider the maps $\phi_i : \mathbb{R}^n \to \mathbb{RP}^n$,

$$\phi_i(x_1, \ldots, x_n) = [x_1, \ldots, x_{i-1}, 1, x_i, \ldots, x_n],$$

for all $i = 1, \ldots, n+1$. The ϕ_i are local parametrizations with $U_i = \mathbb{R}^n$ and

$$V_i = \{[y_1, \ldots, y_i, \ldots, y_{n+1}] \,|\, y_i \neq 0\}.$$

They make $\mathbb{RP}^{n+1}$ into an n-dimensional manifold. The induced topology is the quotient of the natural topology on $\mathbb{R}^{n+1} \setminus \{0\}$ by the equivalence relation. The proof is left as an exercise.

For $n = 2$, this construction gives the real projective plane $\mathbb{RP}^2$. Every point on the southern hemisphere of S^2 is identified with its antipodal

1.3. SMOOTH MANIFOLDS

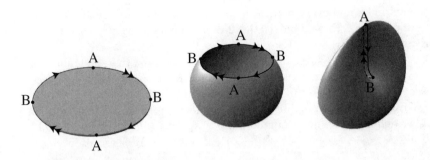

FIGURE 1.3.4
A representation of the real projective plane as a crosscap surface. Glue the opposite points on the boundary circle as indicated by the arrows. Perform the gluing directly through the surface, by allowing self-intersections.

point in the northern hemisphere, so we can disregard the lower hemisphere and consider only points in the upper hemisphere as representatives for pairs of antipodal points. Since we also need to identify pairs of opposite points along the equator, we can think of $\mathbb{RP}^2$ as a disk with opposite points on the boundary circle identified. The representation of the real projective plane as a subset of $\mathbb{R}^3$, shown in Figure 1.3.4, is known as the crosscap surface. The crosscap has a segment of double points which end at two 'pinch points'. The crosscap is not a regular surface in $\mathbb{R}^3$. □

Some manifolds are contained in other manifolds, from which they inherit the smooth structure.

DEFINITION 1.3.5
A subset N of M is an n-dimensional submanifold provided that $n \leq m$ and there exists a collection of local parametrizations of M that gives the smooth structure on M and contains for each point p in N a parametrization $\phi_\alpha : U_\alpha \to V_\alpha$ with $p \in V_\alpha$ and $\phi_\alpha((\mathbb{R}^n \times \{0\}) \cap U_\alpha) = N \cap V_\alpha$.

Here $\mathbb{R}^n \times \{0\}$ is the set of vectors in $\mathbb{R}^m$ with the last $m-n$ coordinates equal to 0.

The restrictions $\phi_\alpha : (\mathbb{R}^n \times \{0\}) \cap U_\alpha \to N \cap V_\alpha$ endow N with an

n-dimensional manifold structure. The codimension of N in M is $m-n$.

Every nonempty open subset of a manifold M is a codimension 0 submanifold, and every point is a codimension m submanifold. Another example of a submanifold is the 2-dimensional torus $\mathbb{T}^2$ described in Example 1.3.3 (i).

A 1-dimensional submanifold of $\mathbb{R}^3$ will be referred to as an embedded curve, and a 2-dimensional submanifold of $\mathbb{R}^3$ as an embedded surface. Every regular surface is a 2-dimensional submanifold of $\mathbb{R}^3$.

Given two smooth manifolds, M of dimension m, and N of dimension n, we define a new smooth manifold $M \times N$, called the product manifold. If $p \in M$, $q \in N$, $\phi_\alpha : U_\alpha \subseteq \mathbb{R}^m \to V_\alpha$ is a local parametrization near p in M, and $\psi_\alpha : U_\beta \subseteq \mathbb{R}^n \to V_\beta$ is a local parametrization near q in N, then
$$\phi_\alpha \times \psi_\beta : U_\alpha \times U_\beta \subseteq \mathbb{R}^{m+n} \to V_\alpha \times V_\beta \subseteq M \times N$$
given by
$$(\phi_\alpha \times \psi_\beta)(x_1, \ldots, x_{m+n}) = (\phi_\alpha(x_1, \ldots, x_m), \psi_\beta(x_{m+1}, \ldots, x_{m+n})),$$
defines a local parametrization near $(p, q) \in M \times N$. The resulting topology of $M \times N$ is the product topology. For any $p \in M$ and $q \in N$, $\{p\} \times N$ and $M \times \{q\}$ are submanifolds of $M \times N$.

Example 1.3.6

We give another description of a 2-dimensional torus as a smooth manifold. Let $S^1 = \mathbb{R}/\mathbb{Z}$ be the unit circle in $\mathbb{R}^2$ with the natural smooth structure (the 1-dimensional analogue of the construction in Example 1.3.3 (ii)). Then $S^1 \times S^1$ is a surface, which can be naturally embedded in $\mathbb{R}^4$. The product manifold $S^1 \times S^1$ is 'differentiably the same' as the 'embedded' torus $\mathbb{T}^2$, defined in Example 1.3.3 (i), which is 'differentiably the same' as the 'abstract' torus $\mathbb{R}^2/\mathbb{Z}^2$, defined in Example 1.3.3 (ii). The precise term for 'differentiably the same' is 'diffeomorphic', which will be defined in the next section. □

1.4 Smooth maps

The natural class of mappings between manifolds is the class of smooth mappings. Since 'smoothness' is a local condition, we are able to express it in local coordinates.

1.4. SMOOTH MAPS

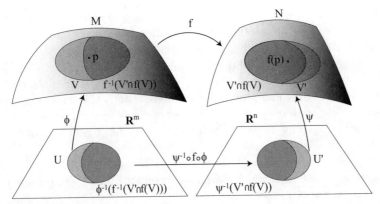

FIGURE 1.4.1
A smooth mapping in local coordinates.

DEFINITION 1.4.1
Let $f : M \to N$ be a map from the m-dimensional manifold M to an n-dimensional manifold N. We say that f is differentiable if for every point $p \in M$ there exists a local parametrization $\phi : U \to V$ for M with $p \in V$, and a local parametrization $\psi : U' \to V'$ for N with $f(p) \in V'$, such that $\psi^{-1} \circ f \circ \phi$ is differentiable at $\phi^{-1}(p)$.

The mapping $\psi^{-1} \circ f \circ \phi$ is defined on $\phi^{-1}(f^{-1}(V' \cap f(V))) \subseteq U \subseteq \mathbb{R}^m$ and takes values in $\psi^{-1}(V' \cap f(V)) \subseteq U' \subseteq \mathbb{R}^n$. See Figure 1.4.1.

The definition is independent of the choice of the charts ϕ and ψ. Suppose that $\bar{\phi} : \bar{U} \to \bar{V}$ and $\bar{\psi} : \bar{U}' \to \bar{V}'$ are other local parametrizations with $p \in \bar{V}$ and $f(p) \in \bar{V}'$. The change of coordinate mappings

$$\theta : \phi^{-1}(V \cap \bar{V}) \to \bar{\phi}^{-1}(V \cap \bar{V}), \quad \text{and}$$
$$\eta : \psi^{-1}(V \cap \bar{V}) \to \bar{\psi}^{-1}(V \cap \bar{V}),$$

are diffeomorphisms between open subsets of $\mathbb{R}^m$ and $\mathbb{R}^n$ respectively, and

$$\bar{\psi}^{-1} \circ f \circ \bar{\phi} = \eta \circ \psi^{-1} \circ f \circ \phi \circ \theta^{-1}$$

on a neighborhood $\bar{\phi}^{-1}\left(f^{-1}(V' \cap \bar{V}' \cap f(V \cap \bar{V}))\right)$ of $\bar{\phi}^{-1}(p)$. It is clear from this formula that $\bar{\psi}^{-1} \circ f \circ \bar{\phi}$ is differentiable at $\bar{\phi}^{-1}(p)$ if and only if $\psi^{-1} \circ f \circ \phi$ is differentiable at $\phi^{-1}(p)$.

We define f to be C^k-differentiable if $\psi^{-1} \circ f \circ \phi$ is C^k-differentiable for any choice of ψ and ϕ. We will consider only C^∞-differentiable maps, unless we specify otherwise. A C^∞-differentiable map will be referred to as a smooth map. In the case when $M = \mathbb{R}^m$ and $N = \mathbb{R}^n$, our definitions reduce to the usual definitions of differentiable and C^k-differentiable maps given in calculus courses.

A diffeomorphism is a bijective smooth map with smooth inverse. Diffeomorphisms exist only between smooth manifolds of the same dimension. Two diffeomorphic manifolds have exactly the same properties from the differentiable point of view. For example, one can show that the 2-dimensional tori defined in Example 1.3.3 (i) and (ii) and in Example 1.3.6 are all diffeomorphic.

The following example illustrates that the analogues of some usual geometric transformations from planar geometry may behave quite differently on smooth manifolds.

Example 1.4.2 *(Translation on the torus)*
Consider a torus $\mathbb{T}^2$ with the smooth structure as in Example 1.3.3 (ii). Consider the planar translation T given by $T(x_1, x_2) = (x_1+t_1, x_2+t_2)$, with $0 \leq t_1, t_2 \leq 1$. Let τ be the map induced by T on $\mathbb{T}^2 = \mathbb{R}^2/\mathbb{Z}^2$. In local coordinates, each $\phi_i^{-1} \circ \tau \circ \phi_j$ is the restriction of T to the appropriate subset of D_j. It is clear that τ is a diffeomorphism of the torus.

This diffeomorphism is quite interesting. If both quantities t_1, t_2 are irrational and rationally independent (i.e. t_2 is not a rational multiple of t_1), then the trajectory $\{\tau^n(p) \,|\, n \in \mathbb{Z}\}$ of any point p of the torus is dense on the surface of the torus. See Figure 1.4.2. If both quantities t_1, t_2 are rational, the trajectory of any point is finite. The remaining case may lead to either a dense set of points on the surface of the torus or to a dense set of points on some finite collection of closed curves on the surface of the torus. □

Two diffeomorphic manifolds are the same from a differentiable point of view. It is however possible to have two distinct smooth structures on the same set, even if they are diffeomorphic. This is illustrated in the following example.

Example 1.4.3
Consider the set of all real numbers $\mathbb{R}$ and the parametrization $\phi : \mathbb{R} \to \mathbb{R}$, $\phi(x) = x$. This satisfies the conditions in Definition 1.3.1, so it gives a smooth structure on $\mathbb{R}$. Then consider the same set $\mathbb{R}$ and the parametrization $\psi : \mathbb{R} \to \mathbb{R}$, $\psi(x) = x^3$, which also gives a smooth structure

1.5. TANGENT VECTORS AND THE TANGENT BUNDLE

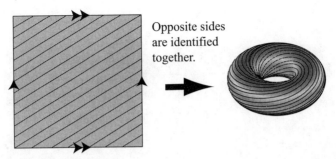

FIGURE 1.4.2
Translation on the torus.

on $\mathbb{R}$.

One the one hand, the two parametrizations are incompatible, since $\psi^{-1} \circ \phi(x) = x^{1/3}$, which is not differentiable at 0, hence the two smooth structures on $\mathbb{R}$ are distinct.

On the other hand, the map $x \in \mathbb{R} \to f(x) = x^3 \in \mathbb{R}$ is a diffeomorphism, since f is bijective, $\psi^{-1} \circ f \circ \phi(x) = x$, and $\phi^{-1} \circ f^{-1} \circ \psi(x) = x$ for all x. □

REMARK 1.4.4 Given a topological space M, there may exist several smooth structures compatible with the topology of M that are not diffeomorphic. This is the case, for example, of the topological n-dimensional sphere S^n. Although for $1 \leq n \leq 6$, S^n can only have one smooth structure, for $n = 7, 8, 9, 10, 11, 12, 13, 14, 15$, the number of possible non-diffeomorphic smooth structures on S^n is 28, 2, 8, 6, 992, 1, 3, 2, 16256, respectively! In the same vein, for $n \neq 4$, all smooth structures on $\mathbb{R}^n$ are diffeomorphic. On $\mathbb{R}^4$, however, there exist infinitely many non-diffeomorphic smooth structures! ■

1.5 Tangent vectors and the tangent bundle

From elementary physics, we know that the velocity vector of a moving particle is always tangent to its trajectory. If the trajectory lies on some surface, then the velocity vector is tangent to that surface. Thus, a tangent vector to the surface at a point is the velocity vector of some

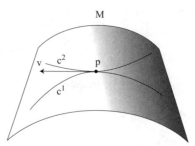

FIGURE 1.5.1
A vector tangent to two trajectories.

trajectory on the surface through that point. Different trajectories passing through the same point may have the same velocity at that point. In fact we can identify a vector with the set of all curves that are tangent to it. See Figure 1.5.1.

Since 'abstract' manifolds are not embedded in an ambient space, we cannot picture tangent vectors as 'sticking out' of a manifold. But we can describe all the curves which should share a given tangent vector.

Let M be an m-dimensional smooth manifold. A smooth curve c in M is a differentiable mapping $c : I \subseteq \mathbb{R} \to M$, with I being some open interval of the real numbers. That is, for every local parametrization $\phi : U \to V$ with V containing some portion of c, the pullback curve $\phi^{-1}(c(t))$ is a differentiable curve in $U \subseteq \mathbb{R}^m$. Let us fix a point $p \in M$ and consider two curves c^1 and c^2 passing through p at $t = 0$. Let $\phi : U \to V$ be a local parametrization near p. Then $\phi^{-1} \circ c^1$ and $\phi^{-1} \circ c^2$ are two curves in $U \subseteq \mathbb{R}^m$, passing through $\phi^{-1}(p)$ at $t = 0$. If they have the same tangent vector at this common point, i.e., if $(\phi^{-1} \circ c^1)'(0) = (\phi^{-1} \circ c^2)'(0)$, then we say that c^1 and c^2 are equivalent at p. This equation makes sense since both sides are vectors in $\mathbb{R}^m$.

DEFINITION 1.5.1
A tangent vector v to a smooth surface M at a point p is an equivalence class of curves c in M, passing through the point p at $t = 0$, and for which $(\phi^{-1} \circ c)'(0)$ represent the same vector $(v_1, \ldots, v_m) \in \mathbb{R}^m$, with respect to some local parametrization $\phi : U \to V$ near p.

We also denote the tangent vector v determined by the equivalence class of c by $\dfrac{dc}{dt}(0)$ or $c'(0)$, and we denote by $(v_1, \ldots, v_m)$ its coordi-

1.5. TANGENT VECTORS AND THE TANGENT BUNDLE

nates with respect to the local parametrization. The vector v uniquely determines its local coordinates

The definition of a tangent vector is independent of the choice of a local parametrization. Suppose that c^1 and c^2 are smooth curves with $c^1(0) = c^2(0) = p$ and $(\phi \circ c^1)'(0) = (\phi \circ c^2)'(0)$ for some local parametrization $\phi : U \to V$ around p. Then $(\bar\phi \circ c^1)'(0) = (\bar\phi \circ c^2)'(0)$ for any local parametrization $\bar\phi : \bar U \to \bar V$ around p. The key point is that the change of coordinates map $\bar\phi^{-1} \circ \phi$ is a diffeomorphism (from an open subset of $U \subseteq \mathbb{R}^m$ to an open subset of $\bar U \subseteq \mathbb{R}^m$), so its derivative at $\phi^{-1}(p)$ is an invertible linear map of $\mathbb{R}^m$ to itself. Using this and the chain rule, we obtain

$$\begin{aligned}(\bar\phi^{-1} \circ c^1)'(0) &= d\left(\bar\phi^{-1} \circ \phi\right)_{\phi^{-1}(p)} (\phi^{-1} \circ c^1)'(0) \\ &= d\left(\bar\phi^{-1} \circ \phi\right)_{\phi^{-1}(p)} (\phi^{-1} \circ c^2)'(0) \\ &= (\bar\phi^{-1} \circ c^2)'(0).\end{aligned}$$

We denote the set of all tangent vectors to M at p by T_pM. Once we choose a local parametrization $\phi : U \to V$ around p, if c is a smooth curve in M with $c(0) = p$, the tangent vector $c'(0) \in T_pM$ uniquely determines its local coordinates, since they are just the components of the vector $(\phi^{-1} \circ c)'(0)$. On the other hand, any $(v_1, \ldots, v_m) \in \mathbb{R}^m$ is the local coordinates of $c'(0)$ for some smooth curve c; the simplest example is obtained by composing ϕ with the straight line in $\mathbb{R}^m$ that passes through $\phi^{-1}(p)$ tangent to $(v_1, \ldots, v_m)$, in other words

$$c(t) = \phi(\phi^{-1}(p) + t(v_1, \ldots, v_m)).$$

Thus the correspondence between a tangent vector at p and its coordinates determined by the local parametrization ϕ is a bijection between T_pM and $\mathbb{R}^m$.

Example 1.5.2
We would like to construct the tangent vectors to M at p that correspond to the standard vector basis of $\mathbb{R}^m$:

$$e_1 = (1, 0, \ldots, 0),\ e_2 = (0, 1, \ldots, 0), \ldots, e_m = (0, 0, \ldots, 1).$$

For this purpose we consider the curve $t \mapsto \phi(\phi^{-1}(p) + t \cdot e_i)$, which is the lift to V by ϕ of the straight line in $\mathbb{R}^m$ through the point $\phi^{-1}(p)$ tangent to e_i. For reasons that we explain in the next section, these vectors are denoted by

$$\frac{\partial}{\partial x_1}\Big|_p, \frac{\partial}{\partial x_2}\Big|_p, \ldots, \frac{\partial}{\partial x_m}\Big|_p,$$

where $x_1, x_2, \ldots, x_m$ are the local coordinates defined by ϕ. □

Another local parametrization around p, $\bar{\phi} : \bar{U} \to \bar{V}$, will determine a different bijection between $T_p M$ and $\mathbb{R}^m$. Suppose c is a smooth curve in M with $c(0) = p$ and the local components of the tangent vector $c'(0)$ determined by ϕ and $\bar{\phi}$ are $(v_1, \ldots, v_m)$ and $(\bar{v}_1, \ldots, \bar{v}_m)$. Then

$$(\bar{v}_1, \ldots, \bar{v}_m) = d\left(\bar{\phi}^{-1} \circ \phi\right)_{\phi^{-1}(p)} (v_1, \ldots, v_m), \qquad (1.5.1)$$

because $(v_1, \ldots, v_m) = (\phi^{-1} \circ c)'(0) =$ and $(\bar{v}_1, \ldots, \bar{v}_m) = (\bar{\phi}^{-1} \circ c)'(0)$. We see that the change in coordinates of a vector when we switch from one local parametrization to another is given by an invertible linear map.

REMARK 1.5.3 There is another, closely related, way to think about tangent vectors at p. Suppose that c is a smooth curve in M with $c(0) = p$ and

$$\left(\phi^{-1} \circ c\right)'(0) = (v_1, \ldots, v_m) \in \mathbb{R}^m$$

for some local parametrization $\phi : U \to V$. It is natural to want to identify the tangent vector $c'(0)$ with $(v_1, \ldots, v_m)$. This is too naive, however, since $(v_1, \ldots, v_m)$ depends on the choice of a local parametrization, as well as on c. It is better to think of $c'(0)$ as being the pair $(\phi : U \to V, (v_1, \ldots, v_m))$. But we should also allow for the possibility of using a different local parametrization $\bar{\phi} : \bar{U} \to \bar{V}$. If

$$\left(\bar{\phi}^{-1} \circ c\right)'(0) = (\bar{v}_1, \ldots, \bar{v}_m),$$

then the pair $(\bar{\phi} : \bar{U} \to \bar{V}, (\bar{v}_1, \ldots, \bar{v}_m))$ also represents $c'(0)$. The vectors $(v_1, \ldots, v_m)$ and $(\bar{v}_1, \ldots, \bar{v}_m)$ are related by equation (1.5.1). We are led to the following equivalence relation on pairs consisting of a local parametrization around p and an element of $\mathbb{R}^m$:

$$(\phi : U \to V, (v_1, \ldots, v_m)) \sim (\bar{\phi} : \bar{U} \to \bar{V}, (w_1, \ldots, w_m))$$

if and only if (1.5.1) holds. It is not difficult to verify that each equivalence class consists precisely of the pairs that represent a single tangent vector at p. Many books define tangent vectors at p as equivalence

1.5. TANGENT VECTORS AND THE TANGENT BUNDLE

classes of pairs under this relation, rather than as equivalence classes of curves. ∎

The set of tangent vectors T_pM has a natural vector space structure. Let v, w be two tangent vectors at p, and c^1, c^2 be two differentiable curves on M representing v and w, respectively. In p were the origin in $\mathbb{R}^m$, then $v+w$ would be tangent to the curve $t \mapsto c^1(t) + c^2(t)$ and $a \cdot v$ would be tangent to $t \mapsto a \cdot c^1(t)$. We can use a local parametrization $\phi : U \to V$ around p to perform analogous operations on curves in M. For this purpose, we want to have $\phi(0) = p$, which can always be achieved by composing ϕ with a translation of $\mathbb{R}^m$. Now we can define

$$(c^1 + c^2)(t) = \phi(\phi^{-1} \circ c^1 + \phi^{-1} \circ c^2)(t), \qquad (1.5.2)$$
$$(a \cdot c^1)(t) = \phi(a \cdot \phi^{-1} \circ c^1)(t). \qquad (1.5.3)$$

The curves $c^1 + c^2$ and $a \cdot c^1$ are differentiable curves through p in M. We define

$$v + w = \text{the equivalence class of } c^1 + c^2 \text{ at } p, \qquad (1.5.4)$$
$$a \cdot v = \text{the equivalence class of } a \cdot c^1 \text{ at } p. \qquad (1.5.5)$$

The sum $v+w$ and the scalar multiple $a \cdot v$ are independent of the choice of a local parametrization even though the curves $c^1 + c^2$ and $a \cdot c^1$ are not. This can be checked using the chain rule.

If $(v_1, \ldots, v_m)$ and $(w_1, \ldots, w_m)$ are the local coordinates of v and w determined by ϕ, the curve $t \mapsto (\phi^{-1} \circ c^1 + \phi^{-1} \circ c^2)(t)$ is tangent to $(v_1+w_1, \ldots, v_m+w_m)$, and hence these are the local coordinates of $v+w$. Similarly the local coordinates of av are $(av_1, \ldots, av_m)$. Thus addition and scalar multiplication correspond to addition and scalar multiplication of their representations in local coordinates. The bijection between T_pM and $\mathbb{R}^m$ given by the local parametrization ϕ is not just a bijection, it is an invertible linear map. Since the vectors

$$\frac{\partial}{\partial x_1}\Big|_p, \ldots, \frac{\partial}{\partial x_m}\Big|_p$$

correspond to the basis $e_1, \ldots, e_m$ of $\mathbb{R}^m$, they must form a basis for T_pM.

When we switch from one local parametrization to another, the coordinates of vectors change linearly. It is clear from this that we obtain the same operations on T_pM no matter which local parametrization we use.

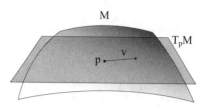

FIGURE 1.5.2
Tangent plane.

We sum up the results we have obtained in the following proposition.

PROPOSITION 1.5.4
The tangent space T_pM, endowed with addition and scalar multiplication operations, is an m-dimensional vector space. A vector basis of this space is $\frac{\partial}{\partial x_1}|_p, \ldots, \frac{\partial}{\partial x_m}|_p$, defined with respect to some local parametrization ϕ near p.

In the case of a regular surface in $\mathbb{R}^3$, the tangent space at a point p is essentially the tangent plane to the surface at p. See Figure 1.5.2. The tangent plane can be identified to a vector subspace of $\mathbb{R}^3$ through a translation that takes p to the origin.

Now we consider the tangent bundle TM, which consists of the all the tangent spaces at points of M glued together in a natural way. Set

$$TM = \bigcup_{p \in M} \{p\} \times T_pM.$$

If $S \subset M$, we define $T_SM = \bigcup_{p \in S}\{p\} \times T_pM$.

PROPOSITION 1.5.5
The tangent bundle TM naturally has the structure of a $2m$-dimensional manifold.

PROOF For each local parametrization $\phi_\alpha : U_\alpha \subseteq \mathbb{R}^m \to V_\alpha \subseteq M$ on M, define $\Phi_\alpha : U_\alpha \times \mathbb{R}^m \to TM$ by

$$\Phi_\alpha(x_1, \ldots, x_m, v_1, \ldots, v_m) = (p, v), \qquad (1.5.6)$$

where p is the point $\phi_\alpha(x_1, \ldots, x_m) \in M$, and v is the vector at p whose local coordinates with respect to ϕ_α are $(v_1, \ldots, v_m)$.

1.5. TANGENT VECTORS AND THE TANGENT BUNDLE

The map Φ_α is a bijection from U_α to $T_{V_\alpha}M$ because ϕ_α is a bijection from U_α to V_α and the local coordinates define a bijection between T_pM and $\mathbb{R}^m$ for each $p \in V_\alpha$. Since the open sets V_α cover M, the sets $T_{V_\alpha}M$ cover TM.

Thus our charts are bijective and their images cover all of TM. It remains to check that the coordinate changes $\Phi_\beta^{-1} \circ \Phi_\alpha$ are diffeomorphisms. Suppose

$$\Phi_\beta(y_1,\ldots,y_m,w_1,\ldots,w_m) = \Phi_\alpha(x_1,\ldots,x_m,v_1,\ldots,v_m).$$

Then

$$(y_1,\ldots,y_m) = (\phi_\beta^{-1} \circ \phi_\alpha)(x_1,\ldots,x_m)$$

and

$$(w_1,\ldots,w_m) = d\left(\phi_\beta^{-1} \circ \phi_\alpha\right)_{\phi_\alpha^{-1}(p)}(v_1,\ldots,v_m).$$

Since $\phi_\beta^{-1} \circ \phi_\alpha$ is a diffeomorphism and $d\left(\phi_\beta^{-1} \circ \phi_\alpha\right)_{\phi_\alpha^{-1}(p)}$ is an invertible linear map of $\mathbb{R}^m$ to itself for each $p \in V_\alpha \cap V_\beta$, it is clear that $\Phi_\beta^{-1} \circ \Phi_\alpha$ is a diffeomorphism. ∎

The local parametrizations of the tangent bundle of a manifold M show that TM locally looks like a product $U_\alpha \times \mathbb{R}^m$. To be more precise, let $\pi : TM \to M$ be the projection that assigns to each bound vector (p,v) its base point p. By definition, this map π is smooth and surjective. Equation (1.5.6) shows that the projection π has the property that $\pi^{-1}(U_\alpha)$ is diffeomorphic to $U_\alpha \times \mathbb{R}^m$. These two properties are used to define the more general notion of a vector bundle, introduced in Section 6.2.

The tangent bundle is a natural concept in mechanics. Consider a system of particles subject to constraints, for example that are bound to a certain surface or they constitute a rigid body. One can always describe simultaneously the relative positions of all particles in the system in terms of some coordinates $(q_1,\ldots,q_n)$, called the generalized coordinates of the system. The set of all possible relative positions of the particles typically forms a manifold, called the configuration space. The dimension of the configuration space equals the number of degrees of freedom of the system. The state of the system evolves in time according to Newton's Second Law of Motion:

$$\frac{d^2q}{dt^2} = F(q,v),$$

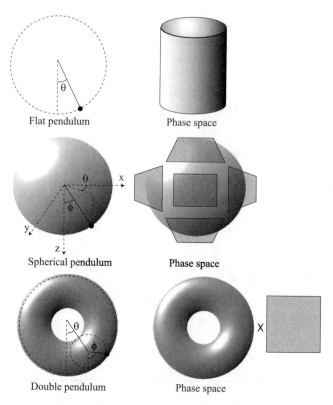

FIGURE 1.5.3
The flat pendulum, the spherical pendulum and the double flat pendulum.

or

$$\frac{dq}{dt} = v, \ \frac{dv}{dt} = F(q,v),$$

where v represents the velocity corresponding to the generalized position vector q and $F(q,v)$ is a force vector. Newton's Second Law of Motion says that the initial position vectors and velocities of the particles of a system determine uniquely its entire evolution in time. The set of all pairs (q,v) is called the phase space of the system, and is represented by the tangent bundle TM.

Example 1.5.6

For a flat pendulum, the position of the bob is determined by the angle made with the vertical direction, so the system has one degree of freedom. The configuration space is a circle. For a spherical pendulum, the position is given by two angles representing the angular spherical coordinates of the position vector, so the system has two degrees of freedom. The configuration space is a sphere. For the double flat pendulum, the position is also given by two angles, so this system has also two degrees of freedom. The configuration space is a torus $\mathbb{T}^2 = S^1 \times S^1$. See Figure 1.5.3.

For the flat pendulum, the phase space is the tangent bundle of a circle, which is diffeomorphic to a cylinder. For the spherical pendulum, the phase space is the tangent bundle of a sphere (which cannot be identified with any well-known shape). For the double flat pendulum the phase space is the tangent bundle of a torus, which is diffeomorphic to the product manifold $S^1 \times S^1 \times \mathbb{R}^2$. See Exercise 1.15.13. □

1.6 Tangent vectors as derivations

We can compute the directional derivative of a function in the direction of a tangent vector. In calculus, the derivative of a function f in the direction of a vector v at a point p is given by $\frac{d}{dt}f(p + t \cdot v)\big|_{t=0}$. More generally, the chain rule tells us that this directional derivative is equal to $\frac{df}{dv} = \frac{d}{dt}f(c(t))\big|_{t=0}$ for any smooth curve c with $c(0) = p$ and $c'(0) = v$.

This idea carries over to a manifold M. Given a smooth mapping $f : M \to \mathbb{R}$, a point p on M, and a tangent vector v to M at p, we want to define the derivative of f in the direction of v at p. Let c be one of the curves with $c(0) = p$ that represents v. The derivative of f in the direction of v is defined by $v(f)(p) = \frac{d}{dt}f(c(t))|_{t=0}$. We leave it as an exercise for the reader to show that this definition is independent of the choice of the curve c representing v.

Example 1.6.1

We now explain the notation $\frac{\partial}{\partial x_i}\big|_p$ introduced in Example 1.5.2. As usual, suppose that $\phi : U \to V$ is a local parametrization around p and $x_1, \ldots, x_m$ are the local coordinate functions induced by ϕ, so that $\phi^{-1}(q) = (x_1(q), \ldots, x_m(q))$ for any point $q \in V$. Recall that $\frac{\partial}{\partial x_i}\big|_p$ is the tangent vector to the curve $t \to \phi(\phi^{-1}(p) + te_i)$. If we move along this curve for time t, the x_i coordinate increases by t and the other coordinates stay constant. Thus $\frac{\partial}{\partial x_i}\big|_p(f)$ is, in a very natural sense, the partial derivative of f in the x_i-direction. □

If $v = (v_1, \ldots, v_m)$ in local coordinates, then

$$v(f) = \sum_{i=1}^{m} v_i \frac{\partial f}{\partial x_i}.$$

The directional derivative $v(f)$ of f at p behaves like the usual derivative operator, meaning that it is linear, i.e., $v(af + bg) = av(f) + bv(g)$, and it satisfies the product rule, i.e., $v(f \cdot g) = f(p)v(g) + g(p)v(f)$. Any linear operator on smooth functions that satisfies these properties is called a derivation at p.

DEFINITION 1.6.2
Let $C^\infty(M)$ be the set of all C^∞ functions from M to $\mathbb{R}$. A derivation at p is a linear map $D : C^\infty(M) \to C^\infty(M)$ satisfying

$$D(fg)(p) = f(p)D(g) + g(p)D(f)$$

for all $f, g \in C^\infty(M)$.

Each tangent vector at p defines a unique derivation at p as the directional derivative in the direction of that vector. We now show that each derivation at p is defined by a unique tangent vector. We need to establish a few simple properties of derivations. First, it is easy to see that $D(1) = 0$, where 1 is the constant function of value 1 on M. This is because $D(1) = D(1 \cdot 1) = 1D(1) + 1D(1) = 2D(1)$. We immediately obtain that $D(c) = 0$ for any constant function c on M.

LEMMA 1.6.3
Suppose that $f = 0$ in a neighborhood of p. Then $Df = 0$.

1.6. TANGENT VECTORS AS DERIVATIONS

PROOF We can choose a C^∞ function g such that $g(p) = 0$ and $g(q) = 1$ whenever $f(q) \neq 0$. Then $f = fg$ and $D(f) = D(fg) = f(p)D(g) + g(p)D(f) = 0 \cdot D(g) + 0 \cdot D(f) = 0$. ∎

COROLLARY 1.6.4
If $f_1 = f_2$ on a neighborhood of p, then $Df_1 = Df_2$.

We can define an equivalence relation on $C^\infty(M)$ at p, by letting $f_1 \sim f_2$ provided $f_1 = f_2$ on some neighborhood of p. The equivalence class of a function f at p is called the germ of f at p. The corollary says that Df is determined by the germ of f at p.

LEMMA 1.6.5
Suppose that $f : M \to \mathbb{R}$ is a C^∞ function on the smooth manifold M, and p is a point in M. Then there are C^∞ functions $g_1, \ldots, g_m$ on M such that $g_i(p) = \partial f / \partial x_i(p)$ for $i = 1, \ldots, m$, and

$$f(q) = f(p) + x_1(q)g_1(q) + \cdots + x_m(q)g_m(q) \qquad (1.6.1)$$

for all q in some small neighborhood of p, where $(x_1, \ldots, x_m)$ represent the local coordinates near p.

PROOF We choose a local coordinate system $\phi : U \to V$ near p, where U is a ball in $\mathbb{R}^m$ with the center at 0, and $\phi(0) = p$. In local coordinates, the map f is described by $f \circ \phi : U \to \mathbb{R}$. Note that $(x_1(p), \ldots, x_m(p)) = 0$.

Using the chain rule, we obtain

$$(f \circ \phi)(x) - (f \circ \phi)(0) = \int_0^1 \frac{\partial (f \circ \phi)}{\partial x_i}(tx)dt = \int_0^1 \sum_{i=1}^m x_i \frac{\partial (f \circ \phi)}{\partial x_i}(tx)dt.$$

Note that for each $x \in U$, $tx \in U$ for all $t \in [0, 1]$, because U was chosen to be a ball. Let

$$h_i(x) = \int_0^1 \frac{\partial (f \circ \phi)}{\partial x_i}(tx)dt,$$

for $i = 1, \ldots, m$. Note that $h_i(0) = \dfrac{\partial (f \circ \phi)}{\partial x_i}(0) = \dfrac{\partial f}{\partial x_i}(p)$.

Then $g_i : V \to \mathbb{R}$ given by $g_i(q) = h \circ \phi^{-1}$ satisfies (1.6.1), for $i = 1, \ldots, m$. Using bump functions, we can extend the functions g_i to M, as in the statement. ∎

Using these properties, we obtain

$$D(f) = D(f(p)) + \sum_{i=1}^{m} D(x_i)g_i(p) + \sum_{i=1}^{m} x_i(p)D(g_i)$$

$$0 + \sum_{i=1}^{m} D(x_i)g_i(p) + \sum_{i=1}^{m} 0 \cdot D(g_i).$$

We conclude

$$D(f) = \sum_{i=1}^{m} D(x_i)\frac{\partial f}{\partial x_i}|_p,$$

that is, $D(f)$ is a linear combination of the derivations $\partial/\partial x_i|_p$, where $i = 1, \ldots, m$. In other words, $D(f)(p)$ is the directional derivative of f at p in the direction of the vector at p with local coordinates $(v_1, \ldots, v_m) = (D(x_1), \ldots, D(x_m))$. We have just shown that any derivation at p corresponds to a tangent vector at p. Clearly, the map that assigns to each tangent vector at p the directional derivative at p in the direction of that vector is linear and onto the space of all derivations at p is a linear isomorphism. In conclusion, we can identify the space of all derivations at p with the tangent space T_pM.

1.7 The derivative of a smooth map

The derivative of a smooth map at a point represents a linear approximation of the map near that point. Let f be a smooth map from an m-dimensional manifold M to an n-dimensional manifold N.

DEFINITION 1.7.1
The derivative of f at a point $p \in M$ is the map $(df)_p : T_pM \to T_{f(p)}N$ defined by

$$(df)_p\left(\frac{dc}{dt}(0)\right) = \frac{d(f \circ c)}{dt}(0),$$

where $c : I \to M$ is a smooth curve in M with $c(0) = p$.

Note that $\frac{dc}{dt}(0)$ represents a tangent vector to M at $p = c(0)$, and

1.7. THE DERIVATIVE OF A SMOOTH MAP

$\frac{d(f \circ c)}{dt}(0)$ represents a tangent vector to N at $f(p)$, since $f \circ c : I \to N$ is a smooth curve in N with $(f \circ c)(0) = f(p)$.

The derivative $(df)_p$ is well defined and linear. To see this, choose local parametrizations ϕ for M around p and ψ for N around $f(p)$. Observe that

$$\left(\psi^{-1} \circ (f \circ c)\right)'(0) = \left(\psi^{-1} \circ f \circ \phi \circ \phi^{-1} \circ c\right)'(0)$$
$$= d\left(\psi^{-1} \circ f \circ \phi\right)_{\phi^{-1}(p)} \left(\phi^{-1} \circ c\right)'(0).$$

This says that the local coordinates for $df(v)$ with respect to ψ are the image under the linear map $d\left(\psi^{-1} \circ f \circ \phi\right)_{\phi^{-1}(p)} : \mathbb{R}^m \to \mathbb{R}^n$ of the local coordinates for v with respect to ϕ. We saw in Section 1.5 that the map which takes a vector to its local coordinates is linear and invertible. By applying this twice, we see that the vector at $f(p)$ whose local coordinates are $\left(\psi^{-1} \circ (f \circ c)\right)'(0)$ is uniquely determined by v and depends linearly on v. Thus $(df)_p(v)$ is well defined and linear.

Applying $(df)_p$ to the vectors of the standard basis $\frac{\partial}{\partial x_1}|_p, \ldots, \frac{\partial}{\partial x_m}|_p$ of $T_p M$, we obtain the vectors $(df)_p \left(\frac{\partial}{\partial x_1}|_p\right), \ldots, (df)_p \left(\frac{\partial}{\partial x_m}|_p\right)$ in $T_{f(p)} N$. We express these vectors as linear combinations of the basis $\frac{\partial}{\partial y_1}|_p, \ldots, \frac{\partial}{\partial y_n}|_p$ of $T_p M$ as

$$(df)_p \left(\frac{\partial}{\partial x_j}|_p\right) = \sum_{i=1}^n a_{ij} \frac{\partial}{\partial y_i}|_{f(p)}$$

for all $j = 1, \ldots, m$. The $n \times m$ matrix made with these coefficients $(Jf)_p = (a_{ij})_{\substack{i=1,\ldots,n \\ j=1,\ldots,m}}$ is called the Jacobi matrix of f at p. The coefficients a_{ij} for j fixed and $i = 1, \ldots, n$ give the j-th column of the matrix. The entries of the matrix depend on the local coordinate systems $(x_1, \ldots, x_m)$ and $(y_1, \ldots, y_n)$ near p and $f(p)$, respectively, but its rank is independent of them. The vectors $(df)_p \left(\frac{\partial}{\partial x_j}|_p\right)$, $j = 1, \ldots, m$, will be a basis of $T_{f(p)} N$ if and only if $m = n$ and the matrix $(Jf)_p$ is invertible.

Thinking of tangent vectors as derivations, the image of a derivation v at p through $(df)_p$ must be a derivation $(df)_p(v)$ at $f(p)$. In this vein, Definition 1.7.1 is equivalent to

$$((df)_p(v))(h)_{f(p)} = v(h \circ f)_p,$$

where h is any smooth real valued function on N.

PROPOSITION 1.7.2
Let $f : M \to N$ and $g : N \to P$ be two smooth maps. Then
$$d(g \circ f)_p = (dg)_{f(p)} \circ (df)_p.$$

PROOF Let ϕ_α, ψ_β and ξ_γ be local parametrizations near p, $f(p)$ and $g(f(p))$, respectively. Our claim reduces to
$$\frac{d}{dt}\left(\xi_\gamma \circ g \circ f \circ c\right) = d\left(\xi_\gamma \circ g \circ \psi_\beta\right) \circ \frac{d}{dt}\left(\psi_\beta^{-1} \circ f \circ c\right),$$
which is the standard chain rule for multi-variable functions. ∎

Example 1.7.3 *(The Gauss map)*

Let M be an embedded surface in $\mathbb{R}^3$. Given a local parametrization $\phi : U_\alpha \to V_\alpha$, for each $p \in V_\alpha$, the tangent vectors $\partial\phi/\partial x_1|_p$, $\partial\phi/\partial x_2|_p$ represent a basis of T_pM. There is a unit normal vector to M at each point of the chart, given by
$$N_p = \frac{\partial\phi/\partial x_1(p) \times \partial\phi/\partial x_2(p)}{\|\partial\phi/\partial x_1(p) \times \partial\phi/\partial x_2(p)\|}. \tag{1.7.1}$$

Here $\times$ denotes the vector product in $\mathbb{R}^3$. Assume that it is possible to extend N_p smoothly to the whole surface M (we will see in Section 1.8 that this not always possible). We define the Gauss map $G : M \to S^2$, which assigns to each point $q \in M$ its unit normal vector N_p,
$$G(p) = N_p.$$

Since N_p varies smoothly with $p \in M$, G is a smooth map on M. The derivative dG of the Gauss map measures the rate of change of the direction of N_p as p moves freely along the surface, so it reflects the amount of curving of the surface at every point. See Figure 1.7.1. An easy linear algebra exercise shows that the determinant of its Jacobi matrix $\det(JG)_p$ is independent of the choice of local coordinates. It is called the Gaussian curvature κ_p of M at p. For a 2-dimensional sphere of radius R, it is clear that the Gauss map is a linear contraction by a factor of $1/R$, so
$$(JG)_p = \begin{pmatrix} 1/R & 0 \\ 0 & 1/R \end{pmatrix}.$$

1.8. ORIENTATION

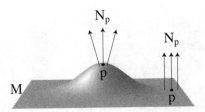

FIGURE 1.7.1
The rate of change of the unit normal vector as a measure of curvature.

Hence $\kappa_p = 1/R^2$ at every point. For a plane, the Gauss map is constant, so $dG = 0$, and $\kappa_p = 0$ at every point. This corresponds to our intuitive perception of curvature, that a larger sphere is less curved than a smaller sphere, while a plane is not curved at all. □

1.8 Orientation

A surface is orientable if it has 'two sides', and is nonorientable if it has only 'one side'. Suppose that M is an embedded surface in $\mathbb{R}^3$. A unit normal vector at p is a vector N_p of length equal to 1, which is perpendicular to T_pM. If ϕ is a local parametrization near a point p, then the tangent vectors $\partial \phi / \partial x_1$, $\partial \phi / \partial x_2$ represent a basis of T_pM, so

$$N_p = \pm \frac{\partial \phi / \partial x_1(p) \times \partial \phi / \partial x_2(p)}{\|\partial \phi / \partial x_1(p) \times \partial \phi / \partial x_2(p)\|}. \tag{1.8.1}$$

Here $\times$ denotes the vector product in $\mathbb{R}^3$. A surface is called orientable if one can continuously assign a unique unit normal vector N_p to each point p in M. For example, the 2-dimensional torus is orientable. The Möbius strip is not orientable (the Möbius strip is defined as the quotient set $[0,1] \times \mathbb{R}/\simeq$ by the equivalence relation $(0,a) \simeq (1,-a)$, with the induced smooth structure). See Figure 1.8.1 (a) and (b).

The two possible choices of unit normal vectors can be obtained one

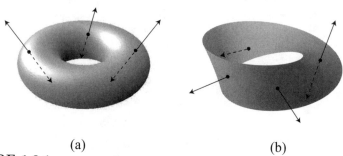

(a)   (b)

FIGURE 1.8.1
The torus and the Möbius strip.

from the other by simply switching the order of the basis vectors in the vector product in (1.8.1). We can choose and fix an ordered basis of T_pM and declare it as positively oriented. Any other ordered basis of T_pM will be assigned a positive orientation provided that the matrix A associated with the change of basis has positive determinant, and it will be assigned a negative orientation otherwise. An embedded surface M is orientable provided that there exists a smooth structure $\{\phi_\alpha\}_\alpha$ on M such that all bases $\{\partial\phi_\alpha/\partial x_1(p), \partial\phi_\alpha/\partial x_2(p)\}$, for all $p \in M$, have the same orientation, whether positive or negative. Note that the change of basis matrix from $\{\partial\phi_\beta/\partial x_1(p), \partial\phi_\beta/\partial x_2(p)\}$ to $\{\partial\phi_\alpha/\partial x_1(p), \partial\phi_\alpha/\partial x_2(p)\}$ is given by the derivative $d\theta_{\alpha\beta}$ of the coordinate change.

We extend the concept of orientability to abstract manifolds.

DEFINITION 1.8.1
A smooth manifold M is said to be orientable if there exists a smooth structure $\{\phi_\alpha\}_\alpha$ on M such that for every pair α, β with $V_\alpha \cap V_\beta \neq \emptyset$ the derivative $d\theta_{\alpha\beta} = d(\phi_\alpha^{-1} \circ \phi_\beta)$ of the smooth coordinate change has a positive determinant. If such a smooth structure does not exist, M is said to be nonorientable.

Orientability is a global property of a manifold. Every manifold is 'locally orientable': for every point $p \in M$ there exists an open neighborhood V of p that is orientable as a submanifold of M.

Every connected, orientable manifold has exactly two distinct orientations. Once we fix a choice of a positive orientation, we call that manifold oriented. The Euclidean space $\mathbb{R}^n$ is orientable, and the standard basis

1.8. ORIENTATION

$\{e_1, \ldots, e_n\}$ of $\mathbb{R}^n$ gives the standard positive orientation.

Orientability is preserved by diffeomorphisms.

PROPOSITION 1.8.2
Let $f : M \to N$ be a diffeomorphism between two smooth manifolds. Then M is orientable if and only if N is orientable.

PROOF   If $\{\phi_\alpha\}_\alpha$ is a smooth structure of M, since f is a diffeomorphism, it results that $\{f \circ \phi_\alpha\}_\alpha$ defines a smooth structure of N. The corresponding coordinate changes $(f \circ \phi_\alpha)^{-1} \circ (f \circ \phi_\beta)$ coincide with $\phi_\alpha^{-1} \circ \phi_\beta$. Hence M is orientable if and only if N is orientable. ∎

Example 1.8.3
The n-dimensional unit sphere S^n is orientable. For each $p \in S^n$, we declare a basis $(v_1, \ldots, v_n)$ of $T_p S^n$ as positively oriented provided $(p, v_1, \ldots, v_n)$ has the same orientation as the standard basis of $\mathbb{R}^{n+1}$. The position vector p represents the outward unit normal vector to S^n.
☐

If M and N are oriented, the diffeomorphism f induces an orientation on N which may or may not agree with the positive orientation on N. If the orientation induced by f is still positive, then we say that f is orientation preserving; otherwise, we say that f is orientation reversing.

Example 1.8.4
We show that the real projective space $\mathbb{RP}^n$ is orientable if and only if n is odd. The antipodal map $A : \mathbb{R}^{n+1} \to \mathbb{R}^{n+1}$, defined by $A(p) = -p$, is orientation preserving if n is odd and orientation reversing if n is even, since the associated change of basis matrix is $-I$, where I is the $n \times n$ identity matrix. The restriction of A to the unit sphere S^n is still orientation preserving if n is odd and orientation reversing if n is even. Indeed, the orientation of $(v_1, \ldots, v_n)$ at p is the same as that of $(p, v_1, \ldots, v_n)$, and so the orientation of $A(p, v_1, \ldots, v_n) = (-p, -v_1, \ldots, -v_n)$ at $-p$ is the same as that of $(-v_1, \ldots, -v_n)$.

Assume by contradiction that n is even and $\mathbb{RP}^n$ has some orientation. Let $\pi : S^n \to \mathbb{RP}^n$ be the canonical projection that assigns to each point $p \in S^n$ its equivalence class $[p] \in \mathbb{RP}^n$. The orientation on $\mathbb{RP}^n$

induces an orientation on S^n by declaring a basis $(v_1, \ldots, v_n)$ of $T_p S^n$ as positively oriented if its image $((d\pi)_p(v_1), \ldots, (d\pi)_p(v_n)) \subseteq T_{\pi(p)}(\mathbb{RP}^n)$ is positively oriented. Since the map induced by A on $\mathbb{RP}^n$ is the identity map, it follows that A itself is orientation preserving on S^n, which is a contradiction.

If n is odd, we induce an orientation on $\mathbb{RP}^n$ by the orientation on S^n. A basis $(w_1, \ldots, w_n)$ of $T_{\pi(p)}(\mathbb{RP}^n)$ is declared as positively oriented provided that there is a positively oriented basis $(v_1, \ldots, v_n)$ of $T_p S^n$ whose image under $(d\pi)_p$ is exactly $(w_1, \ldots, w_n)$. The choice of orientation of $T_{\pi(p)}(\mathbb{RP}^n)$ is unambiguous at $\pi(p) = \pi(-p)$, since the antipodal map is orientation preserving for n odd. $\square$

1.9 Immersions, embeddings and submersions

We will frequently use the the implicit function theorem and the inverse function theorem from calculus. We state them below, and we recommend Spivak (1965) for details.

THEOREM 1.9.1 (Inverse function theorem)

Suppose that $f : U \to \mathbb{R}^n$ is a C^k-smooth map, with U an open subset of $\mathbb{R}^n$. Let p be a point in U such that $(df)_p$ is non-singular. Then f is a local C^k-diffeomorphism at p.

THEOREM 1.9.2 (Implicit function theorem)

Let U be an open subset of $\mathbb{R}^m$ and $g : U \subseteq \mathbb{R}^n \times \mathbb{R}^{m-n} \to \mathbb{R}^n$ be a C^k-smooth function, where $m > n$. Assume that for some $p = (x_0, y_0) \in U$, we have $g(p) = 0$ and the derivative of g at p restricted to the first factor is a linear isomorphism. Then there exist an open neighborhood $V \times W \subseteq U$ of p and a C^k-smooth map $h : W \to V$ satisfying $h(y_0) = x_0$ and $g(h(y), y) = 0$ for all $y \in W$.

PROOF Define the mapping $f : U \subseteq \mathbb{R}^n \times \mathbb{R}^{m-n} \to \mathbb{R}^n \times \mathbb{R}^{m-n}$

1.9. IMMERSIONS, EMBEDDINGS AND SUBMERSIONS

by $f(x,y) = (g(x,y), y)$. Note that $f(x_0, y_0) = (0, y_0)$ and that

$$(df)_p = \begin{pmatrix} \frac{\partial g}{\partial x}(x_0, y_0) & \frac{\partial g}{\partial y}(x_0, y_0) \\ 0 & I_{m-n} \end{pmatrix},$$

where I_{m-n} denotes the $(m-n) \times (m-n)$ identity matrix, and $\partial g/\partial x$ and $\partial g/\partial y$ denote the derivative of g restricted to the first factor, and to the second factor, respectively. Since $(\partial g/\partial x)(x_0, y_0)$ is non-singular, $(df)_p$ is a linear isomorphism, so there exists a smooth inverse f^{-1} of f defined in some small neighborhood $V \times W$ of (x_0, y_0). We can write $f^{-1} = (F^1, F^2)$ for some smooth mappings $F^1 : V \times W \to \mathbb{R}^n$ and $F^2 : V \times W \to \mathbb{R}^{m-n}$, with $F^1(0, y_0) = x_0$ and $F^2(0, y_0) = y_0$. The inverse mapping condition $f(f^{-1}(x,y)) = (x,y)$ translates into

$$\left(g\left(F^1(x,y), F^2(x,y)\right), F^2(x,y)\right) = (x,y),$$

for all $(x,y) \in V \times W$. Hence $F^2(x,y) = y$, so $g(F^1(x,y), y) = x$ for all $x \in V$ and all $y \in W$. Letting $x = 0$, we obtain the smooth function $h(y) = F^1(0,y)$ satisfying $h(y_0) = x_0$ and $g(h(y), y) = 0$ for all $y \in W$.

The implicit function theorem can be interpreted as follows. There is given a set of points (x,y), containing p, on which $g(x,y) = 0$. The linear equation associated to $g(x,y) = 0$ near (x_0, y_0) is

$$g(x_0, y_0) + \frac{\partial g}{\partial x}(x_0, y_0)(x - x_0) + \frac{\partial g}{\partial y}(x_0, y_0)(y - y_0) = 0, \qquad (1.9.1)$$

where $\frac{\partial g}{\partial x}(x_0, y_0)$ is the $n \times n$ matrix of partial derivative of g relative to the first factor, and $\frac{\partial g}{\partial y}(x_0, y_0)$ is the $n \times (m-n)$ matrix of partial derivative of g relative to the second factor. The condition that the derivative of g at p restricted to the first factor is a linear isomorphism means that the above linear equation can be solved for x with respect to y. The theorem says that, near (x_0, y_0), the non-linear equation $g(x,y) = 0$ can be locally solved for x as a smooth function of y. Thus, the level set $g(x,y) = 0$ can be locally represented as the graph of a smooth function $x = h(y)$. Moreover, the equation of the tangent plane to this level set at (x_0, y_0) is given by (1.9.1).

If M and N are two smooth manifolds of the same dimension, a smooth map $f : M \to N$ is called a local diffeomorphism at $p \in M$ if f takes some open neighborhood V of p diffeomorphically onto some open neighborhood W of $f(p)$.

PROPOSITION 1.9.3
Suppose that M and N are two smooth manifolds of the same dimension, $f : M \to N$ is a smooth map, and p is a point in M with $(df)_p$ non-singular. Then f is a local diffeomorphism at p.

PROOF Let ϕ and ψ be local parametrizations near p and $f(p)$, respectively. Since $(df)_p : T_p M \to T_{f(p)} N$ is an invertible linear map, we derive that $\psi^{-1} \circ f \circ \phi$ is a map from an open subset of $\mathbb{R}^m$ to an open subset of $\mathbb{R}^m$ which has invertible derivative at $\phi^{-1}(p)$. Now we use the above version of the inverse function theorem, obtaining that $\psi^{-1} \circ f \circ \phi$ maps some smaller open neighborhood U_1 of $\phi^{-1}(p)$ diffeomorphically onto a smaller neighborhood U_2 of $\psi^{-1}(f^{-1}(p))$. Letting $V = \phi(U_1)$ and $W = \psi(U_2)$, we conclude that f is a diffeomorphism from V to W. ∎

We will now explore the consequences of the injectivity of the derivative and of the surjectivity of the derivative.

DEFINITION 1.9.4
Suppose that M is an m-dimensional smooth manifold, N is an n-dimensional smooth manifold, and $f : M \to N$ is a smooth map.

 (i) *The map f is called an immersion if $(df)_p$ is injective at each point p. That is, the rank of the Jacobian $(Jf)_p$ is equal to m at every point $p \in M$.*

 (ii) *The map f is called an embedding if f is an immersion and f defines a homeomorphism from M to $f(M)$ (with the induced topology from N).*

 (iii) *The map f is a submersion if $(df)_p$ is surjective at each point p. That is, the rank of the Jacobian $(Jf)_p$ is equal to n at every point $p \in M$.*

Note that for an immersion we must have $\dim M \leq \dim N$, for an embedding we must have $\dim M = \dim N$, and for a submersion we must have $\dim M \geq \dim N$.

1.9. IMMERSIONS, EMBEDDINGS AND SUBMERSIONS

As an example, let $M = \mathbb{R}^m$ and $N = \mathbb{R}^n$. If $m \leq n$, the mapping

$$(x_1, \ldots, x_m) \in \mathbb{R}^m \to (x_1, \ldots, x_m, 0, \ldots, 0) \in \mathbb{R}^n$$

is an immersion (referred as the canonical immersion), and also an embedding. If $m \geq n$, the mapping

$$(x_1, \ldots, x_n, x_{n+1}, \ldots, x_m) \in \mathbb{R}^m \to (x_1, \ldots, x_n) \in \mathbb{R}^n$$

is a submersion (referred as the canonical submersion).

The following theorem states that any immersion is locally the same as the canonical immersion.

THEOREM 1.9.5 (*Immersion theorem*)
If $f : M \to N$ is an immersion, then for every point $p \in M$ there exist a local parametrization $\phi : U \to V$ near p and a local parametrization $\psi : \bar{U} \to \bar{V}$ near $f(p)$ such that

$$(\psi^{-1} \circ f \circ \phi)(x_1, \ldots, x_m) = (x_1, \ldots, x_m, 0, \ldots, 0).$$

PROOF The proof is based on the inverse function theorem, and is similar to the proof of the implicit function theorem from above.

We start with a local parametrization $\phi_0 : U_0 \to V_0$ near p with $\phi_0^{-1}(p) = 0$, and a local parametrization $\psi_0 : \bar{U}_0 \to \bar{V}_0$ near $f(p)$ with $\psi_0^{-1}(f(p)) = 0$. For simplicity, we also assume that $f(V_0) \subseteq \bar{V}_0$. When the map $\psi_0^{-1} \circ f \circ \phi_0$ is written in terms of its components,

$$(\psi_0^{-1} \circ f \circ \phi_0)(x_1, \ldots, x_m) = (F_1(x_1, \ldots, x_m), \ldots, F_n(x_1, \ldots, x_m))$$

the immersion condition means that the $n \times m$ Jacobi matrix

$$\frac{\partial(F_1, \ldots, F_m, \ldots, F_n)}{\partial(x_1, \ldots, x_m)}(0),$$

has an invertible $m \times m$ submatrix. By rearranging the order of the components, if necessary, we can assume that

$$\frac{\partial(F_1, \ldots, F_m)}{\partial(x_1, \ldots, x_m)}(0) \text{ is invertible.}$$

In preparation for using the inverse function theorem, we define a new function $G : U_0 \times \mathbb{R}^{n-m} \to \mathbb{R}^n$ by

$$G(x_1, \ldots, x_n) = (F_1, \ldots, F_n)(x_1, \ldots, x_m) + (0, \ldots, 0, x_{m+1}, \ldots, x_n).$$

Note that the dimensions of the domain and codomain are equal. The Jacobi matrix of G at 0

$$(JG)_0 = \begin{pmatrix} \dfrac{\partial(F_1,\ldots,F_m)}{\partial(x_1,\ldots,x_m)}(0) & 0 \\ * & I_{n-m} \end{pmatrix},$$

is invertible. By the inverse function theorem, there exist neighborhoods $U_1 \times B^{n-m}(0,\epsilon)$ and U_2 of 0 in $\mathbb{R}^m$ such that G is a diffeomorphism from $U_1 \times B^{n-m}(0,\epsilon)$ to U_2. Now we define a local parametrization ϕ as the restriction of ϕ_0 to U_1. Since G is a diffeomorphism and $G(0) = (F_1,\ldots,F_n)(0) = \psi_0^{-1}(f(p))$, then we can define a local parametrization ψ near $f(p)$ as the restriction of $\psi_0 \circ G$ to $U_1 \times B^{n-m}(0,\epsilon)$. The local parametrizations ϕ and ψ are compatible to ϕ_0 and ψ_0, respectively. We have

$$(\psi_0^{-1} \circ f \circ \phi_0)(x_1,\ldots,x_m) = (F_1,\ldots,F_n)(x_1,\ldots,x_m)$$
$$= G(x_1,\ldots,x_m,0,\ldots,0).$$

Substituting $\psi = \psi_0 \circ G$, we obtain

$$\psi^{-1} \circ f \circ \phi(x_1,\ldots,x_m) = G^{-1} \circ \left(\psi_0^{-1} \circ f \circ \phi_0\right)(x_1,\ldots,x_m)$$
$$= (x_1,\ldots,x_m,0,\ldots,0).$$

∎

COROLLARY 1.9.6
If $f : M \to N$ is an immersion, then for every point $p \in M$ there exists an open neighborhood U of p such that the restriction of f to U is an embedding in N.

PROOF The previous theorem shows that f is a diffeomorphism (hence homeomorphism) from V onto $f(V)$, whose expression in some local coordinates is the identity $\text{id}_{\mathbb{R}^m}$. ∎

The image $f(M)$ of a manifold M through an immersion $f : M \to N$ needs not be a submanifold of N.

Example 1.9.7
Let f be an immersion which takes the circle S^1 onto the 4-leaf clover $f(M) \subseteq \mathbb{R}^2$, given by $f(\cos(t),\sin(t)) = (\sin(2t)\cos(t),\sin(2t)\sin(t))$.

1.9. IMMERSIONS, EMBEDDINGS AND SUBMERSIONS

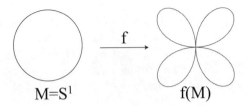

FIGURE 1.9.1
An immersed circle.

FIGURE 1.9.2
An immersed submanifold of the torus.

See Figure 1.9.1. Due to self-intersections, $f(M)$ is not a submanifold of $\mathbb{R}^2$.

□

Even if f is injective, the image of a manifold through an immersion may not be a submanifold.

Example 1.9.8
Let us consider the torus $\mathbb{T}^2 = \mathbb{R}^2/\mathbb{Z}^2$. Let us also consider the map $f: M = \mathbb{R} \to \mathbb{T}^2$ defined by $f(t) = [t, \lambda t]$ with λ irrational, whose image is the line of irrational slope on $\mathbb{T}^2$. The Jacobi matrix of f is given by $(Jf)_p$ is

$$(Jf)_p = (1 \quad \lambda)$$

at each point $p \in \mathbb{R}$. Thus f is an immersion. The image of f is dense because the line wraps around the torus infinitely often. It is possible for $f(t_1)$ and $f(t_2)$ to be very close, even if t_1 and t_2 are far apart in $\mathbb{R}$. See Figure 1.9.2. Thus $f(\mathbb{R})$ is not a submanifold of $\mathbb{T}^2$, since every small neighborhood of any point $q \in f(\mathbb{R})$ in $\mathbb{T}^2$ contains a dense set of arcs.

Note that the set $f(\mathbb{R})$ can be induced with the smooth structure of the real line through the map f. With this induced smooth structure,

$f(\mathbb{R})$ is diffeomorphic to $\mathbb{R}$. This smooth structure is not inherited from the smooth structure of $\mathbb{T}^2$, so $f(\mathbb{R})$ is not a submanifold of $\mathbb{T}^2$.

Consequently, there are two natural topologies on $f(\mathbb{R})$: one induced by $\mathbb{R}$ through f, which is the topology of a smooth structure on $f(\mathbb{R})$, and one inherited from the topology of $\mathbb{T}^2$, which is not the topology of any smooth structure on $f(\mathbb{R})$. □

The image $f(M)$ of an injective immersion into N, together with the smooth structure induced by f, is called an immersed submanifold of N. We emphasize that it is not, in general, a submanifold.

The previous example shows that the injectivity of the derivative of a mapping of a manifold does not ensure that the image of the mapping is a submanifold. In order for the image to be a submanifold, one needs to prevent any dense wrapping, in other words points 'near infinity' should always be mapped into point 'near infinity'. For this, it suffices to require that the inverse image through f of any compact set is also a compact set. A continuous mapping with this property is said to be proper. Equivalently, f has to define a homeomorphism onto its image.

PROPOSITION 1.9.9
Let $f : M \to N$ be an injective immersion. Then f is a homeomorphism from M to $f(M)$ if and only if f is proper.

PROOF If f is a homeomorphism, then f^{-1} is continuous and so $f^{-1}(K)$ is compact for any compact subset K of N.

Conversely, assume that f is proper. In order to show that f is a homeomorphism, it is enough to prove that f maps closed subsets of M to closed subsets of N. Assume, by contradiction, that F is closed in M but $f(F)$ is not closed in N. Then there exists a convergent sequence of points $\{f(x_n)\}_{n \geq 0}$ in $f(F)$ whose limit y is not in $f(F)$. Since f is injective, $f^{-1}(y)$ is a single point x. The set $\{f(x_n) \mid n \geq 0\} \cup \{y\}$ is compact in N, hence its inverse image $\{x_n \mid n \geq 0\} \cup \{x\}$ is compact in M. It follows that $\{x_n\}_{n \geq 0}$ has a convergent subsequence, whose limit is in F. Due to the continuity and injectivity of f, this limit must be x. We obtain $f(x) = y \in f(F)$, which is a contradiction. ■

THEOREM 1.9.10 (Embedding theorem)
If $f : M \to N$ is an embedding, then the image $f(M)$ with the smooth structure induced through f is a submanifold of N.

1.9. IMMERSIONS, EMBEDDINGS AND SUBMERSIONS

PROOF Consider a smooth structure $\{\phi_\alpha\}$ of M. The smooth structure induced on $f(M)$ through f consists of the local parametrizations $f \circ \phi_\alpha : U_\alpha \to f(V_\alpha)$. We want to show that this is the same as the smooth structure induced by N. Pick a point $p \in M$, a local parametrization ϕ_α of M near p, and a local parametrization $\psi_\beta : U_\beta \to V_\beta$ of N near $f(p)$, as in Theorem 1.9.5. We know that f defines a homeomorphism from M to $f(M)$, when $f(M)$ is endowed with the topology induced by N. Thus, the set $f(V_\alpha)$ is relatively open in $f(M)$, so it can be written as $f(M) \cap V$, for some open set V in N. By restricting ψ_β, if necessary, we can assume that $V_\beta \subseteq V$. We have

$$\psi_\beta^{-1} \circ (f \circ \phi_\alpha)(x_1, \ldots, x_m) = (x_1, \ldots, x_m, 0, \ldots, 0),$$

so ψ_β restricted to $U_\beta \cap (\mathbb{R}^m \times 0)$ is a local parametrization near $f(p)$, compatible to ϕ_α. Since we can work this out at every $f(p) \in f(M)$, we conclude that $f(M)$ is a submanifold of N. ∎

In particular, every submanifold is the image of the canonical inclusion, which is clearly an embedding.

Now we show that every submersion is locally the same as the canonical submersion.

THEOREM 1.9.11 *(Submersion theorem)*
If $f : M \to N$ is a submersion, then for every point $p \in M$ there exist a local parametrization $\phi : U \to V$ near p and a local parametrization $\psi : \bar{U} \to \bar{V}$ near $f(p)$ such that

$$(\psi^{-1} \circ f \circ \phi)(x_1, \ldots, x_n, x_{n+1}, \ldots, x_m) = (x_1, \ldots, x_n).$$

PROOF The proof is very similar to the one for immersions. We choose some local parametrizations ϕ_0 near p and ψ_0 near $f(p)$ as in the proof of Theorem 1.9.5, so

$$(\psi_0^{-1} \circ f \circ \phi_0)(x_1, \ldots, x_m) = (F_1(x_1, \ldots, x_m), \ldots, F_n(x_1, \ldots, x_m)).$$

Using the condition on the rank of $(JF)_p$ and rearranging the local coordinates of ψ_0, if necessary, we can assume that

$$\frac{\partial(F_1, \ldots, F_n)}{\partial(x_1, \ldots, x_n)}(0) \text{ is invertible.}$$

In order to apply the inverse function theorem, we add some extra dimensions to the co-domain, so we define

$$G(x_1, \ldots, x_m) = (F_1(x_1, \ldots, x_m), \ldots, F_n(x_1, \ldots, x_m), x_{n+1}, \ldots, x_m).$$

Since $(JG)_0$ is invertible, G is a local diffeomorphism at 0. We define the new local parametrizations ϕ near p and ψ near $f(p)$ as the appropriate restrictions of $\phi_0 \circ G^{-1}$ and ψ_0, respectively. We obtain

$$(\psi^{-1} \circ f \circ \phi)(x_1, \ldots, x_m) = (\psi_0^{-1} \circ f \circ \phi_0) \circ G^{-1}(x_1, \ldots, x_m)$$
$$= (x_1, \ldots, x_n).$$

∎

The statements of the immersion theorem and submersion theorem can be incorporated in one general statement, referred as the rank theorem. See Exercise 1.15.13.

We end this section with a remarkable result. It states that every 'abstract' manifold is an 'embedded' manifold in some Euclidean space.

THEOREM 1.9.12 (Whitney embedding theorem)
Every C^k-smooth ($k \geq 2$) m-dimensional manifold can be embedded in $\mathbb{R}^{2m}$.

A weaker version of the theorem, namely that every compact m-dimensional manifold can be embedded in $\mathbb{R}^{2m+1}$, is proved in Hirsch (1976). A proof of the strong version can be found in Whitney (1944).

Whitney's theorem is difficult, but it is not hard to show the existence of an embedding in some higher dimensional Euclidean space. For simplicity, we assume that M is compact. There exists a finite covering $\{V_1, \ldots, V_n\}$ of M by charts $\phi_i^{-1} : V_i \to B^m(0, 1) \subseteq \mathbb{R}^m$, $i = 1, \ldots, n$, with each V_i homeomorphic to the unit ball in $\mathbb{R}^m$. There exists a smooth partition of unity $\{f_1, \ldots, f_n\}$ with $\text{supp}(f_i) \subseteq V_i$. We extend ϕ_i^{-1} to the whole manifold by making it 0 outside the chart

$$h_i(p) = \begin{cases} f_i(p)\phi_i^{-1}(p), & \text{for } p \in V_i, \\ 0, & \text{for } p \notin V_i. \end{cases}$$

We define $h : M \to \mathbb{R}^n \times (\mathbb{R}^m)^n = \mathbb{R}^{(m+1)n}$ by

$$h(p) = (f_1(p), \ldots, f_n(p), h_1(p), \ldots, h_n(p)).$$

1.10. REGULAR AND CRITICAL POINTS AND VALUES

It is easy to see that h is smooth. We want to check that it is injective. If $h(p) = h(q)$, then $f_i(p) = f_i(q)$ and $\phi_i^{-1}(p) = \phi_i^{-1}(q)$ for all i. There exists j with $f_j(p) = f_j(q) > 0$. Hence p, q are in the same chart V_j, so $h_j(p) = h_j(q)$ implies $\phi_j^{-1}(p) = \phi_j^{-1}(q)$, thus $p = q$. Since any injective continuous map from a compact space to a Hausdorff space is a homeomorphism onto its image (see Munkres (2000)), we obtain that h is an embedding.

Although Whitney's theorem ensures the existence of embeddings of abstract manifolds in Euclidean spaces, some of the embeddings can be too complicated for any practical purposes. See Exercise 1.15.17.

1.10 Regular and critical points and values

Critical points and values are used in calculus in optimization problems. We now start to explore their role in differential geometry. We will continue this exploration in Chapter 8.

DEFINITION 1.10.1
Let f be a smooth map from the m-dimensional manifold M to the n-dimensional manifold N.

(i) A point p in M is called a critical point of f if the derivative $(df)_p : T_pM \to T_{f(p)}N$ is not surjective. That is, the rank of the Jacobi matrix $(Jf)_p$ is less than the dimension n of N. The image $f(p)$ of a critical point is called a critical value of f.

(ii) A point p in M is called a regular point of f if it is not critical. A point q in N is called a regular value of f if its inverse image $f^{-1}(q)$ contains no critical points.

Note that if the dimension of M is smaller than the dimension of N, then all points of M are critical points for f. On the other hand, if $f(M)$ is not all of N, then all points of $N \setminus f(M)$ are regular values, since their inverse images are empty. In other words, non-values are regular values.

Now we would like to investigate whether the inverse image of a point in N is a submanifold of M. In Section 1.13 we will investigate whether the inverse image of a submanifold P of N is a submanifold of M.

THEOREM 1.10.2 (Inverse image of a regular value)
The inverse image $f^{-1}(q)$ of a regular value $q \in N$ of a smooth map $f : M \to N$ is a submanifold of dimension $(m - n)$ (and so of codimension n) of M, unless it is empty.

PROOF For a given point $p \in f^{-1}(q)$, Theorem 1.9.11 allows us to choose local parametrizations ϕ near p and ψ near q such that

$$\psi^{-1} \circ f \circ \phi(x_1, \ldots, x_n, x_{n+1}, \ldots, x_m) = (x_1, \ldots, x_n)$$

and the local coordinates of p and q are 0. The points $p' \in U \cap f^{-1}(q)$ have local coordinates $(0, \ldots, 0, x_{n+1}, \ldots, x_m)$. So we can define a local parametrization $\phi' : U \cap (0 \times \mathbb{R}^{m-n}) \to V \cap f^{-1}(q)$ on $f^{-1}(q)$, near p, by

$$\phi'(x_{n+1}, \ldots, x_m) = \phi(0, \ldots, 0, x_{n+1}, \ldots, x_m).$$

In this way, we obtain a smooth structure on $f^{-1}(q)$, induced by the smooth structure on M, which makes $f^{-1}(q)$ an $(m - n)$-dimensional submanifold. ■

Example 1.10.3
If q is a critical value, $f^{-1}(q)$ does not need to be a submanifold. For $f : \mathbb{R}^2 \to \mathbb{R}$ given by $f(x, y) = x^2 - y^2$, $(df)_{(x_0, y_0)} = (2x_0, 2y_0)$, hence $0 = f(0, 0)$ is the only critical value. The inverse image $f^{-1}(0)$ consists of two intersecting lines $y = \pm x$, so it is not a submanifold of $\mathbb{R}^2$. However, for $q \neq 0$, $f^{-1}(q)$ consists of a pair of hyperbolas. See Figure 1.10.1 □

REMARK 1.10.4 The inverse image of a critical value has, in general, no structure of any kind. One can prove that any closed subset of a manifold is the set of zeroes of some differentiable real valued function. This result is due to Whitney and its proof can be found in Bröcker and Jänich (1982). ■

Example 1.10.5
Theorem 1.10.2 provides us with a convenient method to show that many of the familiar surfaces in $\mathbb{R}^3$ are submanifolds of $\mathbb{R}^3$.

Let $f : \mathbb{R}^3 \to \mathbb{R}$ be the function given by $f(x, y, z) = x^2 + y^2 + z^2 - R^2$. The only critical point of f is $(0, 0, 0)$ so the only critical value of f is 0. Thus, for each $R > 0$, the sphere $f^{-1}(0) = \{(x, y, z) \mid x^2 + y^2 + z^2 = R^2\}$

1.10. REGULAR AND CRITICAL POINTS AND VALUES 47

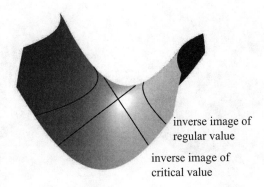

inverse image of regular value

inverse image of critical value

FIGURE 1.10.1
Inverse image of critical and regular values.

is a 2-dimensional submanifold of $\mathbb{R}^3$. Similarly, the function $g(x,y,z) = x^2/a^2 + y^2/b^2 - z^2/c^2 - 1$ has 0 as its only critical value, so the hyperboloid of one sheet $g^{-1}(0) = \{(x,y,z) \mid x^2/a^2 + y^2/b^2 - z^2/c^2 = 1\}$ is a 2-dimensional submanifold of $\mathbb{R}^3$. Also the 2-dimensional torus $\mathbb{T}^2$ can be realized as the inverse image of a regular value. If $h : \mathbb{R}^3 \to \mathbb{R}$ is defined by $h(x,y,z) = (x^2 + y^2 + z^2 - (R^2 + r^2))^2 - 4R^2(r^2 - z^2)$, then 0 is a regular value for h and $\mathbb{T}^2 = h^{-1}(0)$. □

REMARK 1.10.6 A surface of genus g (or a g-holed torus) is an embedded surface homeomorphic to the space obtained from the sphere S^2 by cutting off $2g$ disjoint disks from the surface of the sphere and gluing g cylinders to the boundary circles. See Figure 1.10.2. Thus, a sphere is a surface of genus 0 and a torus is a surface of genus 1. The surfaces of genus g, with $g = 0, 1, 2, \ldots$ represent all compact orientable surfaces. Every surface of genus g is the inverse image of a regular value for some polynomial function on $\mathbb{R}^3$. For example, for $f(x,y,z) = [4x^2(1-x^2) - y^2]^2 + (2z)^2 - 1/5$, we have that 0 is a regular value for f, and $f^{-1}(0)$ represents a surface of genus 2. See Hirsch (1976) for details. ∎

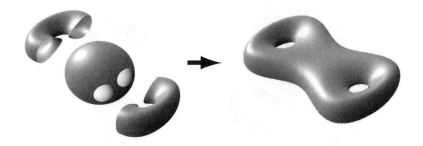

FIGURE 1.10.2
A surface of genus 2.

1.11 Manifolds with boundary

We now discuss manifolds that have boundary points. An example of a manifold with boundary is the closed unit ball in the Euclidean space. The closed unit ball fails to be a manifold in the usual sense since the points lying on the boundary sphere do not have neighborhoods within the ball homeomorphic to open sets in the Euclidean space. However, these boundary points have neighborhoods in the ball homeomorphic to relatively open sets in the half-space $\{(x_1, x_2, x_3) \mid x_3 \geq 0\}$.

DEFINITION 1.11.1
A smooth m-dimensional manifold with boundary is a set M together with a collection of maps $\phi_\alpha : U_\alpha \to M$, with U_α a relatively open subset of the m-half-space $\mathbb{H}^m = \{(x_1, \ldots, x_m) \mid x_m \geq 0\}$, satisfying the conditions in the definition of a smooth manifold — Definition 1.3.1.

See Figure 1.11.1.

A point $p \in M$ that corresponds through a chart to a point with $x_m > 0$ in the m-half-space $\mathbb{H}^m$, is called an interior point of M. The

1.11. MANIFOLDS WITH BOUNDARY

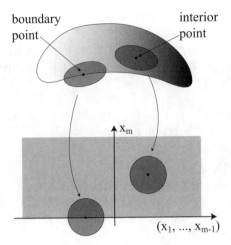

FIGURE 1.11.1
A manifold with boundary.

condition of being an interior point is independent of the chart, as it follows from the inverse function theorem. The interior points form an m-dimensional manifold.

A point $p \in M$ that corresponds through a chart to a point with $x_m = 0$ in $\mathbb{H}^m$ is called a boundary point of M. This condition is also independent of the chart. The collection of all boundary points of M is called the boundary of M and is denoted by ∂M. This is, in general, different from the topological boundary of a set: for example, in the case of an open interval $M = (a,b) \subseteq \mathbb{R}$, the topological boundary $\mathrm{bd}(a,b) = \{a,b\}$ while $\partial M = \emptyset$.

As in the case of manifolds without boundary, we assume that the topology of a manifold with boundary is Hausdorff and has a countable basis.

The definitions of a tangent vector and of a tangent space to a manifold with boundary remain the same, even at boundary points. The definitions of smooth maps between manifolds with boundary, immersions, submersions and embeddings carry over from manifolds without boundary to manifolds with boundary.

THEOREM 1.11.2
The boundary ∂M of an m-dimensional manifold with boundary M is an $(m-1)$-dimensional manifold without boundary.

PROOF   Consider the collection of only those local parametrizations $\phi_\alpha : U_\alpha \to V_\alpha$ on M with $U_\alpha \cap \mathbb{H}^m \neq \emptyset$. By taking restrictions, $\phi_\alpha : U_\alpha \cap \{(x_1, \ldots, x_m) \,|\, x_m = 0\} \to \partial M$, we obtain a collection of local parametrizations on ∂M which are mutually compatible, and thus defines a smooth structure on ∂M. ∎

The following result can be used to generate examples of manifolds with boundary in a manner similar to Example 1.10.5.

PROPOSITION 1.11.3
If $f : M \to \mathbb{R}$ is a smooth function on a smooth manifold without boundary and $a \in \mathbb{R}$ is a regular value for f, then $f^{-1}(-\infty, a]$ is a smooth manifold with boundary equal to $f^{-1}(a)$. See Figure 1.11.2.

PROOF   Since $f^{-1}(-\infty, a)$ is an open set in M, for each $p \in f^{-1}(-\infty, a)$ there is a local coordinate system that maps an open ball in $\mathbb{R}^m$ onto a neighborhood of p in M. Using a translation, if necessary, the open ball in $\mathbb{R}^m$ can be replaced by an open ball in $\mathbb{H}^m$. For each $p \in f^{-1}(a)$, the proof of Theorem 1.10.2 implies that one can find a local parametrization $\phi : U \to V$ such that

$$f \circ \phi(x_1, \ldots, x_m) = x_m,$$

and the points in $f^{-1}(a)$ are mapped into $x_m = 0$. Thus, the points in $V_\alpha \cap f^{-1}(-\infty, a]$ correspond through ϕ to the points in $U \cap \mathbb{H}^m$. ∎

PROPOSITION 1.11.4
Let $f : M \to N$ be a smooth map from the m-manifold with boundary M to the n-manifold without boundary N. If $q \in N$ is a regular value for both f and $f|_{\partial M}$, then $f^{-1}(q)$ is an $(m-n)$-manifold with boundary $\partial(f^{-1}(q))$ equal to $f^{-1}(q) \cap \partial M$.

PROOF   By using local coordinates, we can reduce the problem to the special case when $f : \mathbb{H}^m \to \mathbb{R}^n$ and $q \in f(\partial \mathbb{H}^m)$. If $p \in f^{-1}(q)$ is an interior point of $\mathbb{H}^m$, then Theorem 1.10.2 implies that $f^{-1}(q)$ is a smooth $(n-m)$-manifold near p.

Suppose now that $p \in f^{-1}(q)$ is a boundary point of M. We extend f to a smooth map on an open neighborhood of p in $\mathbb{R}^m$, and we still call

1.11. MANIFOLDS WITH BOUNDARY

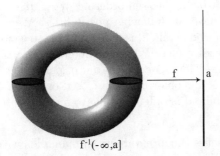

FIGURE 1.11.2
A manifold with boundary as the inverse image of a semi-closed interval.

this extension f. The point p is a regular value for f, and, by restricting the domain of f, if necessary, we can assume that it does not contain any critical points. We have that $f^{-1}(0)$ is a $(m-n)$-manifold.

Let $\pi : f^{-1}(q) \to \mathbb{R}$ be given by

$$\pi(x_1, \ldots, x_m) = x_m.$$

We claim that 0 is a regular value for π. From the facts that (df_p) vanishes precisely on $T_p\left(f^{-1}(q)\right)$ and that p is a regular point for $f|_{\partial \mathbb{H}^m}$, it follows that $T_p\left(f^{-1}(q)\right) \not\subseteq \partial \mathbb{H}^m$. Thus $(d\pi)_p\left(T_p\left(f^{-1}(q)\right)\right) \neq 0$, which proves our claim.

From the fact that 0 is a regular value for h and from Proposition 1.11.3, we conclude that $\pi^{-1}[0, \infty) = f^{-1}(q) \cap \mathbb{H}^m$ is a manifold with boundary equal to $\pi^{-1}(0) = f^{-1}(q) \cap \mathbb{H}^m$. ∎

In the sequel, the term 'manifold' will be reserved for manifolds without boundary. However, it will be possible for a manifold with boundary to have an empty boundary.

REMARK 1.11.5 The 1-dimensional manifolds with boundary can be completely classified. Any smooth, connected, 1-dimensional manifold is diffeomorphic to a circle or to some interval of the real line. There are three types of intervals only: (a, b), $(a, b]$ and $[a, b]$. The real line itself is diffeomorphic to any open interval (a, b). For a proof of this classification result we recommend Milnor (1965). Here we only give the intuitive idea of the proof. Any 1-manifold can locally be given a direction, through the well-ordering of $\mathbb{R}$. If we continue in a given direction, we have the following possibilities. We get back to where we started, and so the manifold is diffeomorphic to a circle. If not, we either keep going for all values of the parameter until we reach an endpoint (in one or both directions), or we run out of parameter. In the former case, the manifold is diffeomorphic to a closed or semi-closed interval, in the latter, the manifold is diffeomorphic to an open interval. ∎

We now discuss the orientability of manifolds with boundary.

PROPOSITION 1.11.6

If a manifold with boundary M is orientable, then so is ∂M.

PROOF Assume that p is a boundary point. If $\phi : U \to V$ is a local parametrization of M near p with $U \subseteq \{(x_1, \ldots, x_m) \mid x_m \geq 0\}$, then $T_p(\partial M)$ coincides with the image of $\{(x_1, \ldots, x_m) \mid x_m = 0\}$ through $(d\phi)_p$. The tangent space $T_p(\partial M)$ is a co-dimension 1 subspace of $T_p M$. It separates $T_p M$ into two kinds of vectors: the images through $(d\phi)_p$ of vectors $(x_1, \ldots, x_m) \in \mathbb{R}^m$ with $x_m > 0$, which we call 'inward' vectors, and the images of vectors $(x_1, \ldots, x_m) \in \mathbb{R}^m$ with $x_m < 0$, which we call 'outward' vectors. This classification of the vectors in $T_p M$ into tangent vectors to ∂M, inward vectors, and outward vectors, is independent of the local parametrization ϕ.

Assume that M is given some orientation and $m > 1$. For each $p \in \partial M$, we choose a positively oriented basis $(v_1, v_2, \ldots, v_m)$ of $T_p M$ so that $(v_2, \ldots, v_m)$ is a basis for $T_p(\partial M)$, and v_1 is an outward vector. We declare the basis $(v_2, \ldots, v_m)$ of $T_p(\partial M)$ as positively oriented. This choice of an oriented basis of $T_p(\partial M)$ can be performed smoothly at all points $p \in \partial M$. Thus ∂M is oriented.

If M is 1-dimensional, i.e., an embedded curve, its boundary is a discrete collection of points. Each boundary point p has a 0-dimensional tangent space. Its orientation is assigned as positive or negative depending whether the outward tangent vector agrees or not with the orientation of $T_p M$. ∎

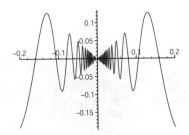

FIGURE 1.12.1
A function with countably many critical values.

1.12 Sard's theorem

In this section we will see that almost every value of a smooth map is a regular value. This can be easily seen in the 1-dimensional case, since a smooth real valued function has at most countably many critical values. See Figure 1.12.1.

In order to make a precise statement, we need to rigorously define the meaning of 'almost every'. There are basically two ways to do this.

The first standard assumes a measure theoretic point of view. A subset A of $\mathbb{R}^m$ is said to have measure zero if for every $\epsilon > 0$, there exists a sequence of cubes $C_i \subseteq \mathbb{R}^m$ such that $A \subseteq \bigcup_i C_i$ and $\sum_i \mathrm{vol}(C_i) < \epsilon$. A subset A of a manifold M is said to have measure zero if $\phi_\alpha^{-1}(A \cap V_\alpha)$ has measure zero for every local parametrization $\phi_\alpha : U_\alpha \to V_\alpha$. The complement of a measure zero set is said to have full measure. From the measure theoretic point of view, a set of points B is regarded as numerous provided that it has full measure.

The second standard assumes a topological point of view. A subset B of a topological space X is said to be residual if it contains the intersection $\bigcap_i U_i$ of a sequence U_i of dense open sets. Since every manifold is locally compact (i.e., each point has at least one compact neighborhood), by the Baire category theorem (see Rudin (1966)) we derive that a residual subset of a manifold M is always dense. From the topological point of view, a set is perceived as numerous if it is residual.

The preference for one standard of numerousness versus the other depends on the context. If a subset A of a manifold M is σ-compact (i.e., it can be written as a countable union of compact sets) and has measure zero, then $B = M \setminus A$ is residual. The converse is not true

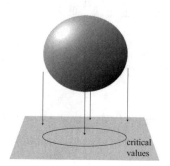

FIGURE 1.12.2
Sard's theorem.

(see Exercise 1.15.18). The set of regular values of a smooth mapping is numerous according to both standards.

THEOREM 1.12.1 *(Sard's theorem)*
Let f be a smooth map from M to N. The set of critical values of f is a measure zero subset of N. The set of regular values is a residual (hence dense) subset of full measure of N.

See Figure 1.12.2. We want to emphasize that this is a theorem about regular values and not about regular points. For example, a constant function of single variable has no regular points (all points are critical) but has only one critical value (so the rest of the points of $\mathbb{R}$ are regular values). A set of one point obviously has measure zero.

The case when $\dim M < \dim N$ is easy (see Exercise 1.15.20). Here we give the proof in the case when $\dim M = \dim N = m$. The remaining case is harder, and the reader is invited to see Sternberg (1964). Also see Milnor (1965).

Through local parametrizations, we can reduce the theorem to the case when M is an open set in $\mathbb{R}^m$ contained in the cube

$$C = \{(x_1, \ldots, x_m) \,|\, 0 \le x_i \le 1 \text{ for each } i\},$$

$N = \mathbb{R}^m$, and $f = (f_1, \ldots, f_m)$ is a smooth function defined on a neighborhood of C. Let $\epsilon > 0$ be fixed.

1.12. SARD'S THEOREM

We first try to measure how the distances between pairs of points change under f in a neighborhood of a critical point $x = (x_1, \ldots, x_m)$. If $y = (y_1, \ldots, y_m)$ is another point in C, by the mean value theorem, for each i there exists $z_i \in C$ on the line segment connecting x with y such that

$$f_i(y) - f_i(x) = \sum_{j=1}^{m} \frac{\partial f_i}{\partial x_j}(z_i)(y_i - x_i).$$

By the uniform continuity of the partial derivatives of f on C, there is a finite number K such that

$$\|f(y) - f(x)\| \leq K\|y - x\|. \tag{1.12.1}$$

Indeed, $K = \sup_{z \in C} \left(\sup_{\|v\|=1} \|(df)_z(v)\| \right)$.

A mapping f with $\|f(y) - f(x)\| \leq K\|y - x\|$ for all x, y is called a Lipschitz mapping, and the constant K is called a Lipschitz constant. Above, we basically proved that any smooth function on a compact domain is Lipschitz.

Let $T_x = ((T_x)_1, \ldots, (T_x)_m)$ be the linearization of f near x given by

$$(T_x)_i(y) = f_i(x) + \sum_{j=1}^{m} \frac{\partial f_i}{\partial x_j}(x)(y_j - x_j).$$

We have

$$\|f(y) - T_x(y)\| \leq \left[\sum_{i,j=1}^{m} \left(\frac{\partial f_i}{\partial x_j}(z_i) - \frac{\partial f_i}{\partial x_j}(x) \right)^2 \right]^{1/2} \|y - x\|.$$

Since each z_i lies in between x and y, by the uniform continuity of the partial derivatives of f on C, there exists $\delta > 0$ such that

$$\|f(y) - T_x(y)\| \leq \epsilon \|y - x\| \tag{1.12.2}$$

provided $\|y - x\| < \delta$. Since x is a critical point, the derivatives of the components $(df_1)_x, \ldots, (df_m)_x$ are not linearly independent, so the image of T_x is contained in some $(m-1)$-dimensional plane P_x in $\mathbb{R}^m$, passing through $T_x(x) = f(x)$. From (1.12.1) and (1.12.2), the image through f of the ball of radius δ centered at x is contained in an m-dimensional parallelepiped centered at $f(x) \in P_x$, with a pair of faces parallel to P_x and at a distance $\epsilon \|y - x\|$ apart from P_x, and the rest

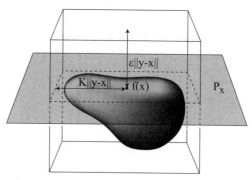

FIGURE 1.12.3
Parallelepiped containing the image through f of a ball of diameter δ centered at x.

of the faces at a distance $K\|y-x\|$ apart from $f(x)$. See Figure 1.12.3. The volume of this parallelepiped is

$$(2\epsilon\|y-x\|)(2K\|y-x\|)^{m-1} = 2^m K^{m-1}\epsilon\|y-x\|^m. \qquad (1.12.3)$$

We divide the whole cube C into k^m small cubes of edges $1/k$. Choose k large enough so that the diameter $\sqrt{m}/k$ of each cube is less than δ. By (1.12.3), each cube containing a critical point x is mapped by f into a parallelepiped of volume $2^m K^{n-1}\epsilon(1/k)^m$. Since there are at most k^m small cubes containing critical points, the set of all critical values is contained in a collection of parallelepipeds of total volume

$$k^m 2^m K^{m-1}\epsilon(1/k)^m = 2^m K^{m-1}\epsilon.$$

As m and K are fixed, this quantity can be made arbitrarily small, which shows that the set of all critical values has measure zero.

REMARK 1.12.2 Sard's theorem remains true provided f is C^k-differentiable, where k is large enough. If the dimension of M is m and the dimension of N is n, then it is sufficient for f to be C^k-differentiable, with $k \geq 1 + \max(m-n, 0)$, in order for Sard's theorem to work. See Hirsch (1976). ∎

Combining Theorem 1.10.2 with Sard's theorem (Theorem 1.12.1), we obtain that $f^{-1}(q)$ is a submanifold of M for almost every point $q \in N$.

1.12. SARD'S THEOREM

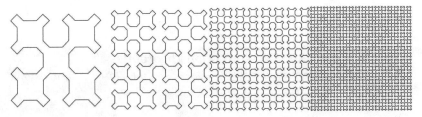

FIGURE 1.12.4
The Sierpinski curve is the limit of the displayed curves.

Example 1.12.3
Is there any continuous and onto mapping from the real line to a square in the plane? In other words, is it possible to fill up a square with a curve? The surprising answer is yes, and an example, due to Sierpinski, is the limit of the curves displayed in Figure 1.12.4. Sierpinski's curve is an example of a so-called fractal set. Sierpinski's curve exhibits infinitely many cusps, so a follow-up question is if there is any smooth and onto mapping from the real line to a square in the plane, or, more generally, if there is any smooth and onto mapping from $\mathbb{R}^n$ to $\mathbb{R}^m$, with $n < m$. This time the answer is negative. Suppose by contradiction that such a mapping $f : \mathbb{R}^n \to \mathbb{R}^m$ exists. Sard's theorem implies that there is at least one regular value. Since $f(\mathbb{R}^n) = \mathbb{R}^m$, the regular value is the image of some regular point. However, by an earlier remark, all points of $\mathbb{R}^n$ are critical points for f, since the dimension of $\mathbb{R}^n$ is smaller than the dimension of $\mathbb{R}^m$. We have reached a contradiction. □

There is a version of Sard's theorem for manifolds with boundary, which follows easily from Sard's theorem for manifolds without boundary.

THEOREM 1.12.4 (Sard's theorem)
The set of critical values of f has measure 0 in N, as does the set of points in N that are critical values of $f|_{\partial M}$. The points in N that are regular values of both f and $f|_{\partial M}$ form a residual (hence dense) subset of full measure in N.

As an application of Sard's theorem, we will prove two important results in topology (following Hirsch (1976)). A continuous map $r : X \to Y$ from a topological space X to a subspace $Y \subseteq X$ is said to be a retraction if the restriction of r to Y is the identity mapping.

THEOREM 1.12.5 (No-retraction theorem)
There exists no retraction $r : \bar{B}^n \to S^{n-1}$ of the closed n-dimensional ball onto its boundary.

PROOF We will first prove that there is no C^∞-smooth retraction. Assume, by contradiction, that there is one, which we call h. By Sard's theorem, the smooth map h has at least one regular value, say y. Then $r^{-1}(y)$ is a compact 1-dimensional manifold with boundary in $\bar{B}^n$. Since $y \in S^{n-1}$ and $r(y) = y$, from Proposition 1.11.4 it follows that

$$\partial(r^{-1}(y)) = r^{-1}(y) \cap S^{n-1},$$

so the boundary of $r^{-1}(y)$ consists of a single point. On the other hand, any connected 1-dimensional manifold with boundary is a diffeomorphic copy of either a circle or of a closed interval, as explained in Remark 1.11.5. Hence $r^{-1}(y)$ is a finite collection of diffeomorphic copies of circles and closed intervals. From here it follows that the boundary of $h^{-1}(y)$ consists of an even number of points (a circle has no boundary points, while an interval has two boundary points). This contradicts the fact that $\partial(r^{-1}(y))$ is a single point. Thus, there is no smooth mapping from a disk onto its boundary which fixes all boundary points.

Next, we assume that there is a continuous map r with this property. We can replace r by another map $\tilde{r}$ with the same property, such that $\tilde{r}$ is constant (hence smooth) in a neighborhood of S^{n-1}, by defining

$$\tilde{r}(x) = \begin{cases} r(x/|x|) & \text{if } 1/2 < |x| \leq 1, \\ r(2x) & \text{if } 0 \leq |x| \leq 1/2. \end{cases}$$

We can approximate $\tilde{r}$ with a C^∞-smooth function $\tilde{\tilde{r}}$ which agrees with $\tilde{r}$ in a neighborhood of S^{n-1}. Such a smooth function $\tilde{\tilde{r}}$ maps the disk onto its boundary and fixes all boundary points. This contradicts the previous step. ∎

We will now discuss Brouwer's fixed point theorem. In 2 dimensions, the theorem can be illustrated experimentally as follows. We take two sheets of graph paper, one lying directly above the other. We crumple the top sheet, and place it on top of the other sheet. Carefully inspecting the grid, we will find that at least one point on the top sheet landed

1.13. TRANSVERSALITY

FIGURE 1.12.5
Brouwer fixed point theorem.

directly above the corresponding point on the bottom sheet. The rectangular shape of the sheet of paper is not relevant; one could use any shape that is homeomorphic to a disk. The same is true in any dimension.

THEOREM 1.12.6 (*Brouwer fixed point theorem*)

Any continuous mapping of the closed unit ball into itself has a fixed point.

PROOF   Let $f : \bar{B}^n \to \bar{B}^n$ and assume $f(x) \neq x$ for all x. Let $g(x)$ be the point where the ray from $f(x)$ through x intersects S^n. See Figure 1.12.5. Then $g : \bar{B}^n \to S^n$ is a retraction, in contradiction with the previous result. ∎

1.13 Transversality

We now investigate whether the inverse image of a submanifold through a smooth map is also a submanifold. We have a positive answer in the particular case of a 0-dimensional submanifold (i.e., a point), provided that the point is a regular value for the map. With this in mind, we define a new differential property, which extends the notion of regularity.

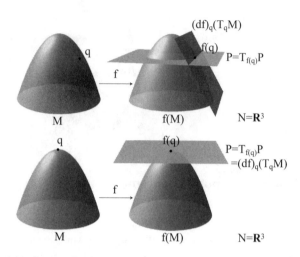

FIGURE 1.13.1
Transverse and non-transverse maps.

DEFINITION 1.13.1
Let f be a smooth mapping from the smooth manifold M to the smooth manifold N, and P be a submanifold of N. We say that f is transverse to P, and we write $f \pitchfork P$, if

$$(df)_q(T_qM) + T_{f(q)}P = T_{f(q)}N \tag{1.13.1}$$

whenever $q \in f^{-1}(P)$.

The above equation says that the image of the derivative of the map at q together with the tangent space to P at $f(q)$ spans the whole tangent space of N at $f(q)$. Examples of transverse and non-transverse maps are shown in Figure 1.13.1. Like regularity, transversality requires conditions only on the points in the inverse image $f^{-1}(P)$. There are some special cases in which transversality follows automatically. For instance, if the image of f is disjoint from P, then (1.13.1) is vacuously satisfied, so $f \pitchfork P$. Also if $P = N$ then every smooth map is transverse to P. On the other hand, transversality places some algebraic restrictions on the dimensions of the manifolds. For instance, if $\dim M < \dim N - \dim P$ then every f is non-transverse to P, unless $f(M) \cap P = \emptyset$. The spaces $(df)_q(T_qM)$ and $T_{f(q)}P$ may have non-trivial intersections, as long as they span $T_{f(q)}N$.

1.13. TRANSVERSALITY

Now we get to one of the main features of transversality.

THEOREM 1.13.2
Let $f : M \to N$ be a smooth map and $P \subseteq N$ be a submanifold of N. If $f \pitchfork P$, then $f^{-1}(P)$ is a submanifold of M, provided $f^{-1}(P)$ is nonempty. The codimension of $f^{-1}(P)$ in M equals the codimension of P in N.

PROOF Using local coordinates, we can assume that $f : \mathbb{R}^m \to \mathbb{R}^n$, and P is a submanifold of $\mathbb{R}^n$ of dimension p. Let $q \in f^{-1}(P)$. There exist local coordinates $(y_1, \ldots, y_p, y_{p+1}, \ldots, y_n)$ near $f(q)$ such that P is given by the points of coordinates $(y_1, \ldots, y_p, 0, \ldots, 0)$.

Let $\pi : \mathbb{R}^n \to \mathbb{R}^{n-p}$ be the canonical submersion given by

$$\pi(y_1, \ldots, y_p, y_{p+1}, \ldots, y_n) = (y_{p+1}, \ldots, y_n).$$

Note that π maps P to $0 \in \mathbb{R}^{n-p}$, and its differential

$$(d\pi)_{f(q)} : T_{f(q)}(\mathbb{R}^n) \to T_0(\mathbb{R}^{n-p})$$

vanishes precisely on $T_{f(q)}P$. Let $h : f^{-1}(P) \to \mathbb{R}^{n-p}$ be defined by $h = \pi \circ f$.

We claim that 0 is a regular value for h. By the chain rule, we have $(dh)_q = (d\pi)_{f(q)} \circ (df)_q$. Note that

$$(d\pi)_{f(q)}\left((df)_q(T_q\mathbb{R}^m)\right) = (d\pi)_{f(q)}\left((df)_q(T_q\mathbb{R}^m) + T_{f(q)}P\right),$$

since the added vectors of $T_{f(q)}P$ do not change the image of $(d\pi)_{f(q)}$. The transversality condition says that $(df)_q(T_q\mathbb{R}^m) + T_{f(q)}P = T_{f(q)}\mathbb{R}^{n-p}$, hence $(dh)_q$ maps the tangent space $T_q\mathbb{R}^m$ onto $T_{h(q)}\mathbb{R}^{n-p}$. Thus $(dh)_q$ is a submersion at every point $q \in f^{-1}(P)$. This proves our claim.

By this claim and by Theorem 1.10.2, it results that $h^{-1}(0)$ is a submanifold near q, and hence $f^{-1}(P)$ is a submanifold of $\mathbb{R}^m$. The codimension of $f^{-1}(P)$ in M is $(n-p)$, equal to the codimension of P in N.
∎

We consider the special case when M is a submanifold of N, and the map from M to N is the canonical embedding $i_M : M \hookrightarrow N$.

DEFINITION 1.13.3
If M, P are two submanifolds of N, then we say that M is transverse to P, and we write $M \pitchfork P$, if
$$T_p(M) + T_p(P) = T_p N$$
whenever $p \in M \cap P$.

This condition is just the translation of $i_M \pitchfork P$. As before, if M and P are disjoint, then they are transverse by vacuous hypothesis. In other words, non-intersecting submanifolds are always transverse. The previous theorem provides sufficient conditions under which the intersection of two submanifolds is itself a submanifold.

COROLLARY 1.13.4
The intersection $M \cap P$ of two transverse submanifolds M and P of N is itself a submanifold of N, unless it is empty. Moreover
$$\operatorname{codim}(M \cap P) = \operatorname{codim} M + \operatorname{codim} P.$$

PROOF The fact that $M \cap P$ is a submanifold is straightforward. If m, p, n and i denote the dimensions of M, P, N and $M \cap P$, respectively, then Theorem 1.13.2 gives us $m - i = n - p$, so $n - i = (n - m) + (n - p)$. We get this much nicer equation for the codimension of $M \cap P$ than we would get for its dimension. ■

Example 1.13.5
Let $N = \mathbb{R}^2$, M be the submanifold given by the graph of $f(x) = x^3 - a$, with a a real parameter, and P be represented by the x-axis. It is easy to see that $M \pitchfork P$ if and only if $a \neq 0$. Although $M \not\pitchfork P$ if $a = 0$, it happens that $M \cap P$ is still a submanifold of N (of dimension 0). See Figure 1.15.1. This is not, in general, the case. See Exercise 1.15.19
◻

1.14 Stability

In experiments, the measurement of physical quantities can only be done within some margin of error. Therefore, only those geometric prop-

1.14. STABILITY

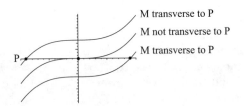

FIGURE 1.13.2
Transverse and non-transverse submanifolds.

erties that persist under small perturbation are considered to be physically relevant. Such properties are said to be stable.

The simplest way to perturb a map is through a homotopy.

If $f : X \to Y$ is a continuous mapping of the topological space X into the topological space Y, a homotopy of f is a continuous mapping $F : X \times [0,1] \to Y$ such that $F_0 = F(\cdot, 0) = f$. If f is a smooth map between two manifolds, it is natural to consider homotopies F which are also smooth. A property of a smooth map $f : M \to N$ is said to be stable provided that for any smooth homotopy F of f, there exists $\epsilon > 0$ such that $F_t = F(\cdot, t)$ has the same property, for all $0 \leq t \leq \epsilon$.

PROPOSITION 1.14.1
The property of a smooth mapping $f : M \to N$ of being an immersion — or a submersion, or an embedding — of a compact manifold M into a manifold N is a stable property.

PROOF Suppose that f is an immersion, that is $(df)_p : T_pM \to T_pN$ is injective at every $p \in M$. This means the matrix representing $(df)_p$ relative to some local coordinate system has rank greater than or equal to the rank m of M. Thus, $(df)_p$ has a $m \times m$ sub-matrix $A(p)$ with $\det(A(p)) \neq 0$. If F is a homotopy of f, the entries of $(dF_t)_p$ vary continuously with both p and t. For q near p, let $A_t(p)$ be the sub-matrix of $(dF_t)_q$ whose entries occupy the same slots as the entries of $A(p)$. For each p in M, there exist $\epsilon_p > 0$ and a neighborhood N_p of p such that $\det A_t(q) \neq 0$ if $q \in N_p$ and $0 \leq t \leq \epsilon_p$. Due to the compactness of M, it follows that F_t is an immersion for all $0 \leq t \leq \epsilon$, provided ϵ is chosen small enough.

The stability of the submersion property of a smooth function f follows along the same line of reasoning, the only difference being that the sub-matrices of interest are now of rank greater than or equal to the dimension n of N.

Since an embedding of a compact manifold is just an injective immersion, we only need to prove that the property of being an injective immersion is stable. Assume the contrary. Then $f = F_0$ is an injective immersion, and there exists a sequence $t_n \to 0$ and two sequences $x_n, y_n \in M$ with $x_n \neq y_n$ such that $F_{t_n}(x_n) = F_{t_n}(y_n)$ for all $n \geq 0$. Since $M \times [0,1]$ is compact, by taking convergent subsequences of (x_n, t_n) and of (y_n, t_n) if necessary, we can assume that $x_n \to x$ and $y_n \to y$. Let $G : M \times [0,1] \to M \times [0,1]$ be given by

$$G(\xi, t) = (F(\xi, t), t).$$

The matrix representing $(df)_0$ relative to some local coordinate system has rank at least m, hence the matrix

$$(dG)_{(x,0)} = \begin{pmatrix} (df)_x & * \\ 0 & 1 \end{pmatrix}$$

contains an $m \times m$ sub-matrix of non-zero determinant. This implies that G is an immersion around $(x, 0)$. By the continuity of G, we have $G(x_n, t_n) \to (f(x), 0)$ and $G(y_n, t_n) \to (f(y), 0)$. Since $f(x)$ must equal $f(y)$ and f is an embedding, it follows that $x = y$, so x_n and y_n both converge to the same limit x. Since G is an immersion around $(x, 0)$, it is injective in some small neighborhood of $(x, 0)$. We can always get (x_n, t_n) and (y_n, t_n) arbitrarily close to $(x, 0)$ by choosing n sufficiently large. Thus, $x_n = y_n$ for all n large enough, which is a contradiction.
∎

COROLLARY 1.14.2

The property of a smooth mapping $f : M \to N$ of being a diffeomorphism is a stable property.

PROPOSITION 1.14.3

The property of a smooth mapping $f : M \to N$ of the compact manifold M into the manifold N of being transverse to a submanifold P of N is a stable property.

PROOF From the proof of Theorem 1.13.2, recall that the transversality condition at a point p translates into a condition that some mapping h is a submersion at every point in P. Since the property of being

1.14. STABILITY

a submersion is stable, it follows that the transversality condition is also stable. ∎

While transversality is a stable property, non-transversality is, on the contrary, unstable, as one can easily infer from Figure 1.13.1.

We will now see that if P is a submanifold of M, then almost every map $f : M \to N$ is transverse to P.

THEOREM 1.14.4 (Transversality theorem)
Suppose that M, N and S are smooth manifolds, and $F : M \times S \to N$ is transverse to a submanifold P of N. Then the set of all s with $F_s = F(\cdot, s) : M \to N$ transverse to P is residual (hence dense) in S.

PROOF By Theorem 1.13.2, $F^{-1}(P)$ is a submanifold of $M \times S$. Let $\pi : F^{-1}(P) \to S$ be the projection of $M \times S$ into the second factor, restricted to $F^{-1}(P)$.

We claim that if s is a regular value for π, then F_s is transverse to P. Choose $q \in M$ such that $F_s(q) = p \in P$. Since $F(q, s) = p \in P$ and F is transverse to P at (p, s), we have

$$(dF)_{(q,s)}(T_{(q,s)}(M \times S)) + T_pP = T_pN.$$

Since s is assumed to be a regular value for π, $(d\pi)_{(q,s)}(T_{(q,s)}(F^{-1}(P))) = T_sS$.

LEMMA 1.14.5
Suppose that L is a linear map and V' is a subspace of V such that $L(V) = L(V')$. Then $V = \ker L + V'$.

PROOF OF THE LEMMA For any $v \in V$, we can find $v' \in V'$ with $L(v') = L(v)$. Then $v - v' \in \ker L$ and the equation $v = (v - v') + v'$ shows that $v \in \ker L + V'$. ∎

Applying the lemma to $(d\pi)_{(q,s)} : T_{(q,s)}(M \times S) \to T_sS$ with $V' = T_{(q,s)}(F^{-1}(P))$ shows that

$$T_{(q,s)}(M \times S) = T_qM \times T_sS = T_{(q,s)}(M \times \{s\}) + T_{(q,s)}(F^{-1}(P)).$$

We then have

$$(dF)_{(q,s)}(T_{(q,s)}(M \times \{s\})) + (dF)_{(q,s)}(T_{(q,s)}(F^{-1}(P))) + T_pP = T_pN,$$

which combined with $(dF)_{(q,s)}(T_{(q,s)}(F^{-1}(P)) \subseteq T_p P$ yields

$$(dF_s)_q(T_q M) + T_p P = T_p N.$$

This proves our claim.

Since F_s is transverse to P whenever s is a regular value of π, the set of all s for which F_s is transverse to P is at least as large as the set of all regular values of π. The latter set, and hence the former set, is a residual in S, due to Sard's theorem — Theorem 1.12.1. ∎

We will now explain how to use this theorem to prove the fact that any mapping $f: M \to N$ can be perturbed by an arbitrarily small amount to a mapping transverse to P. We only consider the special case when $N = \mathbb{R}^n$.

Let $\epsilon > 0$ and $S = B^n(0, \epsilon)$ be the open ball of radius ϵ in $\mathbb{R}^n$. Define $F: M \times S \to N$ by $F(x, s) = f(x) + s$. Since

$$(dF)_{(x,s)} = \bigl((df)_x \ \ I_n\bigr),$$

where I_n is the $n \times n$ identity matrix, then F is a submersion, so F is transverse to P. By the previous theorem, there exists $s \in B^n(0, \epsilon)$ such that the 'perturbation' $f(x) + s$, which is at most ϵ away from $f(x)$, is transverse to P.

For the general case, see Guillemin and Pollack (1974).

1.15 Exercises

1.15.1 Consider a sphere lying on a horizontal plane, and the stereographic projection ϕ_1 from the plane to the sphere, defined as follows: ϕ_1 maps each point Q of the plane to the point P at the intersection of the line through the north pole N and Q with the sphere. Consider the analogous stereographic projection ϕ_2 relative to a horizontal plane lying on the top of the sphere and to the south pole S. Show that ϕ_1 and ϕ_2 induce a smooth structure on the unit sphere.

1.15.2 Consider a bijection between the real line $\mathbb{R}$ and the sphere S^2 (such a bijection exists since these are sets with same cardinality). Show that the composition of the local parametrizations of S^2 from above with

1.15. EXERCISES

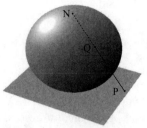

FIGURE 1.15.1
Stereographic projection.

this bijection defines a smooth structure on $\mathbb{R}$. Show that $\mathbb{R}$ endowed with this smooth structure is diffeomorphic to the sphere S^2. With this smooth structure, the real line is a sphere! The point of this exercise is to stress that a manifold is not just a set that can be endowed with some structure, but the set together with that structure.

1.15.3 Show that the n-dimensional real projective plane $\mathbb{R}\mathbb{P}^n$, with the structure from Example 1.3.4, is an n-dimensional smooth manifold.

1.15.4 We define the complex projective n-space $\mathbb{C}\mathbb{P}^n$. On $\mathbb{C}^{n+1}$ we define an equivalence relation $(w_1, \ldots, w_{n+1}) \sim (\lambda w_1, \ldots, \lambda w_{n+1})$ provided λ is a non-zero complex number. Define coordinate charts (U_i, ϕ_i), $0 \leq i \leq n$, by

$$U_i = \mathbb{C}^n = \{(z_1, \ldots, z_n) \,|\, z_1, \ldots, z_n \in \mathbb{C}\},$$
$$V_i = \{[w_1, \ldots, w_i, \ldots, w_{n+1}] \,|\, w_i \neq 0\},$$
$$\phi_i(z_1, \ldots, z_n) = [z_1, \ldots, z_{i-1}, 1, z_i, \ldots, z_n],$$

for all $i = 1, \ldots, n+1$ Show that the collection $(U_i, \phi_i)_{0 \leq i \leq n}$ defines a smooth structure on $\mathbb{C}\mathbb{P}^n$.

1.15.5 Provide the unit cube

$$\{(x_1, \ldots, x_{n+1}) \,|\, \max_{1 \leq i \leq n+1} |x_i| = 1\} \subseteq \mathbb{R}^{n+1}$$

with a smooth structure. The point of this exercise it to illustrate that a smooth manifold may not look smooth! Of course this smooth structure is not compatible with the smooth structure of $\mathbb{R}^{n+1}$.

1.15.6 Using Zorn's lemma, one can show that there exists an uncountable well-ordered set Ω. Consider the set $L^+ = \Omega \times [0,1)$ endowed with the lexicographic order

$$(\alpha, t) < (\beta, s) \quad \text{if and only if } \alpha < \beta \text{ or if } \alpha = \beta \text{ and } t < s.$$

On L^+ we consider the order topology (a sub-basis for this topology consists of sets of the type $\{X \in L^+ \mid X < X_0\}$ and $\{X \in L^+ \mid X > X_0\}$, where $X_0 \in L^+$). This topology is Hausdorff. We can think of L^+ as a really 'long ray', consisting of an uncountable number of copies of $[0,1)$ 'pasted together' end-to-end. Compare this with the ray $[0, \infty)$, which can be viewed as a countable number of copies of $[0,1)$ pasted together end-to-end. We define the 'long line' L as the disjoint union of two copies of the long ray with their origins identified. Show that the topology of L does not have a countable basis of open sets. Also show that is not separable (there is no countable dense set). Conclude that there exists a smooth structure on L, compatible with the order topology, and so fails the countable basis condition in the definition of a manifold. Also conclude that L cannot be embedded in any $\mathbb{R}^n$.

1.15.7 Let F be the 'forked line', constructed as follows. Take the disjoint union of two copies of $\mathbb{R}$ and identify each point $x \neq 0$ in the first copy of $\mathbb{R}$ to the corresponding x in the second copy. Consider the quotient topology on F. We can think of F as the real line $\mathbb{R}$ with two points at the origin instead of one. Show that compact sets in F are not closed. Also show that the topology of F is is not Hausdorff. Conclude that there exists a smooth structure on F, compatible with the quotient topology, and so fails the Hausdorff condition in the definition of a manifold. Also conclude that F cannot be embedded in any $\mathbb{R}^n$.

1.15.8 Show that the tori defined in Example 1.3.3 (i) and (ii) are diffeomorphic. Show that $S^1 \times S^1$ regarded as a product manifold is diffeomorphic to $\mathbb{T}^2$. Define the n-dimensional torus as $\mathbb{R}^n/\mathbb{Z}^n$, as in Example 1.3.3 (ii). Show that $\mathbb{T}^n$ is diffeomorphic to $\underbrace{S^1 \times \cdots \times S^1}_{n \text{ times}}$.

1.15.9 Consider the equivalence relation on $\mathbb{R}^2$: $(x_1, x_2) \sim (x_1', x_2')$ provided that $(x_1', x_2') = (x_1 + n_1, 1 - x_2 + n_2)$, for some $(n_1, n_2) \in \mathbb{Z}^2$. Show that $\mathbb{R}^2$ induces a smooth structure on $K = \mathbb{R}^2/\sim$. The resulting 2-dimensional manifold is known as the Klein bottle. It can be visualized as a cylinder with the circular ends attached one to another, with opposite orientations. See Figure 1.15.2. Show that the Klein bottle K cannot be embedded in $\mathbb{R}^3$.

1.15. EXERCISES

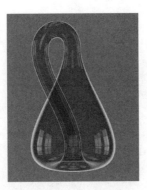

FIGURE 1.15.2
The Klein bottle (reproduced with permission of John M. Sullivan).

1.15.10 Show that any two smooth structures on the circle S^1 are diffeomorphic.

1.15.11 Show that the real projective line $\mathbb{RP}^1$ can be obtained as a submanifold of the real projective plane $\mathbb{RP}^2$ if we start with S^1 viewed as the equator of S^2 and proceed as in Example 1.3.4. Show that there exists no smooth function $f : \mathbb{RP}^2 \to \mathbb{R}$ such that $f^{-1}(q) = \mathbb{RP}^1$ for some regular value q of f. The point of this exercise is to emphasize that it is not true that every submanifold is the inverse image of some regular value.

1.15.12 Give an example to illustrate that a smooth homeomorphism is not necessarily a diffeomorphism, which means that the condition that the inverse mapping is smooth is independent.

1.15.13 Prove that the tangent bundle of a product of manifolds is diffeomorphic to the product of the tangent bundles of the manifolds. Deduce that the tangent bundle of a torus $S^1 \times S^1$ is diffeomorphic to $S^1 \times S^1 \times \mathbb{R}^2$.

1.15.14 Show that any injective immersion of a compact manifold is an embedding.

1.15.15 Show that if an embedded surface S of $\mathbb{R}^3$ is given by $S = f^{-1}(q)$, where $f : \mathbb{R}^3 \to \mathbb{R}$ is a smooth function and q is a regular value for f, then S is orientable.

1.15.16 If $f : M \to N$ is a smooth map, and the rank of the derivative $(df)_p$ at some point $p \in M$ is k, prove that there exist a local parametrization $\phi_\alpha : U_\alpha \to V_\alpha$ near p and a local parametrization $\psi_\beta : U_\beta \to V_\beta$ near $f(p)$ such that

$$(\psi_\beta^{-1} \circ f \circ \phi_\alpha)(x_1, \ldots, x_m) = (x_1, \ldots, x_r, 0, \ldots, 0).$$

1.15.17 Consider the projective plane $\mathbb{RP}^2$, described in Example 1.3.4. Let $f : \mathbb{R}^3 \to \mathbb{R}^4$ defined by

$$f(x, y, z) = (x^2 - y^2, xy, xz, yz).$$

Note that the restriction of f to S^2 satisfies $f(p) = f(-p)$, hence we can define $\tilde{f} : \mathbb{RP}^2 \to \mathbb{R}^4$ by $\tilde{f}([p]) = f(p)$. Prove that $\tilde{f}$ is an embedding.

We point out that that the projective plane has no embedding in $\mathbb{R}^3$ (see Spivak (1999)). This shows that the dimension $2m$ of the embedding Euclidean space in Whitney's embedding theorem is optimal.

1.15.18 Consider a Cantor set of positive measure constructed as follows. Fix $0 < \mu < 1$. From the unit interval $I = [0, 1]$ we first remove an open interval R_1 centered at the midpoint of I, and of length $\mu/2$, at the second step we remove the union R_2 of two open intervals centered at the midpoints of the components of $I \setminus R_1$, and of total length $\mu/4$, and so on. The remaining points of the interval form a Cantor set $C = I \setminus \bigcup_{i=1}^\infty R_i$. Show that the Cantor set C is σ-compact and nowhere dense, and so $\mathbb{R} \setminus C$ is a residual set but not not of full measure in $\mathbb{R}$.

1.15.19 Let C be a Cantor set in $\mathbb{R}^n$. Show that there exists a submanifold M of $\mathbb{R}^{n+1}$ such that $M \cap (\mathbb{R}^n \times \{0\}) = C$. Show that every such an M is not transverse to $\mathbb{R}^n \times \{0\}$. The point of this exercise is to illustrate that if two intersecting manifolds are not transverse, their intersection may fail to be a submanifold.

Hint: Whitney proved that any closed subset of $\mathbb{R}^n$ is the zero set of some smooth function (see Remark 1.10.4).

1.15.20 Show that if $f : M \to N$ is a smooth map of a manifold M into a manifold N with $\dim M < \dim N$, then $f(M)$ has measure 0 in N. Conclude that the critical values of f form a measure zero set in N. This proves Sard's theorem in the special case when $\dim M < \dim N$.

Hint: Use Exercise 1.15.16.

Chapter 2

Vector Fields and Dynamical Systems

2.1 Introduction

Many physical systems are described by differential equations.

As an example, we consider the two-body problem (the Kepler problem). It concerns the motion of two bodies under mutual gravitation. By placing a reference frame at one of the bodies, the problem can be reduced to the motion of a single body in a central gravitational force field. The force is expressed by the universal law of gravitation

$$F = -\mu \frac{x}{\|x\|^3},$$

where the gravitational constant and the total mass of the system are normalized to 1. At every position x in the phase space $\mathbb{R}^3 \setminus \{(0,0,0)\}$ there is a force vector that points towards the origin and has its magnitude inverse proportional to the square of the distance to the origin. The motion of the body is governed by the second order differential equation

$$\frac{d^2 x}{dt^2} = -\frac{x}{\|x\|^3}.$$

This translates into a first order system

$$\frac{dx}{dt} = v, \qquad (2.1.1)$$
$$\frac{dv}{dt} = -\frac{x}{\|x\|^3},$$

with the variables x and v representing the position and the velocity of the body, respectively. The theory of differential equations asserts that the future motion of the body is completely determined by the initial position x_0 and initial velocity v_0 of the body. We should note that many differential systems do not have closed-form solutions. This is the case, for example, of the n-body problem with $n \geq 3$, which studies the general motion of n bodies interacting by mutual gravitation forces (we will discuss this problem in Section 9.2). In such instances, one usually attempts to perform a qualitative study of differential equations.

In the sequel, we will find explicit solutions to the central force problem, and, implicitly, to the two-body problem. We denote by $\cdot$ the dot product and by $\times$ the vector product in $\mathbb{R}^3$. The vector $p = mv$ is called the linear momentum; the vector $I = x \times v$ is called the angular momentum; the quantity $(v \cdot v)/2$ is called the kinetic energy; the quantity $-1/\|x\|$ is called the potential energy; finally

$$E = \frac{v \cdot v}{2} - \frac{1}{\|x\|}$$

represents the total energy of the body. We want to identify scalar quantities that are preserved along the motion (these are called prime integrals). Using the basic properties of the dot product and of the vector product, we can easily check that the total energy and the angular momentum are conserved. Indeed, using (2.1.1) we obtain

$$\frac{dI}{dt} = \frac{dx}{dt} \times v + x \times \frac{dv}{dt} = v \times v + x \times \left(-\frac{x}{\|x\|^3}\right) = 0$$

and

$$\frac{dE}{dt} = v \cdot \frac{dv}{dt} + \frac{dx}{dt} \cdot \frac{x}{\|x\|^3} = v \cdot \left(-\frac{x}{\|x\|^3}\right) + v \cdot \frac{x}{\|x\|^3} = 0.$$

Since $I = x \times v$ is constant, the motion is planar, and we can change the coordinates so that $x_3 = 0$ and $v_3 = 0$. With respect to these new coordinates we have $I = (0, 0, x_1 v_2 - x_2 v_1) = $ constant. If $\|I\| = 0$, the initial velocity of the body is directed towards the origin, so the body will eventually hit the origin. This means that the corresponding solution $(x(t), v(t))$ of (2.1.1) is defined only locally, for a certain interval of time. If $\|I\| \neq 0$, then the solution $(x(t), v(t))$ is defined globally for all time $t \in \mathbb{R}$.

Now we express the conservation equations in polar coordinates $x_1 = r \cos \theta$ and $x_2 = r \sin \theta$. The components of the velocity vector are $v_1 = \frac{dr}{dt} \cos \theta - r \sin \theta \frac{d\theta}{dt}$ and $v_2 = \frac{dr}{dt} \sin \theta + r \cos \theta \frac{d\theta}{dt}$, so

2.1. INTRODUCTION

$$\|I\| = r^2 \frac{d\theta}{dt},$$

$$E = \frac{1}{2}\left[\left(\frac{dr}{dt}\right)^2 + r^2\left(\frac{d\theta}{dt}\right)^2\right] - \frac{1}{r} = \frac{1}{2}\left[\left(\frac{dr}{dt}\right)^2 + \frac{\|I\|^2}{r^2}\right] - \frac{1}{r}.$$

By the chain rule, we have

$$\frac{d\theta}{dt} = \frac{d\theta}{dr}\frac{dr}{dt}. \qquad (2.1.2)$$

We solve for $d\theta/dt$ and dr/dt in the previous two equations and then we substitute in (2.1.2). We obtain

$$\frac{d\theta}{dr} = \frac{\frac{\|I\|}{r^2}}{\sqrt{2\left(E + \frac{1}{r} - \frac{\|I\|^2}{2r^2}\right)}}.$$

Integration by substitution yields

$$\theta = \arccos \frac{\frac{\|I\|}{r} - \frac{1}{r}}{\sqrt{2E + \frac{1}{\|I\|^2}}}.$$

Substituting

$$p = \|I\|^2, \ e = \sqrt{1 + 2E\|I\|^2}, \text{ and } \theta = \arccos\left((p/r - 1)/e\right)$$

we obtain

$$r = \frac{p}{1 + e\cos\theta}.$$

This is the polar equation of a conic: if $e < 1$ (corresponding to $E < 0$), the trajectory is an ellipse; if $e = 1$ (corresponding to $E = 0$), the trajectory is a parabola; if $e > 1$ (corresponding to $E > 0$), the trajectory is a hyperbola. In the case of the planetary motion, the total energy E is negative. This proves Kepler's First Law that the trajectories of the planets around the Sun are ellipses. The previously derived equation $\|I\| = r^2 d\theta/dt = $ constant proves Kepler's Second Law: the area swept out by the position vector per unit of time $r^2 d\theta/dt$ is constant. Kepler's Third Law says that the period T of the rotation depends only on the semi-major axis $T = 2\pi a^{3/2}$, and can be easily deduced from the above. The reader is invited to see Pollard (1976) for more details.

2.2 Vector fields

The force field F on $\mathbb{R}^3\setminus\{(0,0,0)\}$ in the previous section is an example of a vector field, a smooth choice of a tangent vector at each point of the manifold.

DEFINITION 2.2.1
A smooth vector field X on a manifold M is a smooth map that assigns to every point $p \in M$ a vector X_p in the tangent space T_pM at p.

Equivalently, a smooth vector field is a smooth map $X : M \to TM$ satisfying $X(p) \in T_pM$ for each $p \in M$. Since every tangent vector at p can be uniquely expressed as a linear combination of a vector basis of the tangent space, we can write a smooth vector field as

$$X = \sum_{i=1}^m v_i \frac{\partial}{\partial x_i}, \tag{2.2.1}$$

where $(x_1, \ldots, x_m)$ are the corresponding local coordinates, and the vector components $v_i : U_\alpha \to \mathbb{R}$ are smooth real valued functions for all $i = 1, \ldots, m$.

If we think of tangent vectors as derivations, then we can define the real valued function $X(f) : M \to \mathbb{R}$ by

$$X(f)(p) = X_p(f).$$

We will sometimes write Xf instead of $X(f)$. This is a smooth function since all quantities involved in (2.2.1) depend smoothly on $p \in M$. Thus, a smooth vector field can be regarded as a smoothly defined derivation, uniquely determined by the properties

$$(X + Y)_p(f) = X_p(f) + Y_p(f),$$
$$X_p(fg) = f(p)X_p(g) + g(p)X_p(f),$$

for all smooth real valued functions f on M.

We emphasize that a smooth vector field is a globally defined object, unlike tangent vectors, which are separately defined at each point. It turns out that the topology of the manifold restricts the way one can construct smooth (or even continuous) vector fields. This will be discussed in greater detail in Section 7.6 and Section 8.6; for the moment

2.2. VECTOR FIELDS

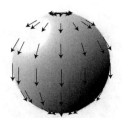

FIGURE 2.2.1
A vector field on the sphere with two zeroes at the poles.

we only show that any continuous vector field on a sphere must vanish somewhere. See Figure 2.2.1. In other words 'one cannot comb the hair of a sphere without making a cowlick'. We start with the smooth version of the theorem, followed by the continuous version.

THEOREM 2.2.2
An even dimensional sphere S^n does not admit any smooth vector field of non-zero tangent vectors.

PROOF We follow Milnor (1978). Assume that the sphere $S^n \subseteq \mathbb{R}^{n+1}$ has a smooth vector field $X(x)$ with $X(x) \neq 0$ at all $x \in S^n$. By replacing $X(x)$ with $X(x)/\|X(x)\|$, we can always assume that the $X(x)$ are unit vectors. We want to prove that n must be odd.

Consider the smooth map $f : S^n \times [0,1] \to \mathbb{R}^{n+1}$ defined by

$$f(x,t) = x + tX(x).$$

Let S_r^n denote the sphere of radius r centered at the origin. For each t, $f_t = f(\cdot, t)$ maps $S^n = S_1^n$ into $S_{\sqrt{1+t^2}}^n$. Since f_0 is the inclusion of S^n into $\mathbb{R}^{n+1}$, it follows immediately from Proposition 1.14.1 that f_t is an embedding of S^n into $\mathbb{R}^{n+1}$ for any t close enough to 0. The map f_t is thus a diffeomorphism from S_r^n to $S_{\sqrt{1+t^2}}^n$ for any t close enough to 0.

We now extend f_t to a smooth map $F_t : \mathbb{R}^{n+1} \setminus \{0\} \to \mathbb{R}^{n+1}$ by setting

$$F_t(x) = \|x\| f_t(x/\|x\|) = x + t\|x\| X(x/\|x\|).$$

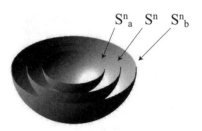

FIGURE 2.2.2
A section through a spherical shell.

If t is close enough to 0 so that f_t is a diffeomorphism from S^n to $S^n_{\sqrt{1+t^2}}$, then F_t will be a diffeomorphism from S^n_r to $S^n_{r\sqrt{1+t^2}}$ for every $r > 0$.

Now fix two radii $b > a > 0$ and consider the spherical shell Σ that lies between S^n_a and S^n_b. It is clear that F_t is a diffeomorphism from Σ to the spherical shell Σ' that lies between $S_{a\sqrt{1+t^2}}$ and $S_{b\sqrt{1+t^2}}$. We compute the volume of Σ' in two different ways. See Figure 2.2.2 in which we have $a < 1 < b$.

Calculation 1. Since Σ' is just Σ scaled by the factor $\sqrt{1+t^2}$, the volume of Σ' is
$$(\sqrt{1+t^2})^{n+1} \cdot \text{Vol}(\Sigma).$$

Calculation 2. We integrate $\det(dF_t)$ over Σ. It is clear from the definition of F_t that dF_t is the identity plus t times the derivative of the map $x \mapsto \|x\|X(x/\|x\|)$. Hence dF_t is an $(n+1) \times (n+1)$ matrix, each entry of which is a linear polynomial in t, whose coefficients are smooth functions of x. Consequently $\det(dF_t)$ is a polynomial of degree $n+1$ in t. By integrating this polynomial over Σ, we see that the volume of Σ' is a polynomial of degree $n+1$ in t.

But Calculation 1 shows that the volume of Σ' is not a polynomial in t unless n is odd. This completes the proof. ∎

We now prove the continuous version of the result by approximating continuous mappings by smooth mappings. This general result is known

2.2. VECTOR FIELDS

as 'the hairy ball theorem', even though it concerns a sphere rather than a ball.

THEOREM 2.2.3 (The hairy ball theorem)
An even dimensional sphere S^n does not admit any continuous vector field of non-zero tangent vectors.

PROOF Suppose by contradiction that there exists a nowhere vanishing continuous vector field $Y(x)$, with $x \in S^n$. By compactness, $\delta = \min_{x \in S^n} \|Y(x)\|$ is positive. Since $x \in S^n \to X(x) \in \mathbb{R}^{n+1}$ is a continuous mapping on a compact domain, we can approximate it by a smooth function (a polynomial) $x \in S^n \to F(x) \in \mathbb{R}^{n+1}$ with

$$\|F(x) - Y(x)\| < \delta, \qquad (2.2.2)$$

for all $x \in S^n$ — this is due to the Weierstrass approximation theorem (see Rudin (1966)). We define

$$X(x) = F(x) - (F(x) \cdot x)x$$

for $x \in S^n$, where $\cdot$ denotes the dot product. Clearly X is a smooth vector field, and is tangent to S^n. We claim that $X(x) \neq 0$ for all $x \in S^n$. If $X(x) = 0$, then $F(x) = (F(x) \cdot x)x$, so $F(x)$ is perpendicular to $Y(x)$. By the Pythagorean theorem, we have

$$\|F(x) - Y(x)\|^2 = \|F(x)\|^2 + \|Y(x)\|^2 \geq \|Y(x)\|^2 \geq \delta^2,$$

which disagrees with (2.2.2).

Thus $X(x)$ is a smooth field of non-zero tangent vectors on S^n, which is a contradiction with Theorem 2.2.3. ∎

For the 2-dimensional sphere, the hairy ball theorem can be interpreted as follows: on the surface of the earth, at any moment, there exists a place where there is no wind at all.

REMARK 2.2.4 Any odd dimensional sphere $S^{2k-1} \subset \mathbb{R}^{2k}$ carries at least one smooth vector field of non-zero tangent vectors, for example

$$X(x) = (x_2, -x_1, x_4, -x_3, \ldots, x_{2k}, -x_{2k-1})$$

for $x = (x_1, x_2, \ldots, x_{2k}, x_{2k-1})$. The fact that the ambient space $\mathbb{R}^{2k}$ is even dimensional makes it possible to write the components of the vector field on S^{2k-1} as a collection of pairs of the form $(x_{2i}, -x_{2i-1})$, with the result that the dot product $x \cdot X(x)$ is 0 at every point $x \in S^{2k-1}$. ∎

Based on the hairy ball theorem, we can give another proof of the Brouwer fixed point theorem.

THEOREM 2.2.5 (Brouwer fixed point theorem)
Suppose that $f : \bar{B}^n \to \bar{B}^n$ is a continuous mapping of the closed unit ball into itself. Then f has a fixed point.

PROOF We again follow Milnor (1978), though with some differences in the details. It is enough to prove the theorem when n is even. If $f : \bar{B}^{2k-1} \to \bar{B}^{2k-1}$ were a continuous map with no fixed points, then
$$(x_1, \ldots, x_{2k}) \mapsto (f(x_1, \ldots, x_{2k-1}), 0)$$
would be a continuous map of B^{2k-1} with no fixed points.

We now suppose that $f : \bar{B}^{2k} \to \bar{B}^{2k}$ is a continuous map with no fixed points, and work towards a contradiction. We think of $\bar{B}^{2k}$ as the unit disk in $\mathbb{R}^{2k}$, bounded by the unit sphere S^{2k-1}. Define a vector field X on $\bar{B}^{2k}$ by
$$X(x) = x - f(x).$$
The properties of f mean that X is continuous and nowhere vanishing.

Claim. $X(x) \cdot x > 0$ if $x \in S^{2k-1}$.

Indeed, if $x \in S^{2k-1}$, $X(x) \cdot x = x \cdot x - f(x) \cdot x = 1 - f(x) \cdot x > 0$. This is because, if $x, y \in \bar{B}^{2k}$, then $x \cdot y \leq 1$, with equality if and only if $\|x\| = 1$ and $y = x$.

The claim says that X points out of $\bar{B}^{2k}$ everywhere on the boundary, but we will need only the weaker consequence that $X(x)$ is not tangent to S^{2k-1} for any $x \in S^{2k-1}$.

Now let $Y(x) = \overline{X(x)}$, where
$$\overline{(v_1, \ldots, v_{2k})} = (-v_2, v_1, \ldots, -v_{2k}, v_{2k-1}).$$
Then $\|Y\| = \|X\|$ and $Y \cdot X = 0$ everywhere on $\bar{B}^{2k}$. Hence Y is continuous and nowhere vanishing. Also $Y(x)$ can be a multiple of x only if $X(x)$ is orthogonal to x, which is impossible on S^{2k-1}.

We now think of S^{2k} as the unit sphere in $\mathbb{R}^{2k+1}$ and $\bar{B}^{2k}$ as the unit disc in $\mathbb{R}^{2k}$. We lift Y to a vector field $\hat{Y}$ on S^{2k} by setting
$$\hat{Y}(y) = \hat{Y}(y_1, \ldots, y_{2k+1}) = (Y(y_1, \ldots, y_{2k}), 0).$$

2.2. VECTOR FIELDS

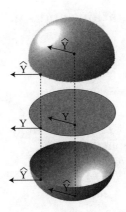

FIGURE 2.2.3
Projections of the disc onto hemispheres.

Then $\widehat{Y}$ is a continuous nowhere vanishing vector field on S^{2k}. Note that $\widehat{Y}(y)$ cannot be a multiple of y unless $y_{2k+1} = 0$. But then $(y_1, \ldots, y_{2k}) \in S^{2k-1}$ and $Y(y_1, \ldots, y_{2k})$ is not a multiple of $(y_1, \ldots, y_{2k})$. Consequently $\widehat{Y}(y)$ is never a multiple of y. Geometrically this means that $\widehat{Y}$ is nowhere perpendicular to S^{2k}. See Figure 2.2.3.

Finally we define the vector field Z to be the component of $\widehat{Y}$ tangent to S^{2k}. It is clear that Z is a continuous nowhere vanishing vector field on S^{2k}. This contradicts the hairy ball theorem and completes the proof. ∎

Based on the hairy ball theorem, we can prove the no-retraction theorem.

THEOREM 2.2.6 (No-retraction theorem)
There exists no continuous retraction $r : \bar{B}^n \to S^{n-1}$.

PROOF Assume that such a retraction exists. The antipodal mapping $S(x) = -x$ of S^{n-1} leaves no point fixed. Then $S \circ r : \bar{B}^n \to S^{n-1} \subseteq \bar{B}^n$ is a continuous mapping on the closed ball with no fixed point. This contradicts the Brouwer fixed point theorem. ∎

2.3 Differential equations and smooth dynamical systems

In this section we consider differential equations defined by vector fields on manifolds and their solutions. We first recall a fundamental result of ordinary differential equations.

THEOREM 2.3.1 *(Existence and uniqueness of solutions of ordinary differential equations)*

Let U be an open subset of $\mathbb{R}^n$ and $f : U \to \mathbb{R}^n$ be a locally Lipschitz function, i.e., for each point $x \in U$ there exist a neighborhood V_x of x in U and a constant $K_x > 0$ such that

$$\|f(x_1) - f(x_2)\| \leq K_x \|x_1 - x_2\| \text{ for all } x_1, x_2 \in V_x.$$

Let x_0 be a point in U, and t_0 be an instant of time in $\mathbb{R}$. Then there exist an $\epsilon > 0$ and a differentiable function

$$\phi_{t_0,x_0} : (t_0 - \epsilon, t_0 + \epsilon) \to U,$$

satisfying

$$\frac{d}{dt}\phi_{t_0,x_0}(t) = f(\phi_{t_0,x_0}(t)), \qquad (2.3.1)$$

$$\phi_{t_0,x_0}(t_0) = x_0. \qquad (2.3.2)$$

This function is unique in the sense that if $\psi_{t_0,x_0} : (t_0 - \epsilon', t_0 + \epsilon') \to U$ also satisfies (2.3.1) and (2.3.2) for some $\epsilon' > 0$, then $\phi_{t_0,x_0}(t) = \psi_{t_0,x_0}(t)$ on their common interval of definition about t_0.

Moreover, $\phi_{t_0,x_0}(t)$ depends continuously on the initial condition x_0, in the sense that for every $\epsilon > 0$, there exists $\delta > 0$ such that $\|x_0 - y_0\| < \delta$ implies $\|\phi_{t_0,x_0}(t) - \phi_{t_0,y_0}(t)\| < \epsilon$.

We say that the function ϕ_{t_0,x_0} is the solution of the differential equation

$$\frac{dx}{dt} = f(x),$$

with initial condition $x(t_0) = x_0$. If f is C^k-differentiable with $k \geq 1$, then it is automatically locally Lipschitz, and the corresponding solution ϕ_{t_0,x_0} is C^{k+1}-differentiable. If we fix $t_0 = 0$, then the solution depends

2.3. SMOOTH DYNAMICAL SYSTEMS

only on its initial value x_0 and is denoted by ϕ_{x_0}. One can extend ϕ_{x_0} to the largest interval I_{x_0} for which the conditions (2.3.1) and (2.3.2) are satisfied. Such an interval exists and it must be open, since otherwise one could extend the solution even more. The mapping ϕ_{x_0} on I_{x_0} defines a curve in U, called a trajectory of f (or flow line). Note that the vector $f(\phi_{x_0}(t_0))$ is tangent to this curve at the point corresponding to $t = t_0$. Then we can define the mapping $\phi : \{(t, x) \,|\, x \in U \text{ and } t \in I_x\} \to U$ by

$$\phi(t, x) = \phi_x(t).$$

Note that the domain $\Omega = \{(t, x) \,|\, x \in U \text{ and } t \in I_x\}$ of ϕ is an open neighborhood of (t_0, x_0). Since ϕ_x is a solution of (2.3.1) with initial condition (2.3.2) at $t_0 = 0$, the mapping ϕ satisfies

$$\phi(0, x) = x. \qquad (2.3.3)$$

The uniqueness of the solution of the differential equation implies that if t_1, t_2 are real numbers such that t_1, t_2 and $t_1 + t_2$ are all in the same interval I_x, then

$$\phi(t_1 + t_2, x) = \phi(t_1, \phi(t_2, x)). \qquad (2.3.4)$$

The mapping ϕ is called a local flow, and is as smooth as f is. The word 'local' expresses the fact that for a fixed $x \in U$ the solution ϕ_x may only be defined on some open interval about 0 and not for all t.

If the function f has compact support, then one can prove that the maximal interval of definition of each solution ϕ_x is the whole of $\mathbb{R}$. Also if the function f is linear, each solution is defined for all t. The resulting flow is thus globally defined. If we let $t_2 = -t_1$ in (2.3.4), we obtain that the mapping $x \to \phi(t_1, x)$ is invertible for each $t_1 \in \mathbb{R}$ and its inverse is given by

$$x \to \phi(-t_1, x), \qquad (2.3.5)$$

so the mapping

$$x \in U \to \phi^t(x) = \phi(t, x) \in U$$

is a diffeomorphism of U for each $t \in \mathbb{R}$. In particular, if $t = 0$ the mapping ϕ^0 is the identity map of M, according to (2.3.3). This together with (2.3.4) and (2.3.5) shows that $\{\phi^t\}_t$ constitutes a 1-parameter group of diffeomorphisms of M. The proofs for all of the above facts can be found in Coddington and Levinson (1955).

Since we will later study flows not necessarily derived from differential equations, it is worth having the following general definition.

DEFINITION 2.3.2
A flow on a topological space M is a continuous function $\phi : \mathbb{R} \times M \to M$ satisfying the following conditions

$$\phi(0, x) = x, \tag{2.3.6}$$

$$\phi(t_1 + t_2, x) = \phi(t_1, \phi(t_2, x)), \tag{2.3.7}$$

for all $t_1, t_2 \in \mathbb{R}$ and $x \in M$. We say that a flow on a topological space defines a continuous-time dynamical system.

We will adopt the following notation convention. Given a flow ϕ, by ϕ_x we mean the trajectory $\phi_x : \mathbb{R} \to M$ given by $\phi_x(t) = \phi(t, x)$, and by ϕ^t we mean the homeomorphism $\phi^t : M \to M$ (or diffeomorphism, in the smooth case) given by $\phi^t(x) = \phi(t, x)$. Of course, we have $\phi^t(x) = \phi_x(t)$. Sometimes we write $\phi^t x$ instead of $\phi^t(x)$.

Now we consider differential equations on smooth manifolds. The substitute for f in the ordinary differential equation is given by a smooth vector field; this is perfectly reasonable, since $f(x)$ can be thought as a tangent vector to U at x. Given a smooth vector field X on a smooth manifold on M, we set the differential equation

$$\frac{dx}{dt} = X(x).$$

The above results translate to manifolds word by word.

THEOREM 2.3.3 (Existence and uniqueness of solutions of differential equations on manifolds)
Let M be a smooth manifold and X be a smooth vector field on M. For each $p \in M$ there exists an open neighborhood $(-\epsilon, \epsilon)$ of 0 and a smooth curve $\phi_p : (-\epsilon, \epsilon) \to M$ satisfying

$$\frac{d}{dt}\phi_p(t) = X_{\phi_p(t)},$$
$$\phi_p(0) = x. \tag{2.3.8}$$

The solution ϕ_p is unique in the sense that if $\psi_p : (-\epsilon', \epsilon') \to M$ satisfies (2.3.8) and (2.3.8), then $\phi_p(t) = \psi_p(t)$ on their common interval of definition.

The solution $\phi_p(t)$ depends continuously on the initial condition p.

There exists a largest open interval I_p on which the solution ϕ_p is defined. The mapping $\phi : \{(t, p) \,|\, p \in M \text{ and } t \in I_p\} \to M$ is a local

2.3. SMOOTH DYNAMICAL SYSTEMS

flow, that is $\phi(0,p) = p$ and $\phi(t_1 + t_2, p) = \phi(t_1, \phi(t_2, p))$ whenever $t_1, t_2, t_1 + t_2 \in I_p$ for some p. If X has compact support (for example, if the manifold M is compact), then the flow is defined globally on $\mathbb{R} \times M$.

PROOF We transport the vector field from the M to $\mathbb{R}^m$ (where $m = \dim M$), solve the corresponding differential in $\mathbb{R}^m$, and then transport it back to M. Since X takes values into the tangent bundle, we need to use its local coordinate systems. Let p be an arbitrary point of the manifold and X_p be the corresponding tangent vector. With respect to some local parametrization $\phi_\alpha : U_\alpha \to V_\alpha$, $(p, X_p) \in TM$ has local coordinates $(x_1, \ldots, x_m, v_1(p), \ldots, v_m(p)) \in U_\alpha \times \mathbb{R}^m$. Define the smooth function $f : U_\alpha \to \mathbb{R}^m$ by $f(q) = (v_1(q), \ldots, v_m(q))$. Using Theorem 2.3.1 and the subsequent discussion, there exists a local flow Φ on U_α. We lift this flow to a local flow $\phi = \phi_\alpha \circ \Phi$ on V_α. It is clear that it is independent of the local parametrization, it is smooth and it satisfies the equations (2.3.6) and (2.3.7). It is also clear that at each $q \in V_\alpha$,

$$\frac{d\phi_q}{dt}(t) = X_{\phi_q(t)}$$

for all t sufficiently close to 0. The solution is unique. The uniqueness allows us to extend the local flow off V_α: if ϕ_α and ϕ_β are two local parametrizations and $V_\alpha \cap V_\beta$ contains a point p, then the solution of the differential equation (2.3.8) is uniquely defined within $V_\alpha \cap V_\beta$, so the trajectories through q, obtained within V_α, and respectively within V_β, coincide on their common domain of definition.

We now need to show that the flow is globally defined provided that the support $\text{supp}(X) = \text{cl}\{p \in M \,|\, X_p \neq 0\}$ is compact. We cannot derive this directly from the corresponding property in $\mathbb{R}^m$, since we work only locally. For a point p with $X_p = 0$, the constant function $\phi_p(t) = p$ satisfies (2.3.8) and (2.3.8), so it represents a globally defined solution. Now let p be a point of M for which $X_p \neq 0$. Suppose, by contradiction, that the open maximal interval of definition $I_p = (a_p, b_p)$ of the solution $\phi_p(t)$ has finite right endpoint $b_p < \infty$. Let t_n be an increasing sequence of positive numbers approaching b_p. Then the sequence of points $p_n = \phi_p(t_n)$ belongs to the compact set $\text{cl}\{p \in M \,|\, X_p \neq 0\}$, so it has a convergent subsequence. We may assume for simplicity that the sequence p_n itself converges to q, for some $q \in M$. There is an open neighborhood $\Omega = (-\epsilon, \epsilon) \times U$ of $(0, q)$ in $\mathbb{R} \times M$ for which the local flow is defined. Then $p_n = \phi_p(t_n) \in \Omega$ for all n sufficiently large, and $\phi_{p_n}(t)$ is defined for all $t \in (-\epsilon, \epsilon)$. As a consequence, $\phi_p(t_n + t)$ is defined for all $t \in (-\epsilon, \epsilon)$. Since $t_n + \epsilon \to b_p + \epsilon$, we conclude that ϕ_p can be extended for t beyond b_p, which is a contradiction. A similar argument works in the case when $a_p > -\infty$. Therefore, the flow ϕ is defined for all t. ∎

The points where a vector field vanishes are remarkable: the corresponding trajectories reduce to points. Of special interest are also the points whose trajectories return to their original position after a certain period of time.

DEFINITION 2.3.4
A point p is called a critical point of a vector field X if $X_p = 0$. It is called a fixed point for a flow ϕ if $\phi_p(t) = p$ for all t. A point p is called a periodic point of the flow ϕ if $\phi_p(t_0) = p$, for some $t_0 > 0$. The trajectory of a periodic points is called a closed trajectory or a periodic orbit. The smallest t_0 with this property is called its principal period.

Example 2.3.5 (Constant vector field of the torus)
Let $\mathbb{T}^2 = \mathbb{R}^2/\mathbb{Z}^2$ be the 2-dimensional torus. Let (v_1, v_2) be a constant vector field on $\mathbb{R}^2$; this induces a constant vector field

$$X_p = v_1 \frac{\partial}{\partial x_1} + v_2 \frac{\partial}{\partial x_2}$$

on the torus, where (x_1, x_2) represent the local coordinates of p. The trajectories in $\mathbb{R}^2$ are straight lines of direction (v_1, v_2). We want to describe the flow induced by the vector field X on $\mathbb{T}^2$. When we factor $\mathbb{R}^2$ by $\mathbb{Z}^2$, the effect on these straight lines can be thought of as follows: when a trajectory hits the upper edge of the square $[0, 1] \times [0, 1]$, then it continues from the corresponding point on the lower edge; when it hits the right hand-side edge, then it continues from the corresponding point on the left hand-side edge, and so on. See Figure 2.3.1. So the trajectories are given by

$$\phi_{(x_1, x_2)}(t) = (x_1 + v_1 t, x_2 + v_2 t) \pmod{1}.$$

We can distinguish two cases. If v_1 and v_2 are rationally independent, then all trajectories hit each edge at a dense set of points. Hence all trajectories wind densely round the torus. If v_1 and v_2 are rationally dependent, then all trajectories hit each edge at a finite number of points. Hence all trajectories are periodic. □

2.3. SMOOTH DYNAMICAL SYSTEMS

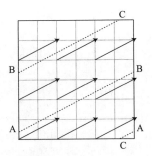

FIGURE 2.3.1
A constant vector field on the torus.

The above example has two interesting features: (i) the vector field has no critical points and (ii) the trajectories in local coordinates are parallel straight lines. It turns out that, in general, condition (i) locally implies condition (ii), if the local coordinates are chosen appropriately.

THEOREM 2.3.6 *(Flow box coordinates)*
Let X be a vector field on a smooth manifold M, such that $X_{p_0} \neq 0$ at some point p_0. Then there exists a local coordinate system near p such that trajectories of the local flow of X in these local coordinates are parallel straight lines.

PROOF Let ϕ be the flow induced by X. Take a relatively compact, embedded submanifold S of codimension 1 in M, passing through p_0 and transverse to X at p_0. Then X is transverse to S at all p in some neighborhood of p_0 in S, that is, $X_p \notin T_pS$ for all p near p_0 in S. We may assume for simplicity that X is transverse to S at all points. Using the Immersion Theorem 1.9.5, we find a local coordinate system $(x_1, \ldots, x_m)$ near p_0 such that $S = \{(x_1, \ldots, x_m) \mid x_m = 0\}$. We can choose the local coordinate system with the additional properties that p_0 has local coordinates $(0, \ldots, 0)$ and X_{p_0} has local coordinates $(0, \ldots, 0, v_m(p_0))$, with $v_m(p_0) > 0$. By shrinking S, if necessary, we can assume that the m-th local coordinate $v_m(p)$ of X_p is positive for all $p \in S$. Since S is relatively compact, there exists $\epsilon > 0$ such that $v_m(\phi^t(p)) > 0$ for all t in the time interval $(-\epsilon, \epsilon)$ and all $p \in S$.

The desired local coordinate system near p_0 is then given by the correspondence $\phi^t(p) \to (x_1, \ldots, x_{m-1}, t)$, with $(x_1, \ldots, x_{m-1})$ representing the first $m-1$ coordinates of a point $p \in S$ and $t \in (-\epsilon, \epsilon)$ the time corresponding to $\phi^t(p)$. We need to check that these new coordinates are compatible to the original ones. The coordinate change $(x_1, \ldots, x_{m-1}, t) \to (x_1, \ldots, x_{m-1}, x_m)$ satisfies

$$\frac{\partial(x_1, \ldots, x_{m-1}, x_m)}{\partial(x_1, \ldots, x_{m-1}, t)} = \begin{pmatrix} \mathrm{id}_{\mathbb{R}^{m-1}} & 0 \\ * & \frac{\partial}{\partial t} x_m(\phi^t(p)) \end{pmatrix}.$$

Since $(\partial/\partial t)(x_m(\phi^t(p))) = v(\phi^t(p))) > 0$, the above matrix is non-singular, so the coordinate change is a diffeomorphism.

It is clear that $t \to (x_1, \ldots, x_{m-1}, t)$ represents the flow ϕ written in these local coordinates, so the trajectories are parallel straight lines.
∎

2.4 Lie derivative, Lie bracket

In this section we are interested in measuring the rate of change of a smooth vector field Y in the direction of another smooth vector field X. We recall that the directional derivative of a smooth function f on M in the direction of X_p of local coordinates $(v_1(p), \ldots, v_m(p))$ is given by

$$X_p(f) = \sum_{i=1}^{m} v_i(p) \frac{\partial f}{\partial x_i}|_p. \tag{2.4.1}$$

The following alternative formula expresses the directional derivative as the rate of change of the function along the local flow line generated by the vector field.

PROPOSITION 2.4.1
Let ϕ be the local flow defined by the smooth vector field X on M. Then

$$X_p(f) = \lim_{t \to 0} \frac{f(\phi^t(p)) - f(p)}{t}. \tag{2.4.2}$$

PROOF We have that $X_p(f) = \frac{d}{dt}(f \circ c)(0)$ for any curve c with $c(0) = p$ and $\frac{dc}{dt}(0) = X_p$. Since the trajectory $\phi_p(t)$ is such a curve, we have

$$X_p(f) = \frac{d}{dt}(f \circ \phi_p)(t)|_{t=0} \tag{2.4.3}$$

2.4. LIE DERIVATIVE, LIE BRACKET

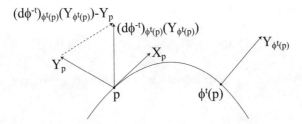

FIGURE 2.4.1
The definition of the Lie derivative.

Using the definition of the derivative and writing $\phi_p(t)$ as $\phi^t(p)$ concludes the proof. ∎

This method of calculating the derivative can be applied to objects more general than smooth functions. The rate of change of a differentiable object along the local flow ϕ generated by a certain vector field is called the Lie derivative of that object in the direction of the vector field. For now, we consider the Lie derivative of a vector field in the direction of another vector field (in Section 6.11 we will consider the Lie derivative of a differential form). In view of (2.4.2), we might guess that we should define the Lie derivative of a vector field Y in the direction of X as the limit of $\left(Y_{\phi^t(p)} - Y_p\right)/t$ as $t \to 0$, where ϕ is the flow determined by X. This, however, does not make sense, since $Y_{\phi^t(p)}$ and Y_p are vectors in different tangent spaces. We can obtain vectors in the same tangent space by transporting $Y_{\phi^t(p)}$ from the base point $\phi^t(p)$ to the base point p through the derivative of ϕ^{-t}. See Figure 2.4.1.

DEFINITION 2.4.2
Let X, Y be two smooth vector fields on a smooth manifold M, p be a point of M, and ϕ be a local flow of X near p. The Lie derivative of Y in the direction of X at p is

$$L_X(Y)(p) = \lim_{t \to 0} \frac{(d\phi^{-t})_{\phi^t(p)}\left(Y_{\phi^t(p)}\right) - Y_p}{t}. \qquad (2.4.4)$$

Since $t \to (d\phi^{-t})_{\phi^t(p)}\left(Y_{\phi^t(p)}\right)$ is a smooth curve in $T_p M$ with value Y_p when $t = 0$, the above limit exists and is a tangent vector at p. We can rewrite the definition

$$L_X(Y)(p) = \frac{d}{dt}\left[(d\phi^{-t})_{\phi^t(p)}\left(Y_{\phi^t(p)}\right)\right]|_{t=0}. \qquad (2.4.5)$$

It is obvious that $L_X(Y)(p)$ depends smoothly on p. We will sometimes write $L_X Y$ instead of $L_X(Y)$.

PROPOSITION 2.4.3
Let X and Y be smooth vector fields on M. Then

$$(L_X(Y))_p(f) = X_p(Y(f)) - Y_p(X(f))$$

for any smooth function f on M.

PROOF First note that the right hand side in the above equation makes sense since $X(f), Y(f)$ are both smooth functions and X_p, Y_p are both derivations.

Let ϕ^t and ψ^t be the local flows defined by X and Y respectively. Using (2.4.5) and the linearity of $(df)_p$ we obtain

$$\begin{aligned}(L_X Y)_p f &= (df)_p(L_X Y)_p \\ &= \frac{d}{dt}\left[(df)_p\left((d\phi^{-t})_{\phi^t p}(Y_{\phi^t p})\right)\right]|_{t=0} \\ &= \frac{d}{dt}\left[d\left(f \circ \phi^{-t}\right)_{\phi^t p}(Y_{\phi^t p})\right]|_{t=0} \\ &= \frac{d}{dt}\left[Y_{\phi^t p}\left(f \circ \phi^{-t}\right)\right]|_{t=0}.\end{aligned}$$

For small $a, b, c \in \mathbb{R}$, let

$$F(a, b, c) = f(\phi^c \circ \psi^b \circ \phi^a p).$$

This map moves the point p for a time a along the flow ϕ^t, then for a time b along the flow ψ^t, and then for a time c along the flow ϕ^t. Then

$$F(t, s, -t) = (f \circ \phi^{-t})(\psi^s(\phi^t p)).$$

By (2.4.3),

$$\frac{\partial}{\partial s} F(t, s, -t) = Y_{\phi^t p}(f \circ \phi^{-t}),$$

2.4. LIE DERIVATIVE, LIE BRACKET

and hence

$$(L_X Y)_p f = \frac{\partial^2}{\partial t \partial s} F(t, s, -t)|_{s=0, t=0}$$
$$= \frac{\partial^2 F}{\partial a \partial b}(0,0,0) - \frac{\partial^2 F}{\partial c \partial b}(0,0,0). \quad (2.4.6)$$

Since $F(a, b, 0) = f\left(\psi^b\left(\phi^a p\right)\right)$, (2.4.3) yields $\partial F/\partial b(a, 0, 0) = Y_{\phi^a p} f = (Yf)(\phi^a p)$ and hence

$$\frac{\partial^2 F}{\partial a \partial b}(0,0,0) = X_p(Yf).$$

Similarly, $F(0, b, c) = f\left(\phi^c\left(\psi^b\right)\right)$, so $\partial F/\partial b(0, 0, c) = X_{\psi^b p} f = (Xf)(\psi^b)$ and

$$\frac{\partial^2 F}{\partial c \partial b}(0,0,0) = Y_p(Xf).$$

Substituting the last two expressions into (2.4.6), we conclude

$$(L_X Y)_p = X_p(Yf) - Y_p(Xf).$$

∎

The following definition is natural in the light of the previous proposition.

DEFINITION 2.4.4
The Lie bracket of two smooth vector fields X and Y is the smooth vector field denoted $[X, Y]$ defined by

$$[X, Y]_p(f) = X_p(Y(f)) - Y_p(X(f)) \quad (2.4.7)$$

for any smooth function f.

Proposition 2.4.3 tells us that this formula really does define a vector field and that, in fact, $[X, Y] = L_X Y$. This provides us another interpretation of the Lie derivative of a vector field in the direction of another vector field: it measures the extent to which the directional derivatives in the directions of X and Y fail to commute.

We now express $[X, Y]$ in local coordinates. Let $(v_1(p), \ldots, v_m(p))$ and $(w_1(p), \ldots, w_m(p))$ be the local coordinates of X_p and Y_p, respectively.

Using the fact that X_p and Y_p are derivations, we compute

$$X(Yf) = X\left(\sum_{j=1}^m w_j \frac{\partial f}{\partial x_j}\right) = \sum_{i,j=1}^m v_i \frac{\partial w_j}{\partial x_i}\frac{\partial f}{\partial x_j} + \sum_{i,j=1}^m v_i w_j \frac{\partial^2 f}{\partial x_i \partial x_j},$$

$$Y(Xf) = Y\left(\sum_{j=1}^m v_j \frac{\partial f}{\partial x_j}\right) = \sum_{i,j=1}^m w_i \frac{\partial v_j}{\partial x_i}\frac{\partial f}{\partial x_j} + \sum_{i,j=1}^m v_i w_j \frac{\partial^2 f}{\partial x_i \partial x_j}.$$

Subtracting the two equations (and dropping the parentheses) we obtain

$$(XY - YX)f = \sum_{i,j=1}^m \left(v_i \frac{\partial w_j}{\partial x_i} - w_i \frac{\partial v_j}{\partial x_i}\right)\frac{\partial f}{\partial x_j}. \qquad (2.4.8)$$

This also illustrates that the Lie bracket $[X,Y]$ measures the extent to which the directional derivatives fail to commute. An easy consequence of (2.4.8) is that $[\partial/\partial x_i, \partial/\partial x_j] = 0$ for all $1 \le i,j \le m$.

It is clear from (2.4.8) that $[X,Y]_p$ depends not only on X_p and Y_p, but also on the first derivatives of X and Y at p. On the other hand, the fact that the right hand side of (2.4.8) is a linear combination of the partial derivatives $\partial f/\partial x_j$ shows us again that $[X,Y]$ is a vector field, cf. (2.4.1)

Many properties of the Lie bracket which are not evident from Definition 2.4.2 follow from Definition 2.4.4.

PROPOSITION 2.4.5
Let X, Y and Z be smooth vector fields on M. We have:

(i) $[X,Y]$ is antisymmetric: $[Y,X] = -[X,Y]$, so $[X,X] = 0$.

(ii) $[X,Y]$ is bilinear: $[aX + bY, Z] = a[X,Z] + b[Y,Z]$, and $[X, aY + bZ] = a[X,Y] + b[X,Z]$ for all $a,b \in \mathbb{R}$.

(iii) $[fX, gY] = fg[X,Y] + fX(g)Y - gY(f)X$; hence $[X,Y]$ is not linear over smooth functions.

(iv) Jacobi identity: $[[X,Y],Z] + [[Y,Z],X] + [[Z,X],Y] = 0$.

PROOF The proofs for (i) and (ii) are left as exercises.
(iii) First, we notice that a simple computation in local coordinates shows that $X(gY) = X(g)Y + gXY$, for all smooth vector fields X, Y

2.4. LIE DERIVATIVE, LIE BRACKET

and all smooth functions g. Then we calculate

$$[fX, gY] = fX(gY) - gY(fX)$$
$$= fX(g)Y + fgXY - gY(f)X + gfYX$$
$$= fg[X,Y] + fX(g)Y - gY(f)X.$$

(iv) By the definition, $[[X,Y],Z] = [XY - YX, Z] = (XY - YX)Z - Z(XY - YX)$. Permuting the letters, we obtain

$$[[X,Y],Z] = XYZ - YXZ - ZXY + ZYX,$$
$$[[Y,Z],X] = YZX - ZYX - XYZ + XZY,$$
$$[[Z,X],Y] = ZXY - XZY - YZX + YXZ.$$

Addition and cancellation lead to (iii). ∎

Lie brackets behave well under diffeomorphisms. In fact, a little more is true. Let $f : M \to N$ be differentiable. Two vector fields X on M and $\bar{X}$ on N are f-related if $(df)_p(X_p) = \bar{X}_{f(p)}$ for all $p \in M$.

PROPOSITION 2.4.6

Suppose that $f : M \to N$ is differentiable and the vector fields X, Y on M are f-related to the vector fields $\bar{X}, \bar{Y}$ on N. Then $[X, Y]$ is f-related to $[\bar{X}, \bar{Y}]$.

PROOF It is clearly enough to show that $\bar{\phi}^t = f \circ \phi^t$ and $\bar{\psi}^t = f \circ \psi^t$, where $\phi^t, \psi^t, \bar{\phi}^t, \bar{\psi}^t$ are the flows generated by $X, Y, \bar{X}, \bar{Y}$, respectively. Suppose now that $\phi_x(t)$ is a flow line associated to X in M, that is, $\phi'_x(t) = X_{\phi_x(t)}$ for all t. Then $(f \circ \phi_x)'(t) = (df)_{\phi_x(t)}(\phi'_x(t)) = (df)_{\phi_x(t)}(X_{\phi_x(t)}) = \bar{X}_{(f \circ \phi_x)(t)}$. Thus $f \circ \phi^t = \bar{\phi}^t$. The analogous statement for ψ^t follows similarly. ∎

We now present a geometric interpretation of $L_X(Y) = [X, Y]$ in terms of the local flows ϕ and ψ generated by X and Y, respectively. We say that the flows ϕ^t and ψ^s commute at a point p if $(\phi^t \circ \psi^s)(p) = (\psi^s \circ \phi^t)(p)$ for all small enough t and s. In general, these flows do not commute. That is, if one departs from p and travels along a flow trajectory of X for a time interval of t, and then travels along a flow trajectory of Y for a time interval of s, the arrival point will not be the same as in the case when one follows the flow trajectories in the reversed

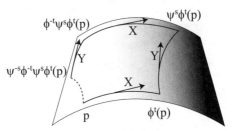

FIGURE 2.4.2
A parallelogram described by the flows.

order. Equivalently, the itinerary $\psi^{-s} \circ \phi^{-t} \circ \psi^s \circ \phi^t(p)$ fails to describe a closed 'parallelogram', since it does not close at p. See Figure 2.4.2.

PROPOSITION 2.4.7
The flows ϕ^t and ψ^s, associated to the vector fields X and Y respectively, commute at all points close enough to p if and only if $[X, Y] = 0$ in a neighborhood of p.

PROOF Let $V(q)$ be the tangent vector at $s = 0$ to the curve $c_{q,t}(s) = (\phi^{-t} \circ \psi^s \circ \phi^t)(q)$. Since $Y_{\phi^t(q)}$ is the tangent vector at $s = 0$ to the curve $s \to (\psi^s \circ \phi^t)(q)$, we have

$$V_q(t) = (d\phi^{-t})_{\phi^t(q)} Y_{\phi^t(q)}. \tag{2.4.9}$$

It follows from Definition 2.4.2 and Definition 2.4.4 that $V_q'(0) = [X, Y]_q$. More generally, we have

$$V_q'(t) = (d\phi^{-t})_{\phi^t(q)}([X, Y]_{\phi^t(q)}).$$

To see this, observe that $c_{q,t+\tau}(s) = \phi^{-t}(c_{\phi^t(q),\tau}(s))$ and hence $V_q(t + \tau) = (d\phi^{-t})_{\phi^t(q)}(V_{\phi^t(q)}(\tau))$. Differentiating with respect to τ at $\tau = 0$ gives the desired equation.

Now suppose that the flows ϕ^t and ψ^s commute at the point q. Then $c_{q,t}(s) = c_{q,0}(s) = \psi^s(q)$ for small enough s and t. Hence $V_q(t) = V_q(0) = Y_q$ for small enough t and $0 = V_q'(0) = [X, Y]_q$.

Conversely, suppose that $[X, Y] = 0$ in a neighborhood of p. Then $V_q'(t) = 0$ for any q close enough to p and any small enough t. It follows that if q is close enough to p, we have $V_q(t) = V_q(0) = Y_q$ for all small enough t. Together with (2.4.9), this says that the vector field Y is is

2.4. LIE DERIVATIVE, LIE BRACKET

ϕ^t-related to itself for small t. But now it follows from the proof of Proposition 2.4.6 that ϕ^{-t} maps the flow line of Y starting at $\phi^q(t)$ to the flow line of Y starting at q. This means that

$$\phi^{-t}(\psi^s(\phi^t(q))) = \psi^s(q)$$

for small s and t, which says that ϕ^t and ψ^s commute at q. ∎

REMARK 2.4.8 One can also prove that $[X,Y]_p$ represents a half of the second derivative of the mapping $t \to \psi^{-t} \circ \phi^{-t} \circ \psi^t \circ \phi^t(p)$. Alternatively, $[X,Y]_p$ represents the tangent vector to the curve given by $t \to \psi^{-\sqrt{t}} \circ \phi^{-\sqrt{t}} \circ \psi^{\sqrt{t}} \circ \phi^{\sqrt{t}}(p)$ at the point p. See Spivak (1999) for details. ∎

The proof of the flow box theorem — Theorem 2.3.6 — is easily adapted to prove the following result.

PROPOSITION 2.4.9
Suppose that $X_1, \ldots, X_k$ are vector fields on M such that $[X_i, X_j] = 0$ for $1 \leq i, j \leq k$ and $X_1(p), \ldots, X_k(p)$ are independent at a point p. Then we can find local coordinates $(x_1, \ldots, x_m)$ near p such that $\partial/\partial x_i = X_i$ for $1 \leq i \leq k$ at all points close enough to p.

The details of the proof are left as an exercise — Exercise 2.7.10.

There is a global version of this proposition, known as Frobenius' theorem, which says that $X_1, \ldots, X_k$ are tangent to a foliation with k-dimensional leaves if $X_1, \ldots, X_k$ are everywhere independent. See, for example Warner (1971).

REMARK 2.4.10 A Lie algebra is a vector space $\mathfrak{g}$ with an antisymmetric bilinear operation $[\cdot, \cdot] : \mathfrak{g} \times \mathfrak{g} \to \mathfrak{g}$ satisfying the Jacobi identity $[[x,y],z] + [[y,z],x] + [[z,x],y] = 0$, for all x, y, z in $\mathfrak{g}$. Proposition 2.4.5 shows that the vector fields on any smooth manifold with the Lie bracket form a Lie algebra.

Lie algebras emerge naturally in the study of Lie groups. A Lie group is a group G with a smooth structure such that the mapping $(x,y) \in G \times G \to xy^{-1} \in G$ is smooth. This ensures that the multiplication operation and the inverse operation are both smooth. It follows that the left translations $L_x : G \to G$, given by $L_x(y) = xy$, and the right translations $R_x : G \to G$, given by $L_x(y) = yx$, are diffeomorphisms. One can show that the vector fields on G that are invariant under all left translations form a vector space, which together with the Lie bracket is a Lie algebra $\mathfrak{g}(G)$. It turns out that this Lie algebra is linearly isomorphic to the tangent space $T_{\text{id}}G$ to G at the identity element id of G. Conversely, every Lie algebra is the Lie algebra of a unique simply connected Lie group.

A fundamental example of a Lie group is the matrix group $G = GL_n(\mathbb{R})$ of all $n \times n$ invertible matrices. Its Lie algebra $\mathfrak{gl}(n)$ is the space of all $n \times n$ matrices.

The idea of a Lie group first appeared in the context of differential equations. The existence of a Lie groups of symmetries for an ordinary differential equation allows one to simplify that equation by reducing the number of variables. See Olver (1993) for details. ∎

2.5 Discrete dynamical systems

Differential equations on manifolds define smooth flows. A discrete time dynamical system can be thought as a 'flow' in which the time is measured in discrete time increments, that is $t = nt_0$, with $n \in \mathbb{Z}$. By rescaling the time units, we can assume that $t_0 = 1$. The restriction $\phi : \mathbb{Z} \times M \to M$ of the flow is completely determined by the knowledge of $\phi(1, \cdot)$, since

$$\phi(n, \cdot) = \underbrace{\phi(1, \cdot) \circ \ldots \circ \phi(1, \cdot)}_{n \text{ times}}.$$

DEFINITION 2.5.1
The action of a continuous map $f : M \to M$ on a topological space M is called a discrete time dynamical system.

In a differentiable context, a discrete dynamical system is described by a smooth map acting on a manifold.

We denote by

$$f^n = \underbrace{f \circ \ldots \circ f}_{n \text{ times}}$$

2.5. DISCRETE DYNAMICAL SYSTEMS

the n-th iterate of f. We put $f^0 = \text{id}$. The (forward) orbit of a point x is the set
$$O^+(x) = \{x, f(x), f^2(x), \ldots\}.$$
If f happens to be a homeomorphism, one can also define the backward orbit of x
$$O^-(x) = \{\ldots, f^{-2}(x), f^{-1}(x), x\},$$
where $f^{-j} = (f^{-1})^j$, for all $j \geq 0$. The total orbit of x
$$O(x) = \{\ldots, f^{-2}(x), f^{-1}(x), x, f(x), f^2(x), \ldots\ldots\}.$$

In the study of a dynamical system, objects of primary importance are its invariant sets. A nonempty subset $S \subseteq M$ is called invariant for f provided $f(S) = S$. Key examples of invariant sets are fixed points and periodic orbits. A point x is called periodic for f if $f^n(x) = x$ for some $n \geq 1$; the smallest n with this property is called the principal period. If $n = 0$ the point p is said to be a fixed point.

It is possible to reduce a continuous time dynamical system to a discrete time dynamical system. One way to do this is to replace the flow $\phi : \mathbb{R} \times M \to M$ by a time discretization $\phi^{t_0} : M \to M$, for some conveniently small time increment $t_0 > 0$. However, periodic orbits of the flow may not correspond to periodic orbits of the map.

Another way in which a continuous time dynamical system can be reduced, in a neighborhood of a periodic orbit, to a discrete time dynamical system, is by taking the first return to a certain surface of section. Let p be a periodic point for a smooth flow ϕ on a manifold M. Let T be the smallest positive number with $\phi(T, p) = p$. The flow line through p is a smooth curve in M. Let S be a smooth submanifold of codimension one in M, transverse to the flow line at p. That is, $(d\phi/dt)(0) \notin T_p S$. The submanifold S is called a local cross-section to the flow. By the implicit function theorem, there exists a neighborhood of p such that, for each q in that neighborhood, the flow line through q is still transverse to S, and moreover, there exists $t(q)$ close to T such that $\phi(t(q), q) \in S$. This defines a smooth map P_S, called the Poincaré map, defined on some neighborhood of p in S by $P_S(q) = \phi(t(q), q)$. See Figure 2.5.1. The discrete time dynamical system induced by P_S on this neighborhood of p reflects quite accurately the behavior of the flow near p. Periodic orbits of the flow that are close to the orbit of p correspond to periodic points of the Poincaré map.

REMARK 2.5.2 An important class of objects similar to orbits consists of the so-called ϵ-pseudo-orbits or ϵ-chains: they differ from

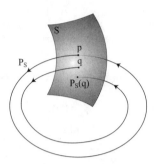

FIGURE 2.5.1
The Poincaré map.

orbits by the fact that at each step of the iteration a jump of size less than some ϵ is allowed. When a point is iterated numerically on a computer, the machine produces a pseudo-orbit rather than a true orbit, due to round-off errors.

If M is a metric space, an ϵ-pseudo-orbit from x to y is a sequence of points $(x_0 = x, x_1, \ldots, x_{n-1}, x_n = y)$ with $d(f(x_{i-1}), x_i) < \epsilon$ for $i = 1, \ldots, n$. A point x is said to be chained to a point y, if for every $\epsilon > 0$, there exists an ϵ-pseudo-orbit from x to y. A point x is called chain recurrent if it is chained to itself.

For example, all periodic points are obviously chain recurrent.

For a rotation of a circle $z \in S^1 \to ze^{i\theta} \in S^1$, where S^1 is thought of as the set of all $z \in \mathbb{C}$ with $|z| = 1$, all points are chain recurrent. In the case when the rotation angle θ is a rational multiple of π, this is true since all points are periodic. In the case when the rotation angle is an irrational multiple of π, this is still true, since any point returns arbitrarily close to its original position after sufficiently many iterations.

The set of all chain recurrent points is closed and invariant. Chain recurrent points that are chained to each other are called chain equivalent. The equivalence classes are called chain components. If any two points of an invariant set are chained one to another, that invariant set is said to be chain transitive.

The chain components are the basic structures of the dynamics. Any dynamical system can be broken into chain components and simple orbits connecting these components. To make this more precise, we introduce some terminology. The chain recurrent set $\mathcal{R}(f)$ is the set of all chain recurrent points. Conley's fundamental theorem of dynamical systems says that for any homeomorphism $f : M \to M$ of a compact metric space M, there exists a real valued function $\Gamma : M \to \mathbb{R}$ such

2.6. HYPERBOLIC FIXED POINTS AND PERIODIC ORBITS

that $\Gamma(x) > \Gamma(f(x))$ for all $x \notin \mathcal{R}(f)$, and the image of M under Γ is a nowhere dense set in $\mathbb{R}$. Furthermore, all chain components of f are level sets for Γ. This result says that one can represent the dynamics on M by a directed graph. The vertices of the graph correspond to the chain components $\Gamma^{-1}(c)$ of f. Two vertices corresponding to two values $c > c'$ in $\mathbb{R}$ are connected by a directed edge if and only if there exists a chain from $\Gamma^{-1}(c')$ to $\Gamma^{-1}(c'')$. This means that there are no orbits starting at one chain component, visiting another chain component, and then going back to the original component. In other words, complicated dynamics can only happen inside the chain components.

An analogue of this result in the case of flows will be discussed in Section 8.3. See Conley (1978) for details. ∎

2.6 Hyperbolic fixed points and periodic orbits

Fixed points and periodic orbits play an important role in the study of dynamical systems and their applications. The conventional wisdom is that only those fixed points and periodic orbits that survive under small perturbations are detectable through experiments. We now briefly discuss a special class of fixed points and periodic orbits that have this property.

DEFINITION 2.6.1
A fixed point p of a diffeomorphism $f : M \to M$ of a manifold M is said to be hyperbolic provided that all the eigenvalues λ of df_p have absolute value different from 1. If all eigenvalues are less than 1 in absolute value, then p is called attracting; if all eigenvalues are greater than 1 in absolute value, then p is called repelling.

The linear mapping $(df)_p : T_pM \to T_pM$ is represented by an $m \times m$ matrix A. In general, the eigenspaces of A do not span the whole of $T_pM \simeq \mathbb{R}^m$. We would like to obtain a natural decomposition of $\mathbb{R}^m$ into vector spaces invariant under A. For this reason, we consider the generalized eigenspaces of A. Let $\lambda_1, \ldots, \lambda_k$ be the distinct eigenvalues of A and $m_1, \ldots, m_k$ be their corresponding multiplicities, where $m_1 + \cdots + m_k = m$. For each real eigenvalue λ_i of multiplicity m_i, the roots of $(A - \lambda_i I)^{m_i} v = 0$ are called generalized eigenvectors. They form a vector subspace E_{λ_i} of dimension m_i in $\mathbb{R}^m$. A basis for this subspace can be obtained by successively solving the matrix equations

$$(A - \lambda_i I)v_1 = 0, (A - \lambda_i I)v_2 = v_1, \ldots, (A - \lambda_i I)^{m_i-1}v_{m_i} = v_{m_i-1}.$$

For each pair of complex conjugate eigenvalues $\lambda_i = \alpha_i + i\beta_i$ and $\bar{\lambda}_i = \alpha_i - i\beta_i$, we consider the solutions $v \in \mathbb{C}^m$ of the equation $(A - \lambda_i I)^{m_i} v = 0$. The real parts $\mathrm{Re}(v)$ and imaginary parts $\mathrm{Im}(v)$ of these vectors span a vector subspace $E_{\lambda_i, \bar{\lambda}_i}$ of dimension $2m_i$ in $\mathbb{R}^m$. A basis for this subspace can be obtained by successively solving the matrix equations

$$(A - \lambda_i I)v_1 = 0, (A - \lambda_i I)v_2 = v_1, \ldots, (A - \lambda_i I)^{m_i-1}v_{m_i} = v_{m_i-1},$$

and taking the real parts $\mathrm{Re}(v_j)$ and imaginary part $\mathrm{Im}(v_j)$ of the vectors $v_1, \ldots, v_{m_i}$. We obtain a decomposition of $\mathbb{R}^m$ into subspaces invariant under A:

$$\mathbb{R}^m = \bigoplus_{\lambda \in \mathbb{R}} E_\lambda \oplus \bigoplus_{\lambda, \bar{\lambda} \in \mathbb{C}} E_{\lambda, \bar{\lambda}}.$$

We can group the generalized eigenspaces into the invariant subspaces, according to their corresponding eigenvalues

$$E_p^s = \bigoplus_{|\lambda|<1} E_\lambda \oplus \bigoplus_{|\lambda|<1} E_{\lambda, \bar{\lambda}},$$

and

$$E_p^u = \bigoplus_{|\lambda|>1} E_\lambda \oplus \bigoplus_{|\lambda|>1} E_{\lambda, \bar{\lambda}},$$

so

$$\mathbb{R}^m = E_p^s \oplus E_p^u.$$

The space E_p^s is called the stable space; the restriction of A to this subspace has only eigenvalues of absolute value less than one. The space E_p^u is called the unstable space; the restriction of A to this subspace has only eigenvalues of absolute value greater than one. The subspaces E_p^s and E_p^u have no vector in common besides the zero vector.

PROPOSITION 2.6.2
There exists a norm $\|\cdot\|_*$ on $T_pM \simeq \mathbb{R}^m$, equivalent to the Euclidean norm $\|\cdot\|$, such that $A_{|E_p^s}$ is a linear contraction and $A_{|E_p^u}$ is a linear expansion relative to this new norm.

2.6. HYPERBOLIC FIXED POINTS AND PERIODIC ORBITS

PROOF There exists $\rho < 1$ such that all eigenvalues of $A_{|E_p^s}$ are less than ρ, and all eigenvalues of $A_{|E_p^u}$ are greater than ρ^{-1}. For a linear mapping L of $\mathbb{R}^n$, we have the relationship

$$\sup\{|\lambda| \,|\, \lambda \text{ is an eigenvalue of } L\} = \lim_{n \to \infty} \|L^n\|^{1/n}.$$

Here the operator norm $\|L\|$ is defined by

$$\|L\| = \sup_{\|v\|=1} \frac{\|L(v)\|}{\|v\|}.$$

We thus have

$$\lim_{n \to \infty} \|A_{|E_p^s}^n\|^{1/n} < \rho, \text{ and } \lim_{n \to \infty} \|A_{|E_p^u}^{-n}\|^{1/n} < \rho.$$

Define the norms $\|\cdot\|_s$ on E_p^s and $\|\cdot\|_u$ on E_p^u, by

$$\|x\|_s = \sum_{n=0}^{\infty} \frac{1}{\rho^n} \|A_{|E_p^s}^n x\|,$$

$$\|y\|_u = \sum_{n=0}^{\infty} \frac{1}{\rho^n} \|A_{|E_p^u}^{-n} y\|,$$

for $x \in E_p^s$ and $y \in E_p^u$. It is not hard to show that these define norms which are equivalent to the Euclidean norm. Since each $v \in \mathbb{R}^n$ can be uniquely written as $v = x + y$ for some $x \in E_p^s$ and $y \in E_p^u$, we can define $\|\cdot\|_*$ on $\mathbb{R}^m$ by

$$\|v\|_* = \max\{\|x\|_s, \|y\|_u\}.$$

This norm has the required properties. Indeed,

$$\|Ax\|_s = \sum_{n=0}^{\infty} \frac{1}{\rho^n} \|A_{|E_p^s}^{n+1} x\|$$

$$= \rho \left(\sum_{n=0}^{\infty} \frac{1}{\rho^{n+1}} \|A_{|E_p^s}^{n+1} x\| \right)$$

$$\leq \rho \left(\|x\| + \sum_{n=0}^{\infty} \frac{1}{\rho^{n+1}} \|A_{|E_p^s}^{n+1} x\| \right)$$

$$\leq \rho \|x\|_s,$$

for any $x \in E_p^s$. Similarly, we obtain $\|A^{-1}y\|_u \leq \rho\|y\|_u$ for any $y \in E_p^u$. ∎

The above result implies that the subspaces E_p^s and E_p^u are uniquely characterized by

$$E_p^s = \{x \in \mathbb{R}^n \mid \lim_{n \to \infty} A^n x = 0\}, \qquad (2.6.1)$$

$$E_p^u = \{y \in \mathbb{R}^n \mid \lim_{n \to \infty} A^{-n} y \to 0\}. \qquad (2.6.2)$$

The following theorem says that the geometric picture of the dynamics near a hyperbolic fixed point is essentially the same as for the linearized system.

THEOREM 2.6.3 (Hartman-Grobman theorem)
If p is a hyperbolic fixed point of f, then f near p is topologically conjugate to $(df)_p$ near 0. This means that there exist neighborhoods U of p in M and V of 0 in $T_p M$, and a homeomorphism $h : V \to U$ which maps 0 to p, such that $f \circ h = h \circ (df)_p$.

The condition $f \circ h = h \circ (df)_p$ means that h maps orbits of f into orbits of $(df)_p$. Since h is a homeomorphism, it also means that h^{-1} maps orbits of $(df)_p$ into orbits of f. For a proof of this result, we recommend Robinson (1998).

This theorem is similar in spirit to the flow box theorem: away from critical points, and near a hyperbolic fixed point, the geometry of the flow lines is simple. The flow is essentially a translation in the first case, and the map is essentially linear, in the second case.

REMARK 2.6.4
The mapping h is not, in general, smooth. If f is C^k-differentiable, a sufficient condition for h to be C^k-smooth is that the eigenvalues $\lambda_1, \ldots, \lambda_m$ of $(df)_p$ (counted with multiplicity) verify

$$\lambda_i \neq \lambda_1^{k_1} \ldots \lambda_m^{k_m},$$

for all powers $k_1, \ldots, k_m$ with $2 \geq k_1 + \ldots + k_m \leq k$. See Sternberg (1958). ∎

Example 2.6.5
The hyperbolic fixed points of diffeomorphisms of the plane can be classified according to the nature of the eigenvalues of the derivative map.

2.6. HYPERBOLIC FIXED POINTS AND PERIODIC ORBITS

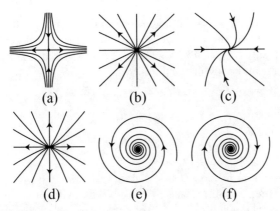

FIGURE 2.6.1
Hyperbolic fixed points.

- Saddle point, with one real eigenvalue less than 1 in absolute value, and one real eigenvalue greater than 1 in absolute value. The nearby orbits follow trajectories as shown in Figure 2.6.1 (a).

- Repelling/attracting fixed point, with equal, real eigenvalues and two independent eigenvectors. Trajectories near such a repelling fixed point are shown in Figure 2.6.1 (b).

- Repelling/attracting fixed point, with equal, real eigenvalues and no two independent eigenvectors. Trajectories near such an attractive fixed point are shown in Figure 2.6.1 (c).

- Repelling/attracting fixed point, with unequal, real eigenvalues. Trajectories near such a repelling fixed point are shown in Figure 2.6.1 (d).

- Repelling/attracting fixed point, with complex, conjugate eigenvalues. Trajectories near such an attractive fixed point are shown in Figures 2.6.1 (e) and (f): the sign of the imaginary part determines the clockwise or the counterclockwise direction of travel along the orbits.

☐

The Hartman-Grobman theorem implies that the subspaces E_p^s and E_p^u of $T_p M$ correspond to objects in M. Indeed, if we choose the domain V of the conjugacy h to be some small disk, then the images of $E_p^s \cap V$

and $E_p^u \cap V$ through h are topological disks in M, along which the orbits behave similarly to the linearized orbits, in a manner described by the equations (2.6.3). These disks are called the local stable and unstable manifolds of p and they are defined by

$$W^s(p, U) = \{x \in U \mid \lim_{n \to \infty} f^n(x) = p\}, \qquad (2.6.3)$$

$$W^u(p, U) = \{y \in U \mid \lim_{n \to \infty} f^{-n}(x) = p\}. \qquad (2.6.4)$$

When the neighborhood U is a disk of radius ϵ relative to some given Riemannian metric on M, these manifolds will be denoted by $W_\epsilon^u(p)$ and $W_\epsilon^s(p)$. When the neighborhood U of p is not specified, we write $W_{loc}^u(p)$ and $W_{loc}^s(p)$. The next theorem says that $W_{loc}^u(p)$ and $W_{loc}^s(p)$ are, as their name suggests, smooth submanifolds of M. This does not follow from the Hartman-Grobman theorem, which only provides topological information near the hyperbolic fixed point.

THEOREM 2.6.6 (Stable manifold theorem)
Assume that $f : M \to M$ is a C^k-diffeomorphism of a C^k-smooth manifold M, and p is a hyperbolic fixed point for f. There exists a neighborhood U of p such that the local stable and unstable manifolds $W^s(p, U)$ and $W^u(p, U)$ are C^k embedded submanifolds of M. Moreover, $T_p W^s(p, U) = E_p^s$ and $T_p W^u(p, U) = E_p^u$. The stable and unstable manifolds are locally unique.

For a proof, see Robinson (1998). Iterating the local stable manifold in backward time, and the local unstable manifold in forward time, we obtain the global stable and unstable manifolds of p:

$$W^s(p) = \bigcup_{n \geq 0} f^{-n} W_{loc}^s(p),$$

$$W^u(p) = \bigcup_{n \geq 0} f^n W_{loc}^u(p).$$

In topological terms, these manifolds are characterized by

$$W^s(p) = \{x \in M \mid \lim_{n \to \infty} f^n(x) = p\},$$

$$W^u(p) = \{y \in M \mid \lim_{n \to \infty} f^{-n}(x) \to p\}.$$

2.6. HYPERBOLIC FIXED POINTS AND PERIODIC ORBITS

The global stable and unstable manifolds are still differentiable. However, they are no longer embedded submanifolds, but only immersed submanifolds. This will be illustrated in the following example.

Example 2.6.7
Consider the Hénon mapping $f_{a,b} : \mathbb{R}^2 \to \mathbb{R}^2$ given by

$$f_{a,b}(x,y) = (1 + y - ax^2, by),$$

where $a = 1.4$ and $b = 0.3$. This map is a diffeomorphism of the plane. A direct computation shows that $f_{a,b}$ has a fixed point p of coordinates

$$x_p = \frac{1}{2a}(-1 + b + \sqrt{(1-b)^2 + 4a}), \quad y_p = bx_p.$$

One eigenvalue of $(df)_p$ is close to -2, and the other one is close to 0, so p is a hyperbolic fixed point. Eigenvectors corresponding to these eigenvalues, together with the global unstable and stable manifolds $W^u(p)$ and $W^s(p)$, are shown in Figure 2.6.2. Notice that the unstable manifold (or the stable manifold) is not an embedded submanifold of $\mathbb{R}^2$. The closure of $W^u(p)$ is a special invariant set, called the Hénon attractor. The dynamics of the map $f_{a,b}$ restricted to this set is 'chaotic'. That is, the trajectories of any two nearby points of the Hénon attractor eventually become uncorrelated. □

We now define hyperbolic periodic orbits.

DEFINITION 2.6.8
A periodic point p of period n of a smooth diffeomorphism $f : M \to M$ is said to be hyperbolic provided that p is a hyperbolic fixed point for f^n. The orbit of p is called a hyperbolic periodic orbit.

Since p is a hyperbolic fixed point of f^n, there exist local (global) stable and unstable manifolds $W^s(p)$ and $W^u(p)$ of p, relative to the iterate f^n. Then we can define the stable and unstable manifolds of the periodic orbit $O(p)$ of p by

$$W^s(O(p)) = \bigcup_{i=0}^{n-1} f^i W^s(p),$$

$$W^u(O(p)) = \bigcup_{i=0}^{n-1} f^i W^u(p).$$

104   2. VECTOR FIELDS AND DYNAMICAL SYSTEMS

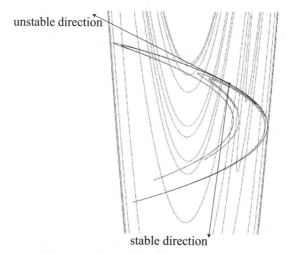

FIGURE 2.6.2
The unstable and stable manifolds of the Hénon map.

The definitions of hyperbolic fixed point and hyperbolic periodic point for continuous time dynamical systems are given in terms of time discretizations of the flow.

DEFINITION 2.6.9
A fixed point p of a smooth flow $\phi : \mathbb{R} \times M \to M$ is hyperbolic if it is hyperbolic as a fixed point of the map $\phi^{t_0} : M \to M$, for some $t_0 \neq 0$. A periodic point p, with $\phi(T, p) = p$ for some $T > 0$, is hyperbolic if $(d\phi^T)_p : T_pM \to T_pM$ has 1 as a simple eigenvalue and no other eigenvalue equal to 1 in absolute value.

In the above definition, the existence of a time t_0 for which ϕ^{t_0} has p as a hyperbolic fixed point guarantees that all ϕ^t with $t \neq 0$ have p as a hyperbolic fixed point (see Exercise 2.7.14).

In the case of a periodic orbit, it is clear that ϕ^T always has 1 as an eigenvalue, since it maps any tangent vector to the flow line through p into itself. If we choose a cross-section S through p, the hyperbolicity condition of the flow at p translates into a hyperbolicity condition for the Poincaré map $P_S : S \to S$ at p as a fixed point.

If the flow ϕ^t is obtained by integrating some smooth vector field X on M, the fixed points of the flow correspond to the zeroes of the vector

2.6. HYPERBOLIC FIXED POINTS AND PERIODIC ORBITS

field. By linearizing the vector field at p, we obtain the differential equation

$$d\xi/dt = (dX)_p(\xi),$$

which gives rise to a linear flow $(t, \xi) \to e^{t(dX)_p}(\xi)$. The original flow is related to the linear flow through the relation

$$(d\phi^{t_0})_p = e^{t_0(dX)_p}.$$

Thus, the hyperbolicity of ϕ^{t_0} at a fixed point p is equivalent to the condition that the real part of any eigenvalue of the linearized vector field $(dX)_p = (\partial X_i/\partial x_j(p))_{i,j}$ is non-zero.

Example 2.6.10
Consider the flat pendulum described in Example 1.5.6. The motion of the pendulum is described by the second order differential equation $ml\theta'' = -mg\sin(\theta)$, where θ is the angle of the pendulum with respect to the vertical, l is the length of the pendulum rod, m is the mass of the bob, and g is the gravitational acceleration. The configuration space is S^1, consisting of all possible values of the angle θ. The second order differential equation is equivalent to the first order system $\theta' = v, v' = -\sin(\theta)$. The phase space is a cylinder of coordinates (θ, v). We can unroll the cylinder into the plane, and visualize the phase portrait of the system, consisting of the vector field $X = (v, -\sin(\theta))$, and the solution curves to X. See Figure 2.6.3. The zeroes of X are the points $(0, k\pi)$, where $k \in \mathbb{Z}$. The linearization of X is given by

$$(dX)_p = \begin{pmatrix} 0 & 1 \\ -\cos(\theta) & 0 \end{pmatrix}.$$

The eigenvalues of $(dX)_p$ are ± 1 at $(0, (2k+1)\pi)$, and $\pm i$ at $(0, 2k\pi)$, for $k \in \mathbb{Z}$. It results that the point p on the cylinder corresponding to $(0, (2k+1)\pi)$ is a hyperbolic saddle. It physically corresponds to the inverted pendulum at rest. The point corresponding to $(0, 2k\pi)$ is not hyperbolic. It corresponds to the pendulum hanging down at rest. The trajectories near the latter fixed point are periodic orbits. □

The Hartman-Grobman theorem remains valid for flows.

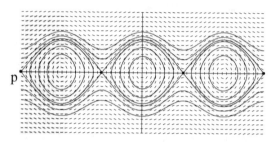

FIGURE 2.6.3
The phase portrait of the pendulum.

THEOREM 2.6.11 *(Hartman-Grobman theorem)*

Assume that X is a smooth vector field on M and ϕ is the (local) flow generated by X. If the p is a hyperbolic fixed point of ϕ^t, then there is a homeomorphism defined in a neighborhood p locally taking orbits of the flow ϕ to orbits of the linearized flow $e^{t(dX)_p}$. The homeomorphism preserves the sense of the orbits but not necessarily the parametrization of the orbits by t.

See Robinson (1998) for a proof. The definitions of the stable and unstable manifolds in the continuous time case are analogous to the corresponding definitions in the discrete time case.

2.7 Exercises

2.7.1 Give an example of a smooth vector field on the sphere with exactly one zero.

2.7.2 Show that the differential equation $\dfrac{dx}{dt} = 3x^{\frac{2}{3}}$ on the real line, with initial condition $x(0) = 0$, has more than one solution. That is, a differential equation that is not Lipschitz can have more than one solution.

2.7. EXERCISES

2.7.3 Give an alternative proof of Theorem 2.3.3 using an embedding of the manifold M in $\mathbb{R}^{2m}$, provided by Whitney's embedding theorem.

2.7.4 Let M be a compact manifold of dimension $m \geq 2$. Show that there exists an injective immersion of $\mathbb{R}$ in M, whose image is not the trajectory of any flow on M.

2.7.5 Give an example of a flow on the projective plane $\mathbb{RP}^2$, which has exactly one fixed point and all other orbits are periodic.

2.7.6 Show that there exists a smooth flow on the Klein bottle (see Exercise 1.15.9) with no fixed points.

2.7.7 Give an example of a smooth vector field on the torus manifold for which no flow trajectory is an embedded submanifold of the torus.

2.7.8 Prove that for any manifold M and any pair of points $p, q \in M$ there exists a diffeomorphism f that takes p to q.

2.7.9 Let ϕ^t be a flow on a manifold M. Define the α- and ω-limit ω-sets of a point x by

$$\alpha(x) = \{y \in M \mid \text{there exists } t_n \to -\infty \text{ such that } \phi^{t_n}(x) \to y\},$$
$$\omega(x) = \{y \in M \mid \text{there exists } t_n \to \infty \text{ such that } \phi^{t_n}(x) \to y\}.$$

Show:

(i) $\alpha(x) = \bigcap_{T \leq 0} \text{cl} \bigcup_{t \leq T} \{\phi^t(x)\}$ and $\omega(x) = \bigcap_{T \geq 0} \text{cl} \bigcup_{t \geq T} \{\phi^t(x)\}$;

(ii) If M is compact and connected, then $\alpha(x)$ and $\omega(x)$ are non-empty, compact and connected sets, invariant under the flow (a set A is said to be invariant under the flow if $\phi^t(A) = A$ for all t).

2.7.10 Let M be a smooth m-dimensional manifold, p be a point in M, and $X_1, \ldots, X_k$ be k linearly independent smooth vector fields defined in a neighborhood of p. Show that there exists a local coordinate system $(x_1, \ldots, x_m)$ near p with $(\partial/\partial x_i)|_p = X_i$, for all $i = 1, \ldots, k$, if and only if $[X_i, X_j] = 0$ for all $1 \leq i, j \leq k$.

2.7.11 The ω-limit set $\omega(x)$ of a point x is the collection of all points $y \in M$ for which there exists an increasing sequence of positive integers n_k such that $f^{n_k}(x) \to y$. If f is a homeomorphism, the α-limit set $\omega(x)$ of x is the collection of all $y \in M$ for which there exists an increasing sequence of positive integers n_k such that $f^{-n_k}(x) \to y$. Show:

(i) $\alpha(x) = \bigcap_{N \leq 0} \text{cl} \bigcup_{n \leq N} \{f^N(x)\}$ and $\omega(x) = \bigcap_{N \geq 0} \text{cl} \bigcup_{n \geq N} \{f^N(x))\}$;

(ii) If M is compact, then $\alpha(x)$ and $\omega(x)$ are non-empty, compact and invariant under f;

(iii) Any point x is chained to any point in $\omega(x)$. Similarly, any point in $\alpha(x)$ is chained to x;

(iv) Any point $x \in M$ is chained to any point in $\omega(x)$, and any point in $\alpha(x)$ is chained to x.

2.7.12 A point $x \in M$ is called a wandering point for f if it has neighborhood U such that $f^n(U)$ is disjoint from U, for all $n \geq 1$. Otherwise, the point is called nonwandering. For example, for a translation, all points in the plane are wandering points, and for a rotation, all points of a circle are nonwandering points. Show:

(i) The wandering points of a map form an open invariant set, and the nonwandering points form a closed invariant set;

(ii) Nonwandering points are chain recurrent;

(iii) Chain recurrent points may be either wandering or nonwandering.

2.7.13 Let $f : \mathbb{R}^2 \to \mathbb{R}^2$ be given by $f(x,y) = (x^2(y^2+1), y)$. Show that the $(0,0)$ is a fixed point of f that is not hyperbolic. Show that there is no linear mapping $A : \mathbb{R}^2 \to \mathbb{R}^2$ such that $f \circ h = h \circ A$ for some local homeomorphism h near $(0,0)$. In other words, the hyperbolicity condition in the Hartman-Grobman theorem is essential.

2.7.14 Show that a fixed point p of a flow $\phi : \mathbb{R} \times M \to M$ is hyperbolic if and only if it is hyperbolic as a fixed point for every map ϕ^t, with $t \neq 0$.

Chapter 3

Riemannian Metrics

3.1 Introduction

In this chapter, we are interested in measuring distances, angles and areas on manifolds.

For an embedded surface S in $\mathbb{R}^3$, the tool that is used to perform such measurements is the dot product

$$u \cdot v = u_1 v_1 + u_2 v_2 + u_3 v_3,$$

where $u = (u_1, u_2, u_3)$ and $v = (v_1, v_2, v_3)$ are vectors in $\mathbb{R}^3$. The components are expressed with respect to the standard basis of $\mathbb{R}^3$. The length of a vector u is given by $\|u\| = \sqrt{u \cdot u}$.

If $c(t)$ is a curve in $S \subseteq \mathbb{R}^3$, the arc length of c between $t = a$ and $t = b$ is obtained by integrating the speed along the curve with respect to time

$$L_{[a,b]}(c) = \int_a^b \left\| \frac{dc}{dt}(t) \right\| dt. \tag{3.1.1}$$

Let $(x_1(t), x_2(t))$ be the equation of the curve with respect to some local coordinate system (x_1, x_2) on S. The velocity vector dc/dt in terms of the basis $\{\partial/\partial x_1, \partial/\partial x_2\}$ of the tangent space of S is given by

$$\frac{dc}{dt} = \frac{dx_1}{dt} \frac{\partial}{\partial x_1} + \frac{dx_2}{dt} \frac{\partial}{\partial x_2},$$

hence the speed can be expressed as

$$\left\| \frac{dc}{dt} \right\| = \left(\frac{dx_1}{dt} \right)^2 \frac{\partial}{\partial x_1} \cdot \frac{\partial}{\partial x_1} + 2 \left(\frac{dx_1}{dt} \frac{dx_2}{dt} \right) \frac{\partial}{\partial x_1} \cdot \frac{\partial}{\partial x_2} + \left(\frac{dx_2}{dt} \right)^2 \frac{\partial}{\partial x_2} \cdot \frac{\partial}{\partial x_2}.$$

The quantities $E = \dfrac{\partial}{\partial x_1} \cdot \dfrac{\partial}{\partial x_1}$, $F = \dfrac{\partial}{\partial x_1} \cdot \dfrac{\partial}{\partial x_2}$ and $G = \dfrac{\partial}{\partial x_2} \cdot \dfrac{\partial}{\partial x_2}$ depend only on the surface and not on the particular curve lying on that surface. Adopting this notation, we obtain

$$\left\| \frac{dc}{dt} \right\|^2 = E \left(\frac{dx_1}{dt} \right)^2 + 2F \left(\frac{dx_1}{dt} \frac{dx_2}{dt} \right) + G \left(\frac{dx_2}{dt} \right)^2. \quad (3.1.2)$$

The quadratic form on the right-hand side is called the first fundamental form of S. The arc length of c is given by

$$L_{[a,b]}(c) = \int_a^b \sqrt{ E \left(\frac{dx_1}{dt} \right)^2 + 2F \left(\frac{dx_1}{dt} \frac{dx_2}{dt} \right) + G \left(\frac{dx_2}{dt} \right)^2 } \, dt. \quad (3.1.3)$$

In the case when S is a plane, we have $E = 1$, $F = 0$ and $G = 1$, so (3.1.2) represents the Pythagorean theorem, and (3.1.3) represents the arc length formula of a parametrized curve from calculus. Thus, the coefficients E, F and G represent the weights by which we have to adjust the Euclidean distance formula in order to apply it in the case of a general surface.

A curve is called regular provided its derivative is never zero. Let c^1 and c^2 be two regular curves that meet at a point $c^1(t_0) = c^2(t_0)$. The angle made by the two curves at this point is the angle made by their velocity vectors. The cosine of this angle is given by

$$\cos(\theta) = \frac{ \frac{dc^1}{dt}(t_0) \cdot \frac{dc^2}{dt}(t_0) }{ \left\| \frac{dc^1}{dt}(t_0) \right\| \left\| \frac{dc^2}{dt}(t_0) \right\| }, \quad (3.1.4)$$

and can be expressed in terms of the first fundamental form. This is easy in the particular case when c^1 and c^2 are the coordinate curves, i.e., $c^1(t) = (x_1(t), 0)$ and $c^2(t) = (0, x_2(t))$, since we have

$$\cos(\theta) = \frac{F}{\sqrt{EG}}. \quad (3.1.5)$$

The area of a bounded domain D is given by

$$\text{Area}(D) = \int_R \left\| \frac{\partial}{\partial x_1} \times \frac{\partial}{\partial x_2} \right\| dx_1 dx_2,$$

where R is the region in the parameter space corresponding to D. Using the linear algebra formula

$$\|u \times v\|^2 + (u \cdot v)^2 = \|u\|^2 \|v\|^2,$$

3.1. INTRODUCTION

for $u = \partial/\partial x_1$ and $v = \partial/\partial x_2$, we obtain

$$\text{Area}(D) = \int_R \sqrt{EG - F^2}\, dx_1 dx_2. \qquad (3.1.6)$$

All of the above measurements rely on the choice of a basis of the tangent space at every point, and on the dot product defined on the ambient space $\mathbb{R}^3$.

In the case of an abstract manifold, there is no ambient space. Although each tangent space to the manifold is linearly isomorphic to a Euclidean space, there is no canonical choice of a smooth inner product on all tangent spaces. Of course, one could locally transport the dot product from the Euclidean space to each tangent space to the manifold through local parametrizations. But these local choices may not agree with each other wherever charts overlap. Hence one has to smooth out these choice on the overlaps. The result is an inner product (a positive definite, symmetric, bilinear form) on each tangent space, smoothly depending on the base point. This is called a Riemannian metric on the manifold. It is not a metric in the sense of a metric space. It is nevertheless called a metric because it does eventually define a distance function between points on the manifold, by the arc length of the shortest path between points.

One of the main motivations for Riemannian metrics is the construction of non-Euclidean geometries. Although axiomatic constructions of non-Euclidean geometries were created before Riemann, the first simple models for these geometries were obtained by defining appropriate Riemannian metrics on familiar spaces.

Perhaps the single most important motivation of Riemannian geometry is the relationship with physics, in particular, with relativity theory. One of the first experimental confirmations of special relativity was the correct prediction of the precession of Mercury. Einstein showed that one can accurately compute the orbit of Mercury if one endows the 'spacetime' with an appropriate 'semi-Riemannian' metric — the Schwarzschild metric — and solves for the paths of minimal action.

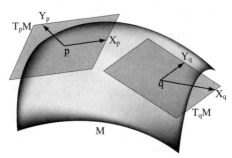

FIGURE 3.2.1
Riemannian metric.

3.2 Riemannian metrics and the first fundamental form

A Riemannian metric is an inner product smoothly defined on each tangent space to a manifold. An inner product on a vector space V is a function $\langle \cdot, \cdot \rangle : V \times V \to \mathbb{R}$ that has the following properties:

(i) bilinearity: $\langle a_1 u_1 + a_2 u_2, v \rangle = a_1 \langle u_1, v \rangle + a_2 \langle u_2, v \rangle$ and $\langle v, a_1 u_1 + a_2 u_2 \rangle = a_1 \langle v, u_1 \rangle + a_2 \langle v, u_2 \rangle$;

(ii) symmetry: $\langle u, v \rangle = \langle v, u \rangle$;

(iii) positive definiteness: if $u \neq 0$ then $\langle u, u \rangle > 0$.

DEFINITION 3.2.1
A Riemannian metric on a smooth manifold M is a mapping g that assigns to every point $p \in M$ an inner product $g_p = \langle \cdot, \cdot \rangle_p$ on the tangent space T_pM, which depends smoothly on $p \in M$, in the sense that, for any two smooth vector fields X and Y on M, the function

$$p \in M \longrightarrow \langle X_p, Y_p \rangle_p \in \mathbb{R}$$

is smooth. A manifold M endowed with a Riemannian metric is called a Riemannian manifold. See Figure 3.2.1.

This definition does not depend on the choice of a local parametrization on M. It is however convenient to express the mapping g in local

3.2. RIEMANNIAN METRICS

coordinates. Let $(x_1, \ldots, x_m)$ be a local coordinate system around a point p. The canonical basis of T_pM is $\partial/\partial x_1|_p, \ldots, \partial/\partial x_m|_p$. Suppose that $X_p = \sum_{i=1}^m v_i(p)\frac{\partial}{\partial x_i}|_p$ and $Y_p = \sum_{i=1}^m w_i(p)\frac{\partial}{\partial x_i}|_p$ are two smooth vector fields. Then

$$\langle X_p, Y_p \rangle = \sum_{i,j=1}^m v_i(p) w_j(p) \left\langle \frac{\partial}{\partial x_i}|_p, \frac{\partial}{\partial x_j}|_p \right\rangle.$$

Hence the Riemannian metric at p can be written in terms of the functions

$$g_{ij}(p) = g_{ij}(x_1, \ldots, x_m) = \left\langle \frac{\partial}{\partial x_i}|_p, \frac{\partial}{\partial x_j}|_p \right\rangle.$$

These functions form an $m \times m$ matrix $(g_{ij})_{i,j=1,\ldots,m}$. The bilinearity of $\langle \cdot, \cdot \rangle_p$ leads to the equation

(i) $\langle X, Y \rangle = (v_1, \ldots, v_m) \begin{pmatrix} g_{11} & \cdots & g_{1m} \\ & \ddots & \\ g_{m1} & \cdots & g_{mm} \end{pmatrix} \begin{pmatrix} w_1 \\ \vdots \\ w_m \end{pmatrix}.$

The symmetry and positive definiteness of $\langle \cdot, \cdot \rangle$ induce the following properties of (g_{ij}):

(ii) symmetry: $g_{ij} = g_{ji}$, for all $i, j = 1, \ldots, m$;

(iii) positive definiteness: $\sum_{i,j=1}^m g_{ij} v_i v_j > 0$ for all $(v_1, \ldots, v_m) \neq 0$.

In particular, since the matrix (g_{ij}) is positive definite, it must also be invertible.

The matrix (g_{ij}) measures the extent to which the basis

$$\frac{\partial}{\partial x_1}|_p, \ldots, \frac{\partial}{\partial x_m}|_p$$

of T_pM differs from orthonormal. We will often denote by g both the Riemannian metric and the matrix associated to the metric. A classical notation for a Riemannian metric is

$$ds^2 = \sum_{i,j=1}^m g_{ij} dx_i dx_j,$$

with s thought of as the 'arc-length' element.

For an embedded surface S in $\mathbb{R}^3$, there exists a canonical Riemannian metric induced by the dot product in $\mathbb{R}^3$. For a given point p, the tangent

plane T_pS is a vector subspace of $\mathbb{R}^3$. The Riemannian metric on S is then defined by
$$\langle X_p, Y_p \rangle = X_p \cdot Y_p,$$
for $X_p, Y_p \in T_pS$.

Similarly, we can induce a natural metric on any immersed submanifold of a Riemannian manifold:

PROPOSITION 3.2.2
If M is a Riemannian manifold with metric $\langle \cdot, \cdot \rangle$ and $f : N \to M$ is an immersion of a manifold N into the manifold M, then the bilinear form $I_p : T_pN \times T_pN \to \mathbb{R}$ defined by
$$I_p(X_p, Y_p) = \langle (df)_p(X_p), (df)_p(Y_p) \rangle_{f(p)},$$
induces a Riemannian metric on N, called the first fundamental form.

PROOF The bilinearity and symmetry of I are immediate. The immersion condition ensures that $X_p \neq 0$ implies $(df)_p(X_p) \neq 0$, so $I_p(X_p, X_p) = \langle (df)_p(X_p), (df)_p(X_p) \rangle \neq 0$. ∎

Historically, the name 'first fundamental form' was given to the quadratic form $v \to I(v,v)$ induced on a surface by the Euclidean metric. To simplify the notation, we will frequently denote the first fundamental form also by $\langle \cdot, \cdot \rangle$.

REMARK 3.2.3 For an abstract manifold, a Riemannian metric is an extra structure on the top of the pre-existing smooth structure. As a physical model, we can think of a system of particles subject to some constraints. The generalized coordinates generate the configuration space, which is a smooth manifold. We can think of a Riemannian metric as describing the kinetic energy
$$K(p) = \frac{1}{2} \langle v, v \rangle_p$$
at each position p in the configuration space, where v denotes the velocity vector. The same trajectories in the configuration space can be travelled with different velocities, and implicitly with different kinetic energies. That is, the same smooth manifold may carry different Riemannian metrics. ∎

3.2. RIEMANNIAN METRICS

Example 3.2.4
We endow the 2-dimensional torus with two distinct Riemannian metrics.

(i) First we consider the embedded torus $\mathbb{T}^2$ given by the equations

$$\phi(x_1, x_2) = ((R + r \cos x_1) \cos x_2, (R + r \cos x_1) \sin x_2, r \sin x_1),$$

where $x_1, x_2 \in \mathbb{R}^2$. We recall that the appropriate restrictions of this equation provide local parametrizations of the torus. Then

$$\frac{\partial}{\partial x_1} = (-r \sin x_1 \cos x_2, -r \sin x_1 \sin x_2, r \cos x_1),$$

and

$$\frac{\partial}{\partial x_2} = (-(R + r \cos x_1) \sin x_2, (R + r \cos x_1) \cos x_2, 0),$$

constitute a basis of $T_p \mathbb{T}^2$. The corresponding first fundamental form is given by

$$g_{11} = E = r^2, \ g_{12} = g_{21} = F = 0, \ g_{22} = G = (R + r \cos x_1)^2.$$

(ii) Now consider the torus $\mathbb{T}^2$ represented as $\mathbb{R}^2/\mathbb{Z}^2$. A local parametrization is given by

$$\phi(x_1, x_2) = [x_1, x_2].$$

A basis of $T_p \mathbb{T}^2$ is

$$\frac{\partial}{\partial x_1} = (1, 0) \text{ and } \frac{\partial}{\partial x_2} = (0, 1).$$

The the first fundamental form is defined by

$$g_{11} = E = 1, \ g_{12} = g_{21} = F = 0, \ g_{22} = G = 1.$$

The torus endowed with this Riemannian structure is locally the same as the Euclidean space $\mathbb{R}^2$, from a metric point of view. For this reason, it is called the flat torus.

It is clear that although the tori defined in (i) and (ii) are diffeomorphic, their metric properties are different. □

Now we will show that any manifold can be endowed with at least one Riemannian metric.

THEOREM 3.2.5
Every smooth manifold has at least one Riemannian metric.

PROOF Let $\{(U_\alpha, \phi_\alpha)\}_\alpha$ be a smooth structure on M. As discussed in Section 1.3, the topology of a manifold is Hausdorff and the family of local parametrizations can be chosen countable. Moreover, by throwing away as many charts as necessary, we can choose the family of open sets $V_\alpha = \phi_\alpha(U_\alpha)$ to be locally finite, which means that for every point $p \in M$, there is an open neighborhood U of p such that $U \cap V_\alpha \neq \emptyset$ for only a finite set of indices α. For each $p \in T_p V_\alpha$ we define the inner product g_p^α on $T_p M$ by taking the dot product

$$g_p^\alpha(X_p, Y_p) = v_1(p)w_1(p) + \cdots + v_m(p)w_m(p),$$

where $X_p = (v_1(p), \ldots, v_m(p))$, and $Y_p = (w_1(p), \ldots, w_m(p))$ in local coordinates. The bilinear form g_p^α is an inner product on $T_p M$, which depends smoothly on $p \in V_\alpha$. In order to glue these inner products together, we use a partition of unity $\{f_\alpha\}$ of smooth functions on M, such that

(i) $0 \leq f_\alpha \leq 1$, for each α;

(ii) the support of each f_α is contained in V_α;

(iii) $\displaystyle\sum_\alpha f_\alpha = 1$.

Define now the form

$$g(X_p, Y_p) = \sum_\alpha f_\alpha g_p^\alpha(X_p, Y_p).$$

Due to the fact that $\{V_\alpha\}$ is locally finite, for each $p \in M$ we have $g_p = f_{\alpha_1} g_p^{\alpha_1} + \cdots + f_{\alpha_m} g_p^{\alpha_m}$, for some $\alpha_1, \ldots, \alpha_m$. It is clear that this defines an inner product on $T_p M$. The mapping g is smooth in the sense required by the definition of a Riemannian metric. ∎

In general, the partition of unity and the local Riemannian metrics are not unique. It is easy to see that we can obtain infinitely many different Riemannian metrics.

3.2. RIEMANNIAN METRICS

DEFINITION 3.2.6

(i) The length of a tangent vector $v \in T_pM$ is defined by

$$\|v\| = \langle v, v \rangle^{1/2};$$

(ii) For a given curve $c : [a, b] \to M$, the arc-length of c between a and b is

$$L_{[a,b]}(c) = \int_a^b \left\| \frac{dc}{dt}(t) \right\| dt;$$

(iii) For a pair of regular curves c^1 and c^2 that meet at a point $p = c^1(t_0) = c^2(t_0)$, the angle between the two curves at p is given by

$$\cos\theta = \frac{\left\langle \frac{dc^1}{dt}(t_0), \frac{dc^2}{dt}(t_0) \right\rangle}{\left\| \frac{dc^1}{dt}(t_0) \right\| \left\| \frac{dc^2}{dt}(t_0) \right\|}.$$

By the change of variable theorem for integrals (see Spivak (1965)), and by the chain rule, it is clear that the arc length of a curve and the angle between two curves are independent of parametrization.

We express arc lengths and angles in local coordinates. If $c(t) = \phi_\alpha(x_1(t), \ldots, x_m(t))$ is a curve in M, where ϕ_α is a local parametrization, then the arc length of the segment corresponding to the closed interval $[a, b]$ is

$$L_{[a,b]} = \int_a^b \sqrt{\sum_{i,j=1}^m g_{ij}(c(t)) \frac{dx_i}{dt} \frac{dx_j}{dt}} \, dt.$$

If $c^i(t) = \phi_\alpha(te_i)$ and $c^j(t) = \phi_\alpha(te_j)$ are two of the coordinate curves, then the angle between them at $\phi_\alpha(0)$ satisfies

$$\cos\theta = \frac{g_{ij}}{g_{ii}g_{jj}}.$$

There is a natural way to define the volume of a domain in an oriented Riemannian manifold. For each point $p \in M$, let $\{E_1(p), \ldots, E_m(p)\}$ be an orthonormal basis of T_pM, with respect to the Riemannian metric.

We define the volume of the 'standard cube' determined by the vectors $E_1(p), \ldots, E_m(p)$ in T_pM as

$$\text{Vol}(E_1(p), \ldots, E_m(p)) = 1.$$

If $X_1(p), \ldots, X_m(p)$ is a collection of tangent vectors at p, then, for each i we have
$$X_i(p) = \sum_{j=1}^{m} a_{ij} E_j(p),$$
for some real numbers a_{ij}, $j = 1, \ldots, m$. By linear algebra, the volume of the parallelepiped determined by the vectors $X_1(p), \ldots, X_m(p)$ is
$$\mathrm{Vol}\,(X_1(p), \ldots, X_m(p)) = |\det(a_{ij})|.$$
This yields the following general definition.

DEFINITION 3.2.7
Let $D = \phi_\alpha(R)$ be an open and relatively compact domain in M, where ϕ_α is some local parametrization. The volume of D (defined by the Riemannian structure) is
$$\mathrm{Vol}(D) = \int_R \mathrm{Vol}\left(\frac{\partial}{\partial x_1}|_p, \ldots, \frac{\partial}{\partial x_m}|_p\right) dx_1 \ldots dx_m.$$

Using the change of variable theorem for multiple integrals, one can easily check that this definition is independent of a local parametrization. We can rewrite this formula in terms of local coordinates. Let
$$\frac{\partial}{\partial x_i}|_p = \sum_{j=1}^{m} a_{ij} E_j(p),$$
for some real numbers a_{ij}, $j = 1, \ldots, m$. Note
$$\left\langle \frac{\partial}{\partial x_i}, \frac{\partial}{\partial x_j} \right\rangle = \left\langle \sum_{k=1}^{m} a_{ik} E_k, \sum_{l=1}^{m} a_{jl} E_l \right\rangle.$$
Using the fact that $\langle E_k, E_l \rangle$ equals 0 provided $k \neq l$ and equals 1 otherwise, we obtain
$$g_{ij} = \sum_{k=1}^{m} a_{ik} a_{jk}.$$
In terms of matrices, we have $(g_{ij}) = (a_{ij})(a_{ij})^T$ (here ^{T} denotes the transpose of a matrix) which implies that $|\det(a_{ij})| = \sqrt{\det(g_{ij})}$. We conclude that
$$\mathrm{Vol}(D) = \int_R \sqrt{\det(g_{ij})}\, dx_1 \ldots dx_m.$$

3.2. RIEMANNIAN METRICS

It is possible to measure volumes in a meaningful way without using the Riemannian structure — see Section 6.5.

From the differentiable point of view, two manifolds are 'the same' if they are diffeomorphic. Any two smooth manifolds are locally the same, since their charts are diffeomorphic copies of $\mathbb{R}^m$. In contrast, two Riemannian manifolds are not necessarily locally the same with respect to their metric properties.

DEFINITION 3.2.8
Let M and N be two Riemannian manifolds.

(i) *We say that M is isometric to N if there exists a diffeomorphism $f : M \to N$ such that*
$$\langle u, v \rangle_p = \langle df_p(u), df_p(v) \rangle_{f(p)}, \text{ for all } u, v \in T_pM.$$
Such a map f is called an isometry.

(ii) *We say that M is locally isometric to N if for every point $p \in M$ there exists an isometry $f : U \to V$ from a neighborhood U of p in M to neighborhood V of $f(p)$ in N.*

These two definitions are not equivalent. The flat torus, for instance, is locally isometric to $\mathbb{R}^2$ but it is not isometric to $\mathbb{R}^2$ (since it is not diffeomorphic to $\mathbb{R}^2$). On the other hand, the embedded torus and the flat torus in Example 3.2.4, although diffeomorphic, are not even locally isometric.

Isometries preserve arc lengths, angles between curves and volumes of domains. The set of all isometries of a manifold forms a group with the composition of functions.

REMARK 3.2.9 We recall Whitney's Theorem 1.9.12 stating that every smooth m-dimensional manifold can be embedded in $\mathbb{R}^{2m}$. In the same vein, a difficult result of John Nash (see Nash (1956)) asserts that for any Riemannian manifold there exists an isometric embedding into some Euclidean space $\mathbb{R}^n$ (endowed with the dot product). A compact m-dimensional manifold can always be isometrically embedded in the $(m/2)(3m + 11)$-dimensional Euclidean space, and a non-compact m-dimensional manifold can always be isometrically embedded in the $(m/2)(m + 1)(3m + 11)$-dimensional Euclidean space. The dimension in the non-compact case is not optimal. The optimal dimension depends on the Riemannian structure. ∎

REMARK 3.2.10 There exist more general ways to perform measurements on a manifold than through a Riemannian metric. In particular, one can drop the positive definiteness condition and ask only for non-degeneracy: if $v \neq 0$, then $g_p(v, v) \neq 0$. Such a metric is called a semi-Riemannian metric. Most definitions and properties from Riemannian geometry carry over to semi-Riemannian geometry. The absence of positive definiteness makes impossible to use the arc length formula of a curve segment (since $g_p(v, v)$ can be negative, so $\sqrt{g_p(v,v)}$ is not defined). This is replaced by the energy of the curve

$$E_{[a,b]}(c) = \int_a^b g_p\left(\frac{dc}{dt}, \frac{dc}{dt}\right) dt.$$

The index of a semi-Riemannian metric is the dimension of the maximal subspace of the tangent space on which the quadratic form $v \to g(v, v)$ is negative definite. A semi-Riemannian metric of index equal to 1 (or equal to $m-1$, where m is the dimension of the manifold) is called a Lorentz metric. See O'Neill (1983) for more details

Semi-Riemannian metrics are the cornerstone of relativity theory. In special relativity, the manifold of interest is the spacetime $\mathbb{R}^4 = \{(t, x, y, z)\}$, resulting from the merger of time t with the Euclidean space coordinates (x, y, z). A point (t, x, y, z) is called an event. The metric of choice is the Minkowski metric

$$ds^2 = -c^2 dt^2 + dx^2 + dy^2 + dz^2$$

where c is the speed of light. The Minkowski metric is an example of a Lorentz metric. In special relativity the measurements of time, length, velocity depend on the observer. The measurements performed by one observer are related to the measurements performed by another observer through a linear mapping, called the Lorentz transformation. The Lorentz transformations are isometries. Based on that, one can derive all transformation laws, including the time dilation and length contraction properties.

In general relativity, one uses Lorentz metrics to quantify the effect of warping of spacetime due to gravity. An example is the Schwarzschild metric, modelling the geometry of spacetime near a spherical star. The manifold of interest is $\{(t, r, \theta, \phi) \mid r \neq 0\}$, where t is time and (r, θ, ϕ)

are spherical coordinates. The Schwarzschild metric is described by

$$ds^2 = -c^2\left(1 - \frac{2M}{r}\right)dt^2 + \left(1 - \frac{2M}{r}\right)^{-1}dr^2 + r^2 d\theta^2 + r^2 \sin^2\theta d\phi^2,$$

where M represents the mass of the star. Note than when $M = 0$ or when r is very large, this reduces to the Minkowski metric written in polar coordinates. See Schutz (1985) for an introduction to relativity.

3.3 Standard geometries on surfaces

One of the most important theorems in geometry is the uniformization theorem. Roughly speaking, this theorem says that every surface admits a canonical geometry. All canonical geometries can be induced from the geometries of three standard surfaces — the Euclidean plane, the sphere, and the hyperbolic plane — through certain group actions.

We start by discussing group actions. Suppose that M is a Riemannian manifold and G is a subgroup of the group of isometric transformations of M. The (left) action of G on M is the mapping

$$(g, p) \in G \times M \to g(p) \in M.$$

We denote $g(p)$ by gp. The characteristic properties of a group action are

$$g_1(g_2 p) = (g_1 g_2)p,$$
$$ep = p,$$

where $g_1, g_2 \in G$, $p \in M$, and e is the identity of G. The orbit of a point p is $Gp = \{gp \mid g \in G\}$. We say that G acts freely if no transformation, besides the identity, has any fixed points, that is, $gp = p$ for some p implies $g = e$. We say that G acts properly discontinuously if every $p \in M$ has a neighborhood V such that $g(V) \cap V = \emptyset$ for all $g \neq e$. The quotient space M/G is defined as the space of all distinct orbits Gp, with $p \in M$. The mapping $\pi : G \to M/G$ that sends each p to its orbit $\tilde{p} = Gp \in M/G$ is called the canonical projection.

Assume that G acts freely and properly discontinuously. Then there exists a canonical smooth structure on M/G that makes π a local diffeomorphism. For $p \in M$, let V be a neighborhood of p such that

$g(V) \cap V = \emptyset$ for all $g \neq e$, and let $\phi_\alpha : U_\alpha \to V_\alpha$ be a local parametrization near p with $V_\alpha \subseteq V$. Then we can define a local parametrization $\widetilde{\phi}_\alpha : U_\alpha \to \pi(V_\alpha)$ on M/G near Gp by $\widetilde{\phi}_\alpha(x) = \pi \circ \phi_\alpha(x)$, for all $x \in U_\alpha$. These local parametrizations define a smooth structure on M/G (the proof is left as an exercise — Exercise 3.4.6).

Moreover, we can define a Riemannian metric on M/G that makes π a local isometry. First note that each point $\widetilde{p} \in M/G$ has a neighborhood $\widetilde{U}$ in M/G such that $\pi^{-1}(\widetilde{U})$ consists of pairwise disjoint diffeomorphic copies of $\widetilde{U}$. For each tangent vector $\widetilde{v}$ at $\widetilde{p}$ and for each $p \in \pi^{-1}(\widetilde{p})$, there is a unique vector $v(p)$ tangent to M at p. Moreover, if $p, p' \in \pi^{-1}(\widetilde{p})$, then $p' = gp$ for some $g \in G$, hence $v(p') = (dg)_p(v(p))$ and so

$$\langle v(p), w(p) \rangle_p = \langle v(p'), w(p') \rangle_{p'},$$

since g is an isometry. We define a Riemannian metric on M/G by

$$\langle \widetilde{v}, \widetilde{w} \rangle_{\widetilde{p}} = \langle (d\pi)_p^{-1}(\widetilde{v}), (d\pi)_p^{-1}(\widetilde{w}) \rangle_p,$$

for some choice of $p \in \widetilde{p}$, and $\widetilde{v}, \widetilde{w} \in T_{\widetilde{p}} M$. The above discussion shows that this metric is well defined.

Every connected surface is diffeomorphic to the quotient of one of the three standard surfaces (described below) by the free and properly discontinuous action of a subgroup of isometries. This a weak version of the uniformization theorem stated in Section 5.3.

The Euclidean plane. The manifold is $\mathbb{R}^2$ and the Riemannian metric is, in classical notation

$$ds^2 = dx^2 + dy^2,$$

that is $g_{ij} = \delta_{ij}$, where δ_{ij}, the Kronecker symbol, is defined by $\delta_{ij} = 1$ if $i = j$ and $\delta_{ij} = 0$ otherwise. An isometry is a translation, a rotation, a reflection, or a glide reflection (a reflection composed with a translation along the axis of the reflection). Only translations and glide reflections act freely on $\mathbb{R}^2$.

There are four possible quotients of the plane: the cylinder, represented as the quotient of $\mathbb{R}^2$ by the group generated by one translation, the flat torus, represented as the quotient of $\mathbb{R}^2$ by the group generated by two translations in independent directions, the Möbius strip, represented as the quotient of $\mathbb{R}^2$ by the group generated by a glide reflection, and the Klein bottle, represented as the quotient of $\mathbb{R}^2$ by the group generated by a translation and a glide reflection in independent directions.

3.3. STANDARD GEOMETRIES ON SURFACES

The sphere. The 2-dimensional sphere S_r^2 can be described through local parametrizations of the type

$$\psi(\theta, \phi) = (r \sin \phi \cos \theta, r \sin \phi \sin \theta, r \cos \phi).$$

A tangent vector basis is given by

$$\frac{\partial \psi}{\partial \theta} = (-r \sin \phi \sin \theta, r \sin \phi \cos \theta, 0),$$

$$\frac{\partial \psi}{\partial \phi} = (r \cos \phi \cos \theta, r \cos \phi \sin \theta, -r \sin \phi),$$

so the Riemannian metric is

$$ds^2 = r^2 \sin^2 \phi \, d\theta^2 + r^2 d\phi^2.$$

The isometries of the spheres are the rotations, the reflections in great circles, and compositions of rotations and reflections perpendicular one to the other, referred to as rotoreflections. Equivalently, the isometries of S_r^2 are given by the 3×3 orthogonal matrices (see Exercise 3.4.7).

There is only one quotient manifold, the real projective plane, induced on S_r^2 by the action of the group generated by the antipodal map (which is a rotoreflection).

The hyperbolic plane. The hyperbolic plane is a model of a non-Euclidean geometry, in which the the parallel postulate fails. There are several possible representations of the hyperbolic plane. We present two of them, and refer to Exercise 3.4.8 for a third one. Higher dimensional analogues are discussed in Exercise 3.4.9.

The Poincaré half-plane. The manifold is the upper half plane

$$\mathbb{H}^2 = \{(x, y) \in \mathbb{R}^2 \mid y > 0\},$$

with the smooth structure induced from $\mathbb{R}^2$. The points on the x-axis, which are excluded from $\mathbb{H}^2$, are thought of as 'points at infinity'. The metric on $\mathbb{H}^2$ is defined so that the points at infinity lie at infinite distance,

$$ds^2 = \frac{1}{y^2} dx^2 + \frac{1}{y^2} dy^2.$$

In complex variables, $\mathbb{H}^2$ is the set of all $z \in \mathbb{C}$ with imaginary part $\Im(z) > 0$, and the metric is given by

$$ds^2 = -\frac{4 dz d\bar{z}}{(z - \bar{z})^2}.$$

The isometries of this model are generated by Möbius transformations of the form

$$z \to \frac{az+b}{cz+d},$$

with $a, b, c, d \in \mathbb{R}$ and $ad - bc = 1$, and by the reflection in the imaginary axis $z \to -\bar{z}$. The above Möbius transformations form a group $PSL_2(\mathbb{R})$, represented by all 2×2 matrices of real entries and of determinants equal to 1.

The Poincaré disk. The manifold is the interior of the open unit disk

$$\mathbb{B}^2 = \{z \in \mathbb{C} \,|\, |z| < 1\},$$

with the smooth structure induced from $\mathbb{R}^2$. The Riemannian metric is defined as

$$ds^2 = \frac{4 dz d\bar{z}}{(1-|z|^2)^2} = \frac{4(dx^2 + dy^2)}{(1-x^2-y^2)^2},$$

making the points on the unit circle behave like points at infinity. The isometries of this model are given by Möbius transformations of the form

$$z \to \frac{az + \bar{c}}{cz + \bar{a}},$$

with $a, c \in \mathbb{C}$ and $|a|^2 - |c|^2 = 1$, and by the reflection in the real axis $z \to \bar{z}$.

The above two representations of the hyperbolic plane are isometric, with an isometry given by

$$z \in \mathbb{H}^2 \to \frac{i-z}{i+z} \in \mathbb{B}^2.$$

The verification that this transformation is an isometry is left as an exercise.

All surfaces of genus $g \geq 1$ can be obtained as quotient spaces of the hyperbolic plane by discrete subgroups of $PSL_2(\mathbb{R})$.

The standard surfaces will be used as examples throughout the book.

REMARK 3.3.1 The analogous statement to the uniformization theorem for 3-dimensional manifolds is known as the geometrization conjecture, and is due to William Thurston. Roughly speaking, this conjecture says that each closed 3-dimensional manifold is decomposable into pieces such that each piece admits exactly one out of eight possible canonical geometries. The geometrization conjecture is much more difficult than the uniformization theorem. It is possible that the recent work of Perelman (2002, 2003a,b) has proved this conjecture (see Remark 5.4.9). In particular, the Geometrization Conjecture implies the Poincaré conjecture: a simply-connected compact 3-dimensional manifold is homeomorphic to the 3-dimensional sphere (a manifold is said to be simply connected if it is connected and any two closed paths at the same base-point are homotopic). ∎

3.4 Exercises

3.4.1 If (M_1, g_1) and (M_2, g_2) are Riemannian manifolds, show that the mapping g defined by $g_{(p1,p2)}((X_1, X_2), (Y_1, Y_2)) = (g_1)_{p_1}(X_1, Y_1) + (g_2)_{p_2}(X_2, Y_2)$, defines a Riemannian metric on $M_1 \times M_2$, called the product metric.

3.4.2 Consider the Riemannian metric induced by $\mathbb{R}^4$ on the torus $S^1 \times S^1$, parametrized as $(\cos(x_1), \sin(x_1), \cos(x_2), \sin(x_2))$. Show that $S^1 \times S^1$ with the induced metric is isometric to the torus $\mathbb{R}^2/\mathbb{Z}^2$ with the metric as in Example 3.2.4 (ii).

3.4.3 Show that the antipodal map $A: S^n \to S^n$ given by $A(p) = -p$ is an isometry of S^n. Show that there is a Riemannian metric on the real projective space $\mathbb{RP}^n$ such that the canonical projection $\pi: S^n \to \mathbb{RP}^n$ is a local isometry.

3.4.4 Suppose that M is compact, N is connected, and M is locally isometric to N. Show that N is locally isometric to M.

3.4.5 Prove that for a connected semi-Riemannian manifold, the index of a metric is the same on every tangent space.

3.4.6 An action of a group G on a manifold M is a mapping $\phi: G \times M \to M$ such that for each $g \in G$, $x \in M \to \phi(g, x) \in M$ is a diffeomorphism, and for each $g_1, g_2 \in G$ and $x \in M$, $\phi(g_1 g_2, x) = \phi(g_1, (\phi(g_2, x)))$ and $\phi(e, x) = x$, where e is the identity of the group. Using the definitions at the beginning of Section 3.3, show that if G acts freely and properly discontinuously on M, then M/G is a smooth manifold.

3.4.7 Show that the isometries of S_r^n are the restrictions to S_r^n of the linear orthogonal maps of $\mathbb{R}^{n+1}$.

3.4.8 Consider the Poincaré half plane $\mathbb{H}^2$ and the Poincaré disk $\mathbb{B}^2$ described in Section 3.3.

(i) Show that the mapping $f : \mathbb{H}^2 \to \mathbb{B}^2$ defined by $f(z) = (i - z)/(i + z)$, is an isometry.

(ii) Consider the manifold $\mathcal{H}^2$ consisting of the upper sheet $(z > 0)$ of the hyperboloid $x^2 + y^2 - z^2 = -1$ with the smooth structure induced by $\mathbb{R}^3$. Endow this manifold with the Minkowski metric
$$ds^2 = dx^2 + dy^2 - dz^2.$$
This is called the hyperboloid model of the hyperbolic plane. Show that the map $g : \mathbb{B}^2 \to \mathcal{H}^2$ given by
$$g(z) = \left(\frac{2\Re(z)}{(1 - |z|^2)}, \frac{2\Im(z)}{(1 - |z|^2)}, \frac{1 + |z|^2}{(1 - |z|^2)} \right),$$
is an isometry.

3.4.9 Consider the following models of the n-dimensional hyperbolic space:

(a) *The Poincaré half-space.* The manifold is the half-space $\mathbb{H}^n = \{(x_1, \ldots, x_n) \mid x_n > 0\}$ with the smooth structure induced from $\mathbb{R}^n$, and the Riemannian metric is
$$ds^2 = \frac{dx_1^2 + \cdots + dx_n^2}{x_n^2}.$$

(b) *The Poincaré ball.* The manifold is the open unit ball $\mathbb{B}^n = \{(x_1, \ldots, x_n) \mid x_1^2 + \cdots + x_n^2 < 1\}$ with the smooth structure induced from $\mathbb{R}^n$, and the Riemannian metric is
$$ds^2 = \frac{dx_1^2 + \cdots + dx_n^2}{1 - (x_1^2 + \cdots + x_n^2)^2}.$$

(c) *The hyperboloid model.* The manifold $\mathcal{H}^n$ is upper sheet $(x_n > 0)$ of the hyperboloid $x_1^2 + \cdots + x_{n-1}^2 - x_n^2 = -1$ with the smooth structure induced from $\mathbb{R}^n$, and the semi-Riemannian metric is
$$ds^2 = dx_1^2 + \cdots + dx_{n-1}^2 - dx_n^2.$$

Show that these three models are mutually isometric.

Chapter 4

Riemannian Connections and Geodesics

4.1 Introduction

In this chapter we introduce the parallel transport of a vector along a curve and geodesics. We start by describing these notions for embedded surfaces and then extend them to abstract manifolds.

In the Euclidean plane, the shortest path between a pair of points is the straight line joining the two points. Let us now consider an embedded surface S in $\mathbb{R}^3$. Given two points on the surface, we would like to find a path between the points that minimizes the arc length. Such a path is not, in general, the straight line between the points, since this line may not lie on the surface. If the surface is compact, one can show that there exists a curve segment lying on the surface, which has the minimum arc length when compared to all other curve segments between the two points. Such a curve segment is called a minimizing geodesic. If the surface is not compact, minimizing geodesics are defined at least locally: any point on the surface has a neighborhood such that there exists an arc length minimizing curve segment from that point to every other point in that neighborhood. A geodesic is a smooth curve whose sufficiently short segments are all minimizing geodesics. For example, the geodesics of a sphere are the great circles.

We now define a notion of parallelism for vector fields along a geodesic. In the Euclidean plane, two lines are parallel provided they make equal angles with a transversal. Consequently, we say that a vector field is parallel provided that any two vectors make equal angles with the straight line joining their base points. Let us consider a geodesic segment on a

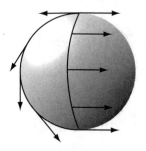

FIGURE 4.1.1
Parallel transport along a broken geodesic.

surface and a tangent vector to the surface, based at one endpoint of the segment. We define the parallel transport of the vector along the geodesic as follows: as the base point of the vector smoothly moves along the geodesic, the tangent vector maintains constant length and constant angle with the geodesic. Then we define the parallel transport of a tangent vector along a broken geodesic consisting of finitely many geodesic segments: we parallel transport the vector along the first geodesic segment from the first vertex to the second, then we parallel transport it along the second geodesic segment from the second vertex to the third, and so on.

For example, consider two orthogonal great circles on the sphere. Take a vector tangent to the first circle at one of the crossings. We parallel transport the vector along the first circle to the other crossing point, and from there we parallel transport it along the second circle back to the initial point. The resulting vector is the opposite of the original one. See Figure 4.1.1.

Finally, the parallel transport of a tangent vector along a smooth curve segment is defined by taking successive approximations of the curve by broken geodesics, performing the parallel transport along each of them, and then taking the limit of the arrival vectors. See Figure 4.1.2.

Using the parallel transport, we can define a new kind of derivative of a smooth vector field along a curve. Let V be a smooth vector field along a curve $c : I \to M$. The curve c is not necessarily a flow line of some vector field on M. We define the covariant derivative of V along c

4.1. INTRODUCTION

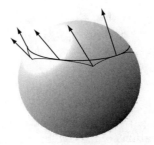

FIGURE 4.1.2
Parallel transport along a curve.

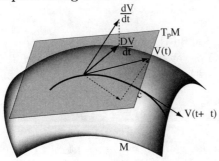

FIGURE 4.1.3
The covariant derivative on a surface.

at the point $c(t)$ by

$$\frac{D_c V}{dt}(c(t)) = \lim_{s \to t} \frac{P_{c(t),c(s)} V_{c(s)} - V_{c(t)}}{s - t},$$

where $P_{c(t),c(s)}$ denotes the parallel transport of the tangent vector $V_{c(s)}$ to a tangent vector at the point $c(t)$.

Geometrically, the covariant derivative $D_c V/dt$ represents the projection of dV/dt onto the tangent plane to the surface. Here dV/dt is the ordinary derivative of a vector valued function $t \in \mathbb{R} \to V(t) \in \mathbb{R}^3$. See Figure 4.1.3. As a physical interpretation, if $c(t)$ represents the trajectory of a particle along the surface, then the covariant derivative $\dfrac{D_c}{dt}\dfrac{dc}{dt}$

4. RIEMANNIAN CONNECTIONS AND GEODESICS

of its velocity vector field $\dfrac{dc}{dt}$ represents the component along the surface of the acceleration of the particle.

In terms of the covariant derivative, we have a much more elegant way to express parallelism. A smooth vector field V along c is parallel if and only if

$$\frac{D_c V}{dt} = 0.$$

This characterization of parallelism is more direct than the method based on successive approximations of a curve by broken geodesics.

One can show that the covariant derivative of a smooth vector field along a curve at a point depends only on the direction dc/dt of the curve at that point, and not on the particular curve. Therefore, the covariant derivative can be naturally extended to a new operation, called an affine connection, that maps a pair of smooth vector fields to a smooth vector field.

The affine connection on the surface assigns to every pair of smooth vector fields X and Y a smooth vector field $\nabla_X Y$ defined as follows: if c is any curve in M with $c(0) = p$ and $dc/dt(0) = X_p$, then

$$\nabla_X Y(p) = \frac{D_c Y}{dt}(p).$$

Geometrically, the affine connection represents the projection of the directional derivative $X_p(Y)$ onto the tangent plane at p.

One can show that the affine connection enjoys the following properties: it is bilinear over the real numbers, it is linear in the variable X over the smooth functions, and it satisfies the product rule in the variable Y with respect to multiplication smooth functions.

This entire process can be reversed. Given an affine connection, that is, a mapping that assigns to every pair of smooth vector fields X and Y another smooth vector field $\nabla_X Y$ satisfying the above properties, one can uniquely define a covariant derivative that agrees with the affine connection, and then a parallel transport that agrees with the covariant derivative. This shows that parallel transport, covariant derivative, and affine connection are equivalent notions.

Although in the surface case the notion of parallel transport is quite intuitive, for an abstract manifold it is more convenient to start with a notion of affine connection. In particular, parallel transport is then described by a single, computable condition in terms of the covariant derivative.

4.2 Affine connections

DEFINITION 4.2.1
(Affine connection) *An affine connection on a manifold M is a smooth map that assigns to every pair of smooth vector fields X and Y on M another smooth vector field $\nabla_X Y$ on M, satisfying the following properties:*

(i) *bilinearity:* $\nabla_{a_1 X_1 + a_2 X_2} Y = a_1 \nabla_{X_1} Y + a_2 \nabla_{X_2} Y$ *and* $\nabla_X(a_1 Y_1 + a_2 Y_2) = a_1 \nabla_X Y_1 + a_2 \nabla_X Y_2$, *for any $a_1, a_2 \in \mathbb{R}$ and any smooth vector fields X_1, X_2, Y_1, Y_2 on M;*

(ii) *linearity in the variable X over the smooth functions:* $\nabla_{f_1 X_1 + f_2 X_2} Y = f_1 \nabla_{X_1} Y + f_2 \nabla_{X_2} Y$, *for any smooth vector fields X_1, X_2 and any smooth functions f_1, f_2 on M;*

(iii) *product rule in the variable Y for multiplication with smooth functions:* $\nabla_X(fY) = f \nabla_X Y + X(f) Y$, *for any smooth function f on M.*

We can express an affine connection in local coordinates $(x_1, \ldots, x_m)$. Since $\nabla_{\partial/\partial x_i} \partial/\partial x_j$ is a vector field, its value at each point can be expressed as a linear combination of the tangent space vector basis

$$\nabla_{\frac{\partial}{\partial x_i}} \frac{\partial}{\partial x_j} = \sum_{k=1}^{m} \Gamma_{ij}^{k} \frac{\partial}{\partial x_k}. \qquad (4.2.1)$$

The n^3 smooth functions Γ_{ij}^k are called the Christoffel symbols and they uniquely determine the affine connection. Indeed, if

$$X = \sum_{i=1}^{m} u_i \frac{\partial}{\partial x_i},$$

$$Y = \sum_{i=1}^{m} v_i \frac{\partial}{\partial x_i},$$

are two smooth vector fields, then

$$\nabla_X Y = \nabla_{\left(\sum_{i=1}^{m} u_i \frac{\partial}{\partial x_i}\right)} \left(\sum_{j=1}^{m} v_j \frac{\partial}{\partial x_j}\right) = \sum_{i=1}^{m} u_i \nabla_{\frac{\partial}{\partial x_i}} \left(\sum_{j=1}^{m} v_j \frac{\partial}{\partial x_j}\right)$$

$$= \sum_{i,j=1}^{m} u_i v_j \nabla_{\frac{\partial}{\partial x_i}} \frac{\partial}{\partial x_j} + \sum_{i,j=1}^{m} u_i \frac{\partial v_j}{\partial x_i} \frac{\partial}{\partial x_j}$$

$$= \sum_{i,j=1}^{m} u_i v_j \left(\sum_{k=1}^{m} \Gamma_{ij}^k \frac{\partial}{\partial x_k} \right) \frac{\partial}{\partial x_j} + \sum_{i,k=1}^{m} u_i \frac{\partial v_k}{\partial x_i} \frac{\partial}{\partial x_k}$$

$$= \sum_{k=1}^{m} \left[\sum_{i,j=1}^{m} u_i v_j \Gamma_{ij}^k + \sum_{i=1}^{m} u_i \frac{\partial v_k}{\partial x_i} \right] \frac{\partial}{\partial x_k}. \qquad (4.2.2)$$

This computation shows that $\nabla_X Y$ at p depends only on the tangent vector X_p at p and on the vector field Y.

It also shows that every smooth manifold admits at least one affine connection. Indeed, on any given chart, we can choose n^3 smooth function Γ_{jk}^i as we please, and then define an affine connection by (4.2.2). Then we glue these locally defined affine connections together into a global affine connection, using a partition of unity, as in the proof of Theorem 3.2.5.

An affine connection uniquely defines a covariant derivative of smooth vector fields along curves.

PROPOSITION 4.2.2 (Covariant derivative)
Given an affine connection ∇ on M, there exists a unique correspondence $V \to D_c V/dt$ that assigns to every smooth vector field V along a curve $c : I \to M$ a smooth vector field $D_c V/dt$ along c, satisfying the properties:

(i) linearity:

$$\frac{D_c(a_1 V_1 + a_2 V_2)}{dt} = a_1 \frac{D_c V_1}{dt} + a_2 \frac{D_c V_2}{dt}, \qquad (4.2.3)$$

for any smooth vector fields V_1, V_2 along c, and any real numbers a_1, a_2;

(ii) product rule:

$$\frac{D_c(fV)}{dt} = f \frac{D_c V}{dt} + \frac{df}{dt} V, \qquad (4.2.4)$$

for any smooth function f defined in some neighborhood of $c(I)$. Here df/dt stands for $d(f \circ c)/dt$;

4.2. AFFINE CONNECTIONS

(iii) *compatibility with the connection:* if there is a smooth vector field Y on M such that $V(t) = Y(c(t))$, then

$$\frac{D_c V}{dt} = \nabla_{\frac{dc}{dt}} Y. \tag{4.2.5}$$

The vector field $D_c V/dt$ is called the *covariant derivative of V along c*. We also denote the covariant derivative by DV/dt, when the curve c is clear from the context.

PROOF We prove the uniqueness part first. We choose a local coordinate system $(x_1, \ldots, x_m)$ of M. The corresponding coordinate tangent vector basis is $\{\partial/\partial x_1, \ldots, \partial/\partial x_m\}$. We can regard $\partial/\partial x_1, \ldots, \partial/\partial x_m$ as smooth vector fields on the coordinate chart. By multiplying these vector fields by a smooth function with the support contained in the coordinate chart, we can extend them to smooth vector fields on the whole manifold. We still call the globally defined vector fields $\partial/\partial x_1, \ldots, \partial/\partial x_m$.

In local coordinates, $c(t) = (x_1(t), \ldots, x_m(t))$,

$$\frac{dc}{dt} = \sum_{j=1}^{m} \frac{dx_j}{dt} \frac{\partial}{\partial x_j}, \quad \text{and}$$

$$V = \sum_{j=1}^{m} v_j \frac{\partial}{\partial x_j},$$

where $v_1, \ldots, v_m$ are smooth functions on the interval I.

Then (4.2.3), (4.2.4) and (4.2.5) lead to

$$\begin{aligned}
\frac{D_c V}{dt} &= \sum_{j=1}^{m} \left(\frac{dv_j}{dt} \frac{\partial}{\partial x_j} + v_j \frac{D}{dt}\left(\frac{\partial}{\partial x_j}\right) \right) \\
&= \sum_{j=1}^{m} \left(\frac{dv_j}{dt} \frac{\partial}{\partial x_j} + v_j \nabla_{\left(\sum_{i=1}^{m} \frac{dx_i}{dt} \frac{\partial}{\partial x_i}\right)} \left(\frac{\partial}{\partial x_j}\right) \right) \\
&= \sum_{k=1}^{m} \left(\frac{dv_k}{dt} + \sum_{i,j=1}^{m} v_j \frac{dx_i}{dt} \Gamma_{ij}^{k} \right) \frac{\partial}{\partial x_k}. \tag{4.2.6}
\end{aligned}$$

This shows that the covariant derivative is uniquely defined through its properties.

For the existence part, we define $D_c V/dt$ on every chart by (4.2.6). It is easy to see that $D_c V/dt$ satisfies all the desired properties. Due to uniqueness, the definitions of $D_c V/dt$ on two overlapping charts agree. In this way, $D_c V/dt$ can be defined on the whole manifold M. ∎

Example 4.2.3
On $\mathbb{R}^3$, there is a standard affine connection $\nabla_X Y$ whose components are the ordinary directional derivative of the components of Y in the direction of X. If $X = \sum u_i \partial/\partial x_i$ and $Y = \sum_{i=1}^m v_i \partial/\partial x_i$, then

$$\nabla_X Y = \sum_{k=1}^m \left[\sum_{i=1}^m u_i \frac{\partial v_k}{\partial x_i} \right] \frac{\partial}{\partial x_k}. \qquad (4.2.7)$$

The Christoffel symbols are all zero.

The induced covariant derivative of a vector field V along a curve $c : I \to \mathbb{R}^3$ is the vector field whose components are the ordinary derivatives of the components of V with respect to t

$$\frac{D_c V}{dt} = \frac{dV}{dt}.$$

Now let M be an embedded surface in $\mathbb{R}^3$. If X and Y are smooth vector fields on M, one can compute $\nabla_X Y$ by (4.2.7). The resulting vector field $\nabla_X Y$ is not necessarily tangent to M. In order to induce an affine connection on M, at each $p \in M$ we take the projection of $(\nabla_X Y)_p$ into the tangent plane $T_p M$. Such a projection can be easily described with respect to a choice of a unit normal vector N_p to M at p. The projection of $\nabla_X Y$ in the direction of N is given by $((\nabla_X Y) \cdot N)N$, hence the projection of $\nabla_X Y$ into the tangent plane is given by

$$\nabla_X^t Y = \nabla_X Y - ((\nabla_X Y) \cdot N)N. \qquad (4.2.8)$$

One can show that the above formula defines an affine connection on M, sometimes referred as the tangential connection. The proof of this fact is left as an exercise. The induced covariant derivative of a vector field V is therefore the projection of dV/dt into the tangent plane. □

The next step concerns the construction of parallel transport.

DEFINITION 4.2.4
A vector field V along a curve c on M is parallel provided that $\dfrac{D_c V}{dt} = 0$.

4.2. AFFINE CONNECTIONS

PROPOSITION 4.2.5
Given an affine connection ∇ on M, a curve $c : I \to M$, and a tangent vector $V_{c(t_0)}$ at a point $c(t_0)$ on the curve, there exists a unique parallel vector field $V_{c(t)}$ along c which extends $V_{c(t_0)}$.

PROOF Let $(x_1, \ldots, x_m)$ be a local coordinate system near $c(t_0)$, and let
$$V_{c(t_0)} = \sum_{k=1}^{m} u_k \frac{\partial}{\partial x_k}$$
be the representation of $V_{c(t_0)}$ in these coordinates. We use (4.2.6) to convert the condition $DV/dt = 0$ into the system of differential equations
$$\begin{cases} \dfrac{dv_k}{dt} + \sum_{i,j=1}^{m} v_j \dfrac{dx_i}{dt} \Gamma_{ij}^k = 0, \\ v_k(t_0) = u_k, \end{cases}$$
for $k = 1, \ldots, m$. By the existence and uniqueness of solutions of ordinary differential equation (Theorem 2.3.1), this system has a unique solution. Since the system is linear, the solution is defined for all t. We obtain a unique vector field V that extends $V_{c(t_0)}$ along the portion of c contained in the domain of the chart. To extend V along the whole of c, we divide c into a finite number of curve segments, each lying in the domain of some chart, and extend V from one curve segment to another. The uniqueness of the solution ensures that the resulting vector field on c is parallel. ∎

DEFINITION 4.2.6
Let $c : I \to M$ be a curve in M and $c(t_0)$ be a point on the curve. The mapping $P_{c(t),c(t_0)} : T_{c(t_0)}M \to T_{c(t)}M$ defined by $P_{c(t),c(t_0)} V_{c(t_0)} = V_{c(t)}$, where $V_{c(t_0)} \in T_{c(t_0)}M$, and $V_{c(t)}$ is the unique extension of $V_{c(t_0)}$ to a parallel vector field along c, is called the parallel transport from $c(t_0)$ to $c(t)$.

PROPOSITION 4.2.7 (Parallel Transport)
The parallel transport $P_{c(t),c(t_0)} : T_{c(t_0)}M \to T_{c(t)}M$ is a linear isomorphisms. If $V(c(t))$ is a vector field along the curve $c(t)$,
$$\lim_{t \to t_0} \frac{P_{c(t_0),c(t)} V_{c(t)} - V_{c(t_0)}}{t - t_0} = \frac{D_c V}{dt}. \tag{4.2.9}$$

PROOF It is easy to verify that $P_{c(t),c(t_0)}$ is a linear isomorphism, whose inverse is $P^{-1}_{c(t),c(t_0)} = P_{c(t),c(t_0)}$. The proof is left as an exercise. ∎

REMARK 4.2.8 Starting with the concept of an affine connection, we constructed a covariant derivative, and then a parallel transport. Conversely, starting with the concept of parallel transport, one can construct a covariant derivative and an affine connection, as outlined in Section 4.1. See Crampin and Pirani (1986) for details. ∎

REMARK 4.2.9 Using the continuous dependence of solutions of ordinary differential equations, one can prove that parallel transport depends nicely on C^1 curves. Assume that c_n is a sequence of smooth parametrized curves that approach a smooth parametrized curve c in the C^1 topology ($c_n(t) \to c(t)$ and $c'_n(t) \to c'(t)$ for all t). If X_n is a parallel vector field along c_n and X_n approaches a vector field X along c, then X is a parallel vector field along c. The statement carries over to the case where c_n and c are piecewise smooth. This agrees with the description of the parallel transport along a curve from Section 4.1. ∎

4.3 Riemannian connections

The machinery introduced in the previous section is independent of a Riemannian structure. The parallel transport is not expected to preserve angles, as in the case of surfaces. In order to abide by the angle preservation property, we need to consider affine connections intimately related to the Riemannian metric.

DEFINITION 4.3.1
Let ∇ be an affine connection and $\langle \cdot, \cdot \rangle$ be a Riemannian metric on a smooth manifold M. We say that the affine connection is compatible with the metric if for every curve $c : I \to M$ and every pair of parallel vector fields V, W along c, the inner product $\langle V, W \rangle$ is constant.

This is equivalent to saying that all vectors of a parallel vector field

4.3. RIEMANNIAN CONNECTIONS

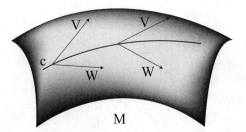

FIGURE 4.3.1
Parallel vector fields for an affine connection compatible with the metric.

have the same length and the angle between any pair of parallel vector fields is constant. See Figure 4.3.1.

PROPOSITION 4.3.2
An affine connection ∇ is compatible to the Riemannian metric $\langle \cdot, \cdot \rangle$ if and only if, for any curve $c : I \to M$, and for any pair of vector fields V, W along c, we have

$$\frac{d}{dt}\langle V, W \rangle = \left\langle \frac{DV}{dt}, W \right\rangle + \left\langle V, \frac{DW}{dt} \right\rangle. \quad (4.3.1)$$

PROOF Assume that (4.3.1) holds true. Let V, W be parallel vector fields along c. Parallelism implies that $DV/dt = DW/dt = 0$, so $\frac{d}{dt}\langle V, W \rangle = 0$ along c, hence $\langle V, W \rangle$ is constant.

Conversely, assume that the affine connection is compatible with the metric. For a point $c(t_0)$ on the curve, we choose an orthonormal basis $\{E_1(c(t_0)), \ldots, E_1(c(t_0))\}$ of the tangent space $T_{c(t_0)}M$. We can extend these vectors to parallel vector fields $\{E_1(c(t)), \ldots, E_1(c(t))\}$ along c, due to Proposition 4.2.5. Since the lengths and the sizes of these vectors are preserved by parallel transport, $\{E_1(c(t)), \ldots, E_1(c(t))\}$ is an orthonormal basis of the tangent space $T_{c(t)}M$, for all $t \in I$. We have $V = \sum_{i=1}^{m} v_i E_i$ and $W = \sum_{i=1}^{m} w_i E_i$ for some smooth functions $v_i, w_i : I \to \mathbb{R}$. Then

$$\langle V, W \rangle = \sum_{i,j=1}^{m} v_i w_j \langle E_i, E_j \rangle = \sum_{i=1}^{m} v_i w_i,$$

hence
$$\frac{d}{dt}\langle V, W\rangle = \sum_{i=1}^{m}\left(\frac{dv_i}{dt}w_i + v_i\frac{dw_i}{dt}\right).$$

On the other side, due to parallelism, we have
$$\frac{DV}{dt} = \sum_{i=1}^{m}\left(\frac{dv_i}{dt}E_i + v_i\frac{DE_i}{dt}\right) = \sum_{i=1}^{m}\frac{dv_i}{dt}E_i,$$

and similarly,
$$\frac{DW}{dt} = \sum_{i=1}^{m}\frac{dw_i}{dt}E_i.$$

We conclude that
$$\left\langle\frac{DV}{dt}, W\right\rangle + \left\langle V, \frac{DW}{dt}\right\rangle = \sum_{i=1}^{m}\left(\frac{dv_i}{dt}w_i + v_i\frac{dw_i}{dt}\right).$$

■

COROLLARY 4.3.3

An affine connection ∇ is compatible to the Riemannian metric $\langle\cdot,\cdot\rangle$ if and only if
$$X\langle Y, Z\rangle = \langle\nabla_X Y, Z\rangle + \langle Y, \nabla_X Z\rangle, \qquad (4.3.2)$$

for all smooth vector fields X, Y, Z on M.

PROOF Consider a curve $c: I \to M$ with $c(0) = p$ and $\frac{dc}{dt}(0) = X_p$. Then
$$X_p\langle Y, Z\rangle = \frac{d}{dt}\langle Y, Z\rangle.$$

On the other side we have
$$\langle\nabla_{X_p}Y, Z\rangle + \langle Y, \nabla_{X_p}Z\rangle = \left\langle\frac{DY}{dt}, Z\right\rangle + \left\langle Y, \frac{DZ}{dt}\right\rangle.$$

Then the previous proposition establishes the equivalence. ■

The above results apply to any affine connection that is compatible with the metric. We now show that there is a unique connection compatible with the metric that also satisfies a certain natural symmetry condition.

4.3. RIEMANNIAN CONNECTIONS

DEFINITION 4.3.4
An affine connection ∇ on a smooth manifold M is called symmetric if

$$\nabla_X Y - \nabla_Y X = [X, Y], \qquad (4.3.3)$$

for all smooth vector fields X, Y, Z on M.

Here $[X, Y]$ is the Lie bracket of the vector fields X and Y. In the case when $X = \partial/\partial x_i$ and $Y = \partial/\partial x_j$, relative to some local coordinate system $(x_1, \ldots, x_m)$, we have $\nabla_X Y - \nabla_Y X = \sum_{i=1}^{m} \left(\Gamma_{ij}^k - \Gamma_{ji}^k \right) \partial/\partial x_k$, while $[X, Y] = 0$. Hence, ∇ is symmetric if and only if $\Gamma_{ij}^k = \Gamma_{ji}^k$, for all i, j and k.

We now give a geometric characterization of symmetry. Consider a parametrized surface in M, i.e. a smooth map $s: U \subseteq \mathbb{R}^2 \to M$ from an open set U in $\mathbb{R}^2$ to M. A smooth vector field along s is a smooth function that assigns to every point $(x, y) \in \mathbb{R}^2$ a tangent vector $V_{s(x,y)} \in T_{s(x,y)} M$. The vector fields $\partial/\partial x$ and $\partial/\partial y$ on $\mathbb{R}^2$ are mapped by the differential ds of s into two smooth vector fields $\partial s/\partial x = ds(\partial/\partial x)$ and $\partial s/\partial y = ds(\partial/\partial y)$ on the surface. Given a point $(x_0, y_0) \in U$, we consider two curves on the surface, $x \to s(x, y_0)$, defined for x within some interval around x_0, and $y \to s(x_0, y)$, defined for y within some interval around y_0. For any smooth vector field V along the surface, we define $\left(\dfrac{DV}{dx} \right)$ as the usual covariant derivative of V along the curve $x \to s(x, y_0)$. The covariant derivative $\left(\dfrac{DV}{dy} \right)$ is defined similarly. See Figure 4.3.2.

PROPOSITION 4.3.5
The affine connection ∇ is symmetric if and only if, for any parametrized surface s in M, we have

$$\frac{D}{dx} \frac{\partial s}{\partial y} = \frac{D}{dy} \frac{\partial s}{\partial x}. \qquad (4.3.4)$$

PROOF Suppose that ∇ is symmetric. Let ϕ_α be a local parametrization on M and $(x_1, \ldots, x_m)$ be the corresponding local coordinate system. We can express the surface s in local coordinates by $(x_1(x, y), \ldots, x_m(x, y))$, with $(x, y) \in U$. The curve $x \to s(x, y_0)$ is represented by $(x_1(x, y_0), \ldots, x_m(x, y_0))$, and the curve $y \to s(x_0, y)$ is

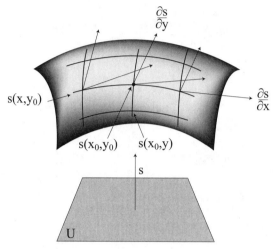

FIGURE 4.3.2
Geometric characterization of symmetry of an affine connection.

represented by $(x_1(x_0, y), \ldots, x_m(x_0, y))$. The restriction of $\partial s/\partial y$ to $x \to s(x, y_0)$ is

$$\frac{\partial s}{\partial y}(x, y_0) = \sum_{i=1}^{m} \frac{\partial x_i}{\partial y}(x, y_0) \frac{\partial}{\partial x_i}\Big|_{s(x,y_0)},$$

and its covariant derivative at $s(x_0, y_0)$, according to (4.2.6), is

$$\frac{D}{dx}\frac{\partial s}{\partial y} = \sum_{k=1}^{m} \left(\frac{\partial^2 x_k}{\partial x \partial y} + \sum_{i,j=1}^{m} \frac{\partial x_i}{\partial x} \frac{\partial x_j}{\partial y} \Gamma_{ij}^k \right) \frac{\partial}{\partial x_k}.$$

Similarly, we obtain

$$\frac{D}{dy}\frac{\partial s}{\partial x} = \sum_{k=1}^{m} \left(\frac{\partial^2 x_k}{\partial x \partial y} + \sum_{i,j=1}^{m} \frac{\partial x_i}{\partial x} \frac{\partial x_j}{\partial y} \Gamma_{ji}^k \right) \frac{\partial}{\partial x_k}.$$

The fact $\Gamma_{ij}^k = \Gamma_{ji}^k$ ends the direct part of the proof.

For the converse, we fix $0 \leq i, j \leq m$, and choose s defined by

$$s(x, y) = \phi_\alpha(0, \ldots, x, \ldots, 0, \ldots, y, \ldots, 0),$$

4.3. RIEMANNIAN CONNECTIONS

where x and y are placed in the i-th and j-th slot, respectively. It follows that

$$\frac{D}{dx}\frac{\partial s}{\partial y} = \sum_{k=1}^{m} \Gamma_{ij}^{k} \frac{\partial}{\partial x_k} \quad \text{and} \quad \frac{D}{dy}\frac{\partial s}{\partial x} = \sum_{k=1}^{m} \Gamma_{ji}^{k} \frac{\partial}{\partial x_k}.$$

Then (4.3.4) implies $\Gamma_{ij}^{k} = \Gamma_{ji}^{k}$. Since i, j were arbitrary, ∇ must be symmetric. ∎

We are ready to prove the main theorem of this section.

THEOREM 4.3.6 *(Fundamental theorem of Riemannian geometry)*
Given a Riemannian manifold, there exists a unique affine connection that is symmetric and compatible to the Riemannian metric. This affine connection is called the Riemannian connection of the manifold (also called the Levi-Civita connection).

PROOF We prove the uniqueness part first. Suppose that such an affine connection exists. Using the compatibility with the metric and the symmetry, we can write

$$\begin{aligned}
X\langle Y, Z\rangle &= \langle \nabla_X Y, Z\rangle + \langle Y, \nabla_X Z\rangle = \\
&= \langle \nabla_X Y, Z\rangle + \langle [X, Z], Y\rangle + \langle \nabla_Z X, Y\rangle, \\
Y\langle Z, X\rangle &= \langle \nabla_Y Z, X\rangle + \langle Z, \nabla_Y X\rangle = \\
&= \langle \nabla_Y Z, X\rangle + \langle [Y, X], Z\rangle + \langle \nabla_X Y, Z\rangle, \\
Z\langle X, Y\rangle &= \langle \nabla_Z X, Y\rangle + \langle X, \nabla_Z Y\rangle = \\
&= \langle \nabla_Z X, Y\rangle + \langle [Z, Y], X\rangle + \langle \nabla_Y Z, X\rangle.
\end{aligned}$$

Adding the the first and the second equation and subtracting the third, we obtain

$$\begin{aligned}
X\langle Y, Z\rangle &+ Y\langle Z, X\rangle - Z\langle X, Y\rangle = \\
&= 2\langle \nabla_X Y, Z\rangle + \langle [X, Z], Y\rangle + \langle [Y, X], Z\rangle - \langle [Z, Y], X\rangle.
\end{aligned}$$

We obtain that $\langle \nabla_X Y, Z\rangle$ is given by the Koszul formula:

$$\langle \nabla_X Y, Z\rangle = \frac{1}{2}\left(X\langle Y, Z\rangle + Y\langle Z, X\rangle - Z\langle X, Y\rangle - \langle [X, Z], Y\rangle - \langle [Y, X], Z\rangle - \langle [Y, Z], X\rangle\right). \quad (4.3.5)$$

Since Z is arbitrary, we conclude that $\nabla_X Y$ is uniquely determined by the Riemannian metric.

For the existence part, we can define $\nabla_X Y$ by (4.3.5) and check that it defines an affine connection that is symmetric and compatible with the metric. This tedious verification is left as an exercise. ∎

For future reference, we will explicitly compute the Christoffel symbols Γ_{ij}^k corresponding to the Riemannian connection, with respect to a local coordinate system $(x_1, \ldots, x_m)$. For this purpose, we choose $X = \partial/\partial x_i$, $Y = \partial/\partial x_j$ and $Z = \partial/\partial x_l$. We plug these in (4.3.5) and obtain

$$\sum_{k=1}^m \Gamma_{ij}^k g_{kl} = \frac{1}{2}\left(\frac{\partial g_{jl}}{\partial x_i} + \frac{\partial g_{il}}{\partial x_j} - \frac{\partial g_{ij}}{\partial x_l}\right).$$

Denote by (g^{kl}) the inverse of the matrix (g_{kl}) representing the metric. We solve for Γ_{ij}^k in the above equation and obtain

$$\Gamma_{ij}^k = \frac{1}{2}\sum_{l=1}^m \left(\frac{\partial g_{jl}}{\partial x_i} + \frac{\partial g_{il}}{\partial x_j} - \frac{\partial g_{ij}}{\partial x_l}\right) g^{lk}. \qquad (4.3.6)$$

This formula shows also that the Riemannian connection is uniquely determined by the metric.

Example 4.3.7

(i) The Riemannian connection on $\mathbb{R}^3$ is the standard affine connection defined in Example 4.2.3. This can be easily seen from (4.3.6), since $(g_{ij}) = I$, implying that all Christoffel symbols equal 0.

(ii) On an embedded surface M is in $\mathbb{R}^3$, the Riemannian connection is the tangential connection $\nabla_X^t Y$ defined in (4.2.8). The proof is left as an exercise. □

4.4 Geodesics

Geodesics are curves on manifolds that generalize the idea of a straight line in a Euclidean space. One characteristic of straight lines is that they locally realize the shortest path between points. Another characteristic

4.4. GEODESICS

is that the tangents at all points are parallel. We will use the latter characteristic as a basis for the definition of a a geodesic.

In the remainder of the chapter, we will always consider a Riemannian manifold M equipped with the Riemannian connection ∇.

DEFINITION 4.4.1
A smooth curve $\gamma : I \to M$ is called a geodesic if its velocity vector field $d\gamma/dt$ is parallel. The restriction of γ to some closed interval $[a,b] \subseteq I$ is called a geodesic segment.

In general, we will only consider geodesics given by regular curves, i.e., curves with $d\gamma/dt \neq 0$ at all points. If $d\gamma/dt$ equals 0 at some point, then it always stays 0, in which case γ reduces to a point.

Note that a curve γ is a geodesic if and only if

$$\frac{D}{dt}\left(\frac{d\gamma}{dt}\right) = 0. \tag{4.4.1}$$

Physically, this means that a particle lying on a surface will keep moving 'in the same direction', and at a constant speed, as long as the tangential component of the acceleration (and hence of force) is zero. This can be thought of as an analogue of Newton's First Law of Motion (every body continues in its state of rest, or of uniform motion in a straight line, unless it is compelled to change that state by forces impressed upon it).

We can rewrite the equation (4.4.1) with respect to a local coordinate system $(x_1, \ldots, x_m)$. If $(x_1(t), \ldots, x_m(t))$ represents a geodesic, then using the covariant derivative equation (4.2.6), we obtain the second order system

$$\frac{d^2 x_k}{dt^2} + \sum_{i,j=1}^{m} \frac{dx_i}{dt}\frac{dx_j}{dt}\Gamma_{ij}^k = 0, \tag{4.4.2}$$

where $k = 1, \ldots, m$. The next theorem says that there exists a unique geodesic through any given point and in any given direction.

THEOREM 4.4.2 (Existence and uniqueness of geodesics)
For every point p in M and every tangent vector $v \in T_p M$, there exists a number $\epsilon > 0$ and a unique geodesic $\gamma_{p,v} : (-\epsilon, \epsilon) \to M$ such that

$$\gamma_{p,v}(0) = p \text{ and } \frac{d\gamma_{p,v}}{dt}(0) = v. \tag{4.4.3}$$

PROOF   Let $(x_1, \ldots, x_m)$ be a local coordinate system near p. Let $(p_1, \ldots, p_n)$ be the local coordinates of p and $(v_1, \ldots, v_n)$ be the local coordinates of the tangent vector v. We can rewrite the second order differential equation (4.4.2) as a linear system by introducing new variables

$$y_i = \frac{dx_i}{dt},$$

obtaining

$$\begin{cases} \dfrac{dx_i}{dt} = y_i, \\ \dfrac{dy_i}{dt} = -\sum_{i,j=1}^{m} \Gamma_{ij}^{k} y_i y_j, \end{cases} \quad (4.4.4)$$

for $i = 1, \ldots, m$. The initial condition is $x_i = p_i$, $y_i = v_i$, for $i = 1, \ldots, m$. By Theorem 2.3.1 on the existence and uniqueness of solutions of ordinary differential equations, there exist an $\epsilon > 0$ and a unique solution $(x_i(t), \ldots, x_m(t), y_1(t), \ldots, y_m(t))$ to (4.4.4), with $t \in (-\epsilon, \epsilon)$. The curve $\gamma_{p,v}$ of local coordinates $(x_1(t), \ldots, x_m(t))$, with $t \in (-\epsilon, \epsilon)$, is the unique geodesic that satisfies the given initial conditions. ∎

Example 4.4.3

We compute the geodesics of the standard surfaces.

(i) In the case of the Euclidean plane $\mathbb{R}^2$, all Γ_{ij} are equal to 0. The geodesic equations are $d^2x/dt^2 = 0$ and $d^2y/dt^2 = 0$, with solutions $x = m_1 t + n_1$ and $y = m_2 t + n_2$. Thus, the geodesics of the Euclidean plane are straight lines. Similarly, the geodesics of $\mathbb{R}^n$ are straight lines.

(ii) We show that the geodesics of the 2-dimensional sphere S_r^2 of radius r in $\mathbb{R}^3$ are great circles. Consider a local coordinate system

$$\psi(\theta, \phi) = (r \sin\phi \cos\theta, r \sin\phi \sin\theta, r \cos\phi)$$

on S_r^2, with (θ, ϕ) restricted to some appropriate domain. The Riemannian metric is given by

$$g_{11} = r^2 \sin^2 \phi, \quad g_{12} = 0, \quad g_{22} = r^2.$$

We use (4.3.6) to compute the Christoffel symbols corresponding to the Riemannian connection

$$\Gamma_{11}^2 = -\sin\phi \cos\phi, \quad \Gamma_{12}^1 = 2\cot\phi,$$

4.4. GEODESICS

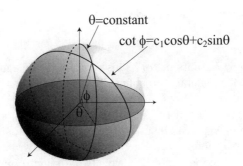

FIGURE 4.4.1
Geodesics on a sphere.

while the rest vanish. According to (4.4.2), the geodesics satisfy the following equations:

$$\begin{cases} \dfrac{d^2\theta}{dt^2} + 2\cot\phi \dfrac{d\theta}{dt}\dfrac{d\phi}{dt} = 0, \\ \dfrac{d^2\phi}{dt^2} - \sin\phi\cos\phi \left(\dfrac{d\theta}{dt}\right)^2 = 0. \end{cases}$$

If $\theta = $ constant on some interval, then the system reduces to $d^2\phi/dt^2 = 0$, with the solution $\phi(t) = at + b$ for some constants a, b. In this case, the geodesic is a vertical great circle of the sphere. See Figure 4.4.1.

If $\theta \neq$ constant, we can (locally) express ϕ as a function of θ, and then apply the chain rule

$$\frac{d\phi}{dt} = \frac{d\phi}{d\theta}\frac{d\theta}{dt}, \quad \frac{d^2\phi}{dt^2} = \frac{d^2\phi}{d\theta^2}\left(\frac{d\theta}{dt}\right)^2 + \frac{d\phi}{d\theta}\frac{d^2\theta}{dt^2}.$$

We substitute these in the above second order system and obtain the second order equation

$$\frac{d^2\phi}{d\theta^2} - 2\cot\phi \left(\frac{d\phi}{d\theta}\right)^2 - \sin\phi\cos\phi = 0.$$

We make the change of variable $w = \cot\phi$ and obtain the second order equation

$$\frac{d^2w}{d\theta^2} + \frac{dw}{d\theta} = 0,$$

which leads to the solution

$$\cot\phi = c_1 \cos\theta + c_2 \sin\theta$$

for some constants c_1, c_2. This represents the equation of a great circle obtained by slicing the sphere by the plane $z = c_1 x + c_2 y$, written in spherical coordinates.

The geometry of the sphere is a model of a non-Euclidean geometry, with 'lines' replaced by great circles. Since any pair of great circles intersect at two points, the Euclidean parallel postulate is invalidated. The 'incidence' axiom, asserting that two distinct points uniquely determine a line also become invalid. We can recover the incidence axiom by identifying each point of the sphere with its antipodal point, obtaining the real projective plane. The real projective plane, with the Riemannian metric induced from the sphere, is another model of non-Euclidean geometry, referred as elliptic geometry. See Greenberg (1994) for details.

(iii) We show that the geodesics of the Poincaré half plane $\mathbb{H}^2$ are vertical lines and semi-circles with the center on the x-axis (see Figure 4.4.2). For the hyperbolic plane, we have $g_{11} = g_{22} = 1/y^2$, and $g_{12} = g_{21} = 0$. The entries of the inverse matrix (g^{kl}) are $g^{11} = g^{22} = y^2$, and $g^{12} = g^{21} = 0$. Using (4.3.6), we compute $\Gamma_{12}^1 = \Gamma_{21}^1 = \Gamma_{22}^2 = -\Gamma_{11}^2 = -1/y$, while the rest equal 0. The geodesic equations are

$$\frac{d^2 x}{dt^2} - \frac{2}{y}\frac{dx}{dt}\frac{dy}{dt} = 0, \text{ and } \frac{d^2 y}{dt^2} + \frac{1}{y}\left(\frac{dx}{dt}\right)^2 - \frac{1}{y}\left(\frac{dy}{dt}\right)^2 = 0.$$

If $dx/dt = 0$ for all t, the first equation is vacuously satisfied, so the vertical lines $x = $ constant are geodesics.

If $dx/dt \neq 0$ at some t, then we can locally solve for x as a function of y. If $u = dx/dy$ then $dx/dt = u\, dy/dt$ and so, by the chain rule,

$$\frac{d^2 x}{dt^2} = \frac{du}{dy}\left(\frac{dy}{dt}\right)^2 + u\frac{d^2 y}{dt^2}.$$

By substituting $d^2 x/dt^2$ from the first equation and $d^2 y/dt^2$ from the second equation, after simplification we obtain

$$\frac{du}{dy} = \frac{u^3 + u}{y}.$$

Integration by partial fraction yields

$$\frac{dx}{dy} = u(y) = \pm\frac{cy}{\sqrt{1 - c^2 y^2}},$$

4.4. GEODESICS

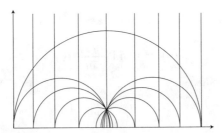

FIGURE 4.4.2
Geodesics of the Poincaré half plane.

for some $c \in \mathbb{R}$, so

$$x = \pm \int \frac{cy}{\sqrt{1-c^2y^2}} dy = \mp \sqrt{\left(\frac{1}{c}\right)^2 - y^2} + d,$$

for some $d \in \mathbb{R}$. This geodesic is a semi-circle of equation $(x-d)^2 + y^2 = (1/c)^2$, centered on the x-axis.

Finding the geodesics of the Poincaré disk and for the hyperboloid model, is left as an exercise.

The hyperbolic plane is a model of a non-Euclidean geometry, with 'lines' replaced by geodesics. The Euclidean parallel postulate fails, since through a given point off a given 'line' one can find infinitely many 'lines' that do not meet the given one. See Figure 4.4.2. The other axioms from Euclidean geometry remain all valid. In particular, one can easily show every pair of points determines a unique 'line'. □

It is important to realize that a geodesic is not simply a geometric image of a curve, but a curve with a certain parametrization. It is however possible to make a change of parameter of the geodesic so that the resulting curve is still a geodesic. For instance, one can increase the speed along a geodesic by proportionally decreasing its interval of definition, and the other way around.

LEMMA 4.4.4 (Homogeneity of geodesics)
Let $p \in M$, $v \in T_pM$, $\epsilon > 0$, and $\gamma_{p,v} : (-\epsilon, \epsilon) \to M$ be the unique geodesic with

$$\gamma_v(0) = p \text{ and } \frac{d\gamma_{p,v}}{dt}(0) = v.$$

Then, for every $a > 0$ there is a unique geodesic $\gamma_{p,av} : (-\epsilon/a, \epsilon/a) \to M$ with
$$\gamma_{p,av}(0) = p \text{ and } \frac{d\gamma_{p,av}}{dt}(0) = av.$$
This geodesic satisfies
$$\gamma_{p,av}(t) = \gamma_{p,v}(at), \qquad (4.4.5)$$
for all $t \in (-\epsilon/a, \epsilon/a)$.

PROOF Define the function $\gamma : (-\epsilon/a, \epsilon/a) \to M$ by $\gamma(t) = \gamma_{p,v}(at)$. It is easy to see that $\gamma(0) = \gamma_{p,v}(0) = p$ and $d\gamma/dt(0) = a \cdot d\gamma_{p,v}/dt(0) = av$. We need to check that γ is still a geodesic.
$$\frac{D}{dt}\left(\frac{d\gamma}{dt}\right) = a \cdot \frac{D}{dt}\left(\frac{d\gamma_{p,v}}{dt}\right) = 0.$$
By the uniqueness of a geodesic with given initial conditions, it follows that $\gamma(t) = \gamma_{p,av}(t)$ for all $t \in (-\epsilon/a, \epsilon/a)$. ∎

Using the smooth dependence of a solution of a differential equation on the initial conditions and the homogeneity property, we can uniformly define geodesics on some large time interval. For technical purposes, we choose this time interval to be $(-2, 2)$, but we could have chosen any open interval containing $[-1, 1]$ instead.

COROLLARY 4.4.5
For every $p \in M$, there exist a neighborhood V of p and a number $\delta > 0$ such that for every $q \in V$ and every $v \in T_q M$ with $\|v\| < \delta$, there exists a unique geodesic $\gamma_{q,v} : (-2, 2) \to M$ satisfying
$$\gamma_{q,v}(0) = q \text{ and } \frac{d\gamma_{q,v}}{dt}(0) = v.$$
Here $\|\cdot\|$ denotes the norm given by the Riemannian metric.

PROOF The system (4.4.3) defines a differential equation on the tangent bundle TM. Through local coordinates, it translates into the system (4.4.4), defined for $(x, y) \in U \times \mathbb{R}^m$, where U is an open set in $\mathbb{R}^m$. From Section 2.3, there is an open set Ω in $\mathbb{R} \times \mathbb{R}^m \times \mathbb{R}^m$,
$$\Omega = \{(t, x, y) \mid t \in I_{(x,y)}, \, x \in U, \, y \in \mathbb{R}^m\},$$

4.5. THE EXPONENTIAL MAP

such that the local flow $\phi : \Omega \to \mathbb{R}^m \times \mathbb{R}^m$ is well defined. Here $I_{(x,y)}$ represents the maximal open interval about 0 for which the solution $\phi_{(x,y)}$ is defined. The set Ω is open and contains $(0, p, 0)$ (the point in the parameter space corresponding to $p \in M$ is still denoted by p). Then there exists a neighborhood of $(0, p, 0)$ in Ω that is a product of open sets of the type

$$\{t \mid |t| < \epsilon_1\} \times U' \times \{y \mid \|y\| < \epsilon_2\},$$

for some $\epsilon_1, \epsilon_2 > 0$ and some open set $U' \subseteq U$. Here $\|y\|$ is the norm induced by the Riemannian metric, expressed in local coordinates.

From local coordinates we go back to TM. It follows that for every $q \in V$ (where V is the image of U' through the appropriate chart), and $v \in T_q M$ with $\|v\| < \epsilon_2$, the geodesic $\gamma_{q,v}$ is defined for $t \in (-\epsilon_1, \epsilon_1)$. Using the above homogeneity property for $a = \epsilon_1/2$, $\gamma_{q,av}$ is defined for $t \in (-2, 2)$. Taking $\delta < a\epsilon_1 = \epsilon_1\epsilon_2/2$, we obtain that the geodesic $\gamma_{q,v}(t)$ is defined for all $q \in V$, $\|v\| < \delta$, and $t \in (-2, 2)$. ∎

We have not yet proved that geodesics represent arc length minimizing curves between sufficiently close points. We will do this in Section 4.6, but we will have to develop a few more tools first.

4.5 The exponential map

The exponential map describes the dependence of geodesics emanating from the same point on their initial velocity. If $v \in T_p M$ and the geodesic $\gamma_{p,v}(t)$ is defined on $[0, 1]$, then we define $\exp_p(v)$ by

$$\exp_p(v) = \gamma_{p,v}(1).$$

By Corollary 4.4.5, there exists $\delta > 0$ depending on p such that $\exp_p(v)$ is defined for all v with $\|v\| < \delta$.

DEFINITION 4.5.1
For $p \in M$, consider the set $V \subseteq T_p M$ of all vectors $v \in T_p M$ such that $\gamma_{p,v}$ is defined on a neighborhood of $[0, 1]$. The exponential map $\exp_p : V \to M$ is defined by

$$\exp_p(v) = \gamma_{p,v}(1). \tag{4.5.1}$$

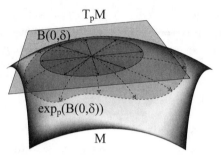

FIGURE 4.5.1
The exponential map.

Roughly speaking, $\exp_p(v)$ is obtained by 'laying' the vector v on the surface, away from p.

In general, we will consider the exponential map $\exp_p$ restricted to a ball $B(0, \delta) \subseteq T_pM$ with $\delta > 0$ as above. See Figure 4.5.1. We emphasize that the ball $B(0, \delta) \subset T_pM$ is defined in terms of the Riemannian metric, and is not a Euclidean ball. However, since any two norms on a Euclidean space are equivalent, we can always squeeze some Euclidean ball centered at 0 inside $B(0, \delta)$, and vice versa.

PROPOSITION 4.5.2
The exponential map $\exp_p$ has the following properties:

(i) $\exp_p(0) = p$;

(ii) $d(\exp_p)_0(v) = v$ for all $v \in T_pM$;

(iii) There exist a neighborhood U of 0 in T_pM and a neighborhood V of p in M such that $\exp_p$ maps U diffeomorphically onto V. Such a neighborhood V is called a geodesic neighborhood of p. In particular, there exists $0 < \rho < \delta$ such that

$$\exp_p : B(0, \rho) \subseteq T_pM \to \exp_p(B(0, \rho)) \subseteq M$$

is a diffeomorphism;

(iv) There exist a neighborhood W of p and a number $\eta > 0$ such that for each $q \in W$, $\exp_q$ is a diffeomorphism from $B(0, \eta)$ to $\exp_q(B(0, \eta))$, and $W \subseteq \exp_q(B(0, \eta))$. Moreover, $\exp_q$ depends smoothly on $q \in W$. Such a neighborhood W is called a uniformly geodesic neighborhood of p.

4.5. THE EXPONENTIAL MAP

PROOF (i) The geodesic of initial point p and initial velocity $v = 0$ degenerates to the point p itself. Hence $\gamma_{p,0}(1) = p$.

(ii) Since $\exp_p : B(0, \delta) \subseteq T_pM \to M$, we have $d(\exp_p)_u : T_u(T_pM) \to T_{\exp_p(u)}M$ for each $u \in B(0, \delta)$. The tangent space to T_pM at u can be identified to T_pM, so we have $d(\exp_p)_0 : T_pM \to T_pM$. Using the homogeneity property, we compute

$$d(\exp_p)_0(v) = \frac{d}{dt}(\exp_p(tv))|_{t=0} = \frac{d}{dt}(\gamma_{p,tv}(1))|_{t=0}$$
$$= \frac{d}{dt}(\gamma_{p,v}(t))|_{t=0} = v.$$

(iii) The previous property says that $d(\exp_p)_0$ is the identity on T_pM. The Inverse Function Theorem implies that $\exp_p$ is a local diffeomorphism, so there exist neighborhoods $U \subseteq B(0, \delta)$ of 0 in T_pM and V of p in M with the prescribed properties.

(iv) The main point here is that one can choose geodesic neighborhoods $\exp_q(B(0, \eta))$ of q uniformly for all q in some neighborhood W of p. See Figure 4.5.2. To prove this, we will regard $\exp_q v$ as a function in both q and v, and apply the inverse function theorem. According to Corollary 4.4.5, there exists an open neighborhood $\mathcal{U}$ of $(p, 0)$ in the tangent bundle TM such that $\exp_q(v)$ is defined for all $(q, v) \in \mathcal{U}$. Define the map $E : \mathcal{U} \to M \times M$ by

$$E(q, v) = (q, \exp_q v).$$

The matrix of $(dE)_{(p,0)}$, with respect to local coordinates on TM and on $M \times M$ induced by a chart around p in M, is given by

$$\begin{pmatrix} \text{id} & 0 \\ * & \text{id} \end{pmatrix},$$

with the upper left block corresponding to the derivative of $q \to q$ at $q = p$ and the lower right block corresponding to the derivative of $v \to \exp_q v$ at $v = 0$.

By the inverse function theorem, there exists a neighborhood $\mathcal{V} \subset \mathcal{U}$ of $(p, 0)$ in TM that is mapped diffeomorphically by E onto a neighborhood of (p, p) in $M \times M$. Since $M \times M$ has the product topology, we can choose a neighborhood $W \subset \mathcal{V}$ of p such that $W \times W \subseteq E(\mathcal{V})$. We now prove that W is a uniform geodesic neighborhood of p.

Step 1. We claim that $\exp_q$ is a diffeomorphism of $B(0, \eta) \subseteq T_qM$ onto its image in M.

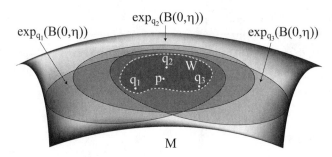

FIGURE 4.5.2
Uniformly geodesic neighborhood.

Since E maps $\mathcal{V}$ diffeomorphically onto its image, the restriction of E to $\mathcal{V} \cap \{(q,v) \mid \|v\| < \eta\}$ is also a diffeomorphism onto its image for each $q \in W$. This implies that $\exp_q$ is a diffeomorphism from $B(0,\eta) \subseteq T_q M$ to $\exp_q(B(0,\eta)) \subseteq M$.

Step 2. We claim that W is contained in each of the sets $\exp_q(B(0,\eta))$, with $q \in W$.

By construction, we know that $\{q\} \times W$ is contained in $E(\mathcal{V})$ and $E(q',v') \in \{q\} \times W$ is possible only if $q' = q$. Hence, for every $q \in W$, we have

$$\{q\} \times W \subseteq E\left(\mathcal{V} \cap \{(q,v) \mid \|v\| < \eta\}\right).$$

This implies that $W \subseteq \exp_q(B(q,\eta))$. ∎

From Proposition 4.5.2, we infer that the image of the closed ball $\bar{B}(0,\rho)$ in $T_p M$ is mapped diffeomorphically into $\exp_p(\bar{B}(0,\rho))$ in M, provided that the radius ρ is chosen small enough. The image $\exp_p(B(0,\rho))$ is called the geodesic ball with center at p and 'radius' ρ, and, with an abuse of notation, is denoted by $B(p,\rho)$. The boundary of this ball is a codimension one submanifold of M, called the geodesic sphere, and denoted by $S(p,\rho)$. The meaning of 'radius' in this context is that the arc length of any geodesic segment joining p to any point of the sphere $S(p,\rho)$ is equal to ρ. The arc length of any geodesic from p to a point q inside the geodesic ball $B(p,\rho)$ is less than ρ.

One of the basic geometric properties of geodesics is that the geodesics emanating from p are orthogonal to the geodesic sphere $S(p,\rho)$. In order to prove this we need a classical lemma.

4.5. THE EXPONENTIAL MAP

LEMMA 4.5.3 *(Gauss' lemma)*
Let $p \in M$. Choose $\delta > 0$ (depending on p) so that $\exp_p$ is defined on $B(0, \delta)$. Let v, w be tangent vectors in $T_p M$ with $\|v\| < \delta$. We identify $T_v(T_p M)$ to $T_p M$. Then

$$\langle d(\exp_p)_v(v), d(\exp_p)_v(w) \rangle = \langle v, w \rangle. \qquad (4.5.2)$$

PROOF Since v is in the open ball $B(0, \delta)$, there exists a number $\epsilon > 0$ such that $t(v + sw) \in B(0, \delta)$ for all $(s, t) \in (-\epsilon, \epsilon) \times (-\epsilon, 1 + \epsilon)$. The smooth map $f : (-\epsilon, \epsilon) \times (-\epsilon, 1 + \epsilon) \to M$, defined by

$$f(t, s) = \exp_p(t(v + sw)),$$

is a parametrized surface in M. Observe that

$$\frac{\partial f}{\partial t}(1, 0) = d(\exp_p)_v(v) \quad \text{and} \quad \frac{\partial f}{\partial s}(1, 0) = d(\exp_p)_v(w),$$

while

$$\frac{\partial f}{\partial t}(0, 0) = 0 = \frac{\partial f}{\partial s}(0, 0).$$

Thus

$$\left\langle \frac{\partial f}{\partial t}, \frac{\partial f}{\partial s} \right\rangle (0, 0) = 0 \text{ and } \left\langle \frac{\partial f}{\partial t}, \frac{\partial f}{\partial s} \right\rangle (1, 0) = \langle d(\exp_p)_v(v), d(\exp_p)_v(w) \rangle.$$

In order to prove the lemma it will suffice to show that, for any t,

$$\frac{\partial}{\partial t} \left\langle \frac{\partial f}{\partial t}, \frac{\partial f}{\partial s} \right\rangle (t, 0) = \langle v, w \rangle.$$

The curve $t \to f(t, s)$ is the geodesic with initial point p and initial velocity $v + sw$, hence $\left\| \frac{\partial f}{\partial t} \right\| = \|v + sw\|$ and $\frac{D}{dt} \frac{\partial f}{\partial t} = 0$. Using this and the properties of ∇, we compute

$$\frac{\partial}{\partial t} \left\langle \frac{\partial f}{\partial t}, \frac{\partial f}{\partial s} \right\rangle = \left\langle \frac{D}{dt} \frac{\partial f}{\partial t}, \frac{\partial f}{\partial s} \right\rangle + \left\langle \frac{\partial f}{\partial t}, \frac{D}{dt} \frac{\partial f}{\partial s} \right\rangle$$

$$= \left\langle \frac{\partial f}{\partial t}, \frac{D}{ds} \frac{\partial f}{\partial t} \right\rangle = \frac{1}{2} \frac{\partial}{\partial s} \left\langle \frac{\partial f}{\partial t}, \frac{\partial f}{\partial t} \right\rangle$$

$$= \frac{1}{2} \frac{\partial}{\partial s} \langle v + sw, v + sw \rangle = \langle v + sw, w \rangle.$$

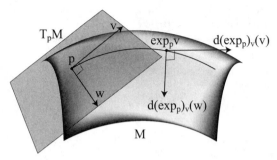

FIGURE 4.5.3
The Gauss lemma.

The last term equals $\langle v, w \rangle$ when $s = 0$. This concludes our proof. ∎

REMARK 4.5.4 The above lemma implies that if v is orthogonal to w, then their image vectors under $d(\exp_p)_v$ are also orthogonal. See Figure 4.5.3. It also shows that $d(\exp_p)_v$ preserves the length of v, as well as of all scalar multiples of v. However, vectors w linearly independent of v may be stretched or shrunk by $d(\exp_p)_v$. Similarly, the inner product of a vector with some scalar multiple of v is preserved by $d(\exp_p)_v$, but the inner product of two vectors, both linearly independent of v, may not be preserved by $d(\exp_p)_v$. ∎

COROLLARY 4.5.5

The geodesics through p are orthogonal to any geodesic sphere $S(p, r)$ centered at p with radius $r < \delta$, where δ is chosen as in Lemma 4.5.3.

PROOF Let v be on the sphere $S(0, \rho) \subseteq T_p M$, and w be a vector tangent to $S(0, \rho)$ at v in $T_p M$, so $\langle v, w \rangle = 0$. The exponential $\exp_p$ takes $S(0, \rho)$ diffeomorphically onto $S(p, \rho)$, so its derivative $d(\exp_p)_v$ takes the vector w tangent to $S(0, \rho)$ at v into the vector $d(\exp_p)_v(w)$ tangent to $S(p, \rho)$ at $\exp_p v$. The geodesic through p and $\exp_p(v)$ is $\gamma_{p,v}(t) = \exp_p(tv)$, so $d(\exp_p)_v(v)$ is a tangent vector to the geodesic at $\exp_p v$. By the Gauss lemma, $\langle d(\exp_p)_v(v), d(\exp_p)_v(w) \rangle = 0$, that is the geodesic is orthogonal to the geodesic sphere at the contact point. See Figure 4.5.1. ∎

REMARK 4.5.6   The reason for the term 'exponential map' lies in the theory of Lie groups (see Remark 2.4.10). On any Lie group G one can define a natural exponential map, even in the absence of a Riemannian structure. Recall that the Lie algebra $\mathfrak{g}(G)$ of the Lie group G is the isomorphic to the tangent space $T_{\mathrm{id}}G$ to G at the identity $\mathrm{id} \in G$. For every $V \in T_{\mathrm{id}}G$, one can show that there exists a unique smooth curve $c_V : \mathbb{R} \to G$ such that $c'(0) = V$, and $c_V(s+t) = c_V(s)c_V(t)$ for all $t, s \in \mathbb{R}$. The later condition says that c_V respects the group operations, so c_V is a smooth group homomorphism from $\mathbb{R}$ to G (also called a 1-parameter subgroup of G). We define the exponential map $\exp : \mathfrak{g}(G) \to G$ by $\exp(V) = c_v(1)$.

In the case of the Lie group $G = GL_n(\mathbb{R})$, the exponential map is the classical exponential:

$$\exp(A) = \sum_{n=0}^{\infty} \frac{A^n}{n!} = e^A,$$

for every $A \in \mathfrak{g}(G)$.

The connection with Riemannian manifolds is the following. We can define a Riemannian structure on a Lie group by choosing an inner product on $T_{\mathrm{id}}G$ and extending it everywhere by left translations. The resulting Riemannian metric is left invariant by left translations. We can now talk about the exponential map in the Riemannian sense, in particular we can talk about the exponential map $\exp_{id}$ at $\mathrm{id} \in G$. If the Riemannian metric is bi-invariant (left invariant and right invariant at the same time), then the geodesics through id in G are precisely the 1-parameter subgroup of G, and so the Riemannian exponential is the same as the Lie exponential. Otherwise the two exponentials are usually different (even if the metric is left invariant). ∎

4.6   Minimizing properties of geodesics

In this section we will show that geodesics locally minimize arc length. If $\gamma_{p,v} : I \to M$ is a geodesic, then the arc length of the segment corresponding to a time interval $[t_0, t_1] \subseteq I$ is

$$L_{[t_0,t_1]}(\gamma_{p,v}) = \int_{t_0}^{t_1} \left\| \frac{d\gamma_{p,v}}{dt} \right\| dt = \|v\|(t_1 - t_0),$$

as the speed along the geodesic is constantly equal to $\|v\|$. Hence the arc length of the geodesic segment is proportional to the time change. If $\|v\| = 1$, the geodesic is called a unit speed geodesic and its arc length is actually equal to the time change. One can always reparametrize a given geodesic to obtain a unit speed geodesic: if $\gamma_{p,v} : [a, b] \to M$ is a geodesic, then $\gamma_{p,(v/\|v\|)} : [\|v\|a, \|v\|b] \to M$ is a unit speed geodesic with the same geometric image as in $\gamma_{p,v}$.

THEOREM 4.6.1 (Geodesics locally minimize arc length)
Let p be a point in M. There exists $\rho > 0$ such that for every point q in the closed geodesic ball $\bar{B}(p, \rho)$, the geodesic $\gamma : [0, 1] \to M$ joining $\gamma(0) = p$ to $\gamma(1) = q$ satisfies

$$L(\gamma) \leq L(c), \tag{4.6.1}$$

for every piecewise smooth curve $c : [0, 1] \to M$ between $c(0) = p$ and $c(1) = q$. Equality holds only if c is a monotone reparametrization of γ.

PROOF We split the proof into two cases.

Case 1. Suppose that the image of c lies entirely within $\bar{B}(p, \rho)$.

We can assume that $c(t) \neq p$ for all $t > 0$, since otherwise we can restrict c to the interval $[t_0, 1]$, where t_0 is the largest t for which $c(t) = p$, and compare the arc lengths of $c_{[t_0,1]}$ and γ. Since $\exp_p$ is a diffeomorphism from $\bar{B}(p, \rho)$ to $\exp_p(\bar{B}(p, \rho))$, we can lift c to the piecewise smooth curve $\alpha = \exp_p^{-1} \circ c$ in T_pM. See Figure 4.6.1. We describe this curve in polar coordinates as $\alpha(t) = r(t)v(t)$, where

$$\begin{cases} r(t) = \|\alpha(t)\|, \\ v(t) = \dfrac{\alpha(t)}{\|\alpha(t)\|}, \end{cases}$$

for $t > 0$. Note that the real valued function function $r(t)$ and the vector valued function $v(t)$ are differentiable wherever $\alpha(t)$ is differentiable. Since $c(t) = \exp_p(r(t)v(t))$, we have

$$\frac{dc}{dt}(t) = \frac{d}{dt}\left(\exp_p(r(t)v(t))\right) = d(\exp_p)_{\alpha(t)}\left(\frac{d}{dt}(r(t)v(t))\right)$$
$$= d(\exp_p)_{\alpha(t)}\left(\frac{dr}{dt}(t)v(t)\right) + d(\exp_p)_{\alpha(t)}\left(r(t)\frac{dv}{dt}(t)\right),$$

4.6. MINIMIZING PROPERTIES OF GEODESICS

wherever $\alpha(t)$ is differentiable. Since $\langle v, v \rangle = 1$ and so $\langle v, dv/dt \rangle = 0$, Gauss' lemma implies $\|d(\exp_p)_{\alpha(t)}(v(t))\| = 1$, and $d(\exp_p)_\alpha ((dr/dt)v)$ and $d(\exp_p)_\alpha (r(dv/dt))$ are orthogonal. By the Pythagorean theorem,

$$\left\|\frac{dc}{dt}\right\|^2 = \left\|d(\exp_p)_{\alpha(t)}\left(\frac{dr}{dt}v\right)\right\|^2 + \left\|d(\exp_p)_{\alpha(t)}\left(r\frac{dv}{dt}\right)\right\|^2$$
$$\geq \|d(\exp_p)_{\alpha(t)}(v)\|^2 \left|\frac{dr}{dt}\right|^2 = \|v\|^2 \left|\frac{dr}{dt}\right|^2 = \left|\frac{dr}{dt}\right|^2.$$

Since these inequalities hold for $t > 0$, we first compute the arc lengths restricted to a time interval $[\epsilon, 1]$, for some small $\epsilon > 0$, and then pass to the limit as $\epsilon \to 0$:

$$\int_\epsilon^1 \left\|\frac{dc}{dt}\right\| dt \geq \int_\epsilon^1 \left|\frac{dr}{dt}\right| dt > \int_\epsilon^1 \frac{dr}{dt} dt = r(1) - r(\epsilon), \quad \text{hence}$$
$$\int_0^1 \left\|\frac{dc}{dt}\right\| dt \geq r(1) - r(0).$$

As $v = d\gamma/dt(0)$ is the initial velocity of the geodesic, from $\exp_p(v) = \gamma(1) = c(1) = q$, we obtain $\alpha(1) = v$, and so $r(1) = \|v\|$. On the other hand, we have $r(0) = \|\exp_p^{-1}(p)\| = 0$. This shows

$$L(c) \geq \|v\| = L(\gamma).$$

In order to have equality, we need $\|d(\exp_p)_{\alpha(t)}(dv/dt)\| = \|dv/dt\| = 0$ and $|dr/dt| = dr/dt \geq 0$ for all except finitely many $t > 0$. This means that $v(t) = v_0$, for some v_0, hence $c(t) = \exp_p(r(t)v_0)$ with r an increasing function in t. Thus c must be a monotone reparametrization of γ.

Case 2. The image of c does not entirely lie within $\bar{B}(p, \rho)$.

Let t_1 be the smallest value of t for which $c(t)$ meets the geodesic sphere $S(p, \rho)$. Then, by Case 1, we have

$$L_{[0,1]}(c) \geq L_{[0,t_1]}(c) \geq \rho \geq L_{[0,1]}(\gamma).$$

Equality holds only if all of the preceding inequalities are, in fact, equalities, so the image of c must lie within the closed geodesic ball, which goes back to Case 1. ∎

This theorem works only to short geodesics between nearby points. It does not say that all geodesic segments minimize length, even if their

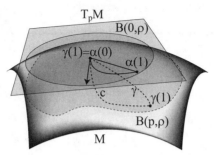

FIGURE 4.6.1
Geodesics locally minimize arc length.

endpoints are close. For example, two non-antipodal points on a sphere are joined by a shorter arc of a great circle and by a longer arc of the same great circle: they are both geodesic segments, but the longer arc is not a minimizing geodesic.

Now we will show that every arc length minimizing curve is, up to reparametrization, a geodesic.

THEOREM 4.6.2 (Length minimizing curves are geodesics)

Suppose that a piecewise smooth curve $c : [a, b] \to M$ is shorter than any other piecewise smooth curve between $c(a)$ and $c(b)$. Then c is a monotone reparametrization of a geodesic.

PROOF For each $t \in [a, b]$, there exists a uniformly geodesic neighborhood $W_{c(t)}$ of $c(t)$. Using the compactness of the interval, and the fact that c is piecewise smooth, there exists a subdivision $a = t_0 < t_1 < \cdots < t_k = b$ of $[a, b]$ such that c is smooth on each subinterval $[t_i, t_{i+1}]$, and $c([t_i, t_{i+1}])$ is contained in $W_{c(t_i)}$, for $i = 0, \ldots, k-1$. By Proposition 4.5.2 (iv), there exists a geodesic ball centered at $c(t_i)$ which contains $W_{c(t_i)}$ and, implicitly, $c(t_{i+1})$. Since the curve segment $c|_{[t_i,t_{i+1}]}$ is arc length minimizing, its length equals the length of the unit speed geodesic segment γ_i from $c(t_i)$ to $c(t_{i+1})$. By Theorem 4.6.1, $c|_{[t_i,t_{i+1}]}$ must be a monotone reparametrization of γ_i. We patch together the parametrizations of all γ_i's and obtain that c is a monotone parametrization of a arc length minimizing, possibly broken, geodesic γ between $c(a)$ and $c(b)$.

We only need to show that γ is regular at the points $c(t_i)$ where the geodesic segments γ_{i-1} and γ_i meet. Let W be a uniformly geodesic neighborhood of $c(t_i)$ and $\eta > 0$ be the radius of a geodesic ball as in

4.6. MINIMIZING PROPERTIES OF GEODESICS

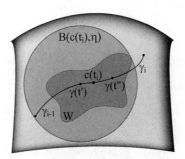

FIGURE 4.6.2
Length minimizing curves are geodesics.

Proposition 4.5.2 (iv). If $t' < t_i$ is sufficiently close to t_i, then $\gamma(t') \in W$ and $B(\gamma(t'), \eta)$ contains $c(t_i)$. Let $t'' > t_i$ be such that the $\gamma(t'') \in B(\gamma(t'), \eta)$ and $\gamma(t'') \neq c(t_i)$. See Figure 4.6.2. Then the curve segment $\gamma|_{[t',t'']}$ is arc length minimizing, so its length has to be equal to the length of a geodesic γ' between $\gamma(t')$ and $\gamma(t'')$ in $B(\gamma(t'), \eta)$. But γ' coincides with γ on the open intervals (t', t_i) and (t_i, t''), hence γ is regular and satisfies the geodesic equation at $c(t_i)$. ∎

The above statement implies that every arc length minimizing, piecewise smooth parametrized curve, with parameter proportional to arc length, is a geodesic. In particular, the curve is actually smooth.

The following result can be used as a practical method to find geodesics in many instances.

PROPOSITION 4.6.3
Every local isometry $f : M \to N$ takes the geodesics of M into geodesics of N. In particular, every isometry defines a bijective correspondence between the geodesics on M and the geodesics on N.

PROOF The main point is that if $c : [0, 1] \to M$ is a piecewise smooth curve, then the piecewise smooth curve $f \circ c : [0, 1] \to N$ has the same length as c, because

$$\|(f \circ c)'(t)\| = \|(df)_{c(t)} c'(t)\| = \|c'(t)\|$$

for all t at which c is differentiable. It follows easily that if c is a geodesic, then the length minimizing property given in Theorem 4.6.2 will carry over to $f \circ c$, and hence $f \circ c$ is a geodesic. ∎

FIGURE 4.6.3
The geodesics of the flat torus.

FIGURE 4.6.4
The geodesics of the flat cylinder.

Example 4.6.4

(i) The flat torus $\mathbb{T}^2 = \mathbb{R}^2/\mathbb{Z}^2$ is locally isometric to $\mathbb{R}^2$, through the local isometry $f(x_1, x_2) = [x_1, x_2]$. The geodesics of $\mathbb{R}^2$ are straight lines. The vertical lines are mapped by f into meridian circles on the torus. The horizontal lines are mapped into parallel circles. All slanted lines are mapped into curves that wind around the torus: the ones of rational slope are mapped into closed curves, while the ones of irrational slope are mapped into dense curves on the torus. Since through every point of the torus and in every direction there is a geodesic obtained as the image of a straight line, we obtained all of the geodesics of the flat torus. See Figure 4.6.3.

(ii) The flat cylinder $C^2 = \mathbb{R}^2/(\mathbb{Z} \times \{0\})$ is locally isometric to $\mathbb{R}^2$. Therefore, the geodesics of the cylinder are the generating lines, the circles cut by planes perpendicular to the generating lines, and helices. See Figure 4.6.4.

4.6. MINIMIZING PROPERTIES OF GEODESICS

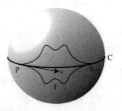

FIGURE 4.6.5
The geodesics of the sphere.

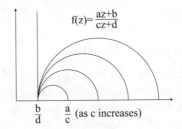

FIGURE 4.6.6
Geodesics of the Poincaré half plane as images of Möbius transformations.

(iii) We provide an alternative way of showing that the geodesics of a sphere are the great circles. Consider the n-dimensional embedded sphere $S^n \subseteq \mathbb{R}^{n+1}$. Let γ be a unit speed geodesic on S^n. Let p be a point on γ and v a tangent vector to γ at p. The plane $E \subseteq \mathbb{R}^{n+1}$ through p and through the center of the sphere and parallel to v slices the sphere by a great circle C. Let $f : S^n \to S^n$ be the reflection of S^n through the plane E. This map is an isometry whose only fixed points are those of C. The curve $f \circ \gamma$ is also a unit speed geodesic through p, and its tangent vector at p is also v. By the uniqueness of the geodesic with respect to initial conditions, we must have $\gamma = f \circ \gamma$ near p. This implies that a piece of γ near p is contained in the great circle C. Since p was arbitrary, it follows that the whole image of γ is contained in C. Since there is a great circle through every point and in every direction, the great circles must be all of the geodesics. See Figure 4.6.5. (iv) We show that the geodesics of the Poincaré half plane are vertical lines and semi-circles centered on the horizontal axis. The geodesic equation easily shows that the vertical axis is a geodesic. The isometries of the Poincaré half plane are the transformations $z \to (az+b)/(cz+d)$, with $a, b, c, d \in \mathbb{R}$ and $ad - bc = 1$ (see Section 3.3). A direct computation shows that the image of the y-axis under such an isometry is either a vertical line $x = b/d$ (if $c = 0$), or a semi-circle of diameter $[b/d, a/c]$ (if $c \neq 0$). By varying the parameters a, b, c, d one can obtain all vertical lines and all semi-circles centered on the horizontal axis. See Figure 4.6.6. Using elementary geometry, at any given point and in any given direction one can draw either a vertical line or a semi-circle. We conclude that these are all of the geodesics. □

4.7 The Riemannian distance

A Riemannian metric was defined as a smoothly varying inner product on tangent spaces. In this section we show that every Riemannian metric induces a distance function between pairs of points. If the manifolds consists of several separated components, it is not possible to measure intrinsically the distance between points located on different components. Therefore, in this section, we will restrict our attention to connected manifolds.

DEFINITION 4.7.1
The Riemannian distance on M is the function $d : M \times M \to \mathbb{R}$ defined by

$$d(p, q) = \inf\{L(c) \,|\, c : [0, 1] \to M \text{ is a piecewise smooth curve between } p \text{ and } q\}.$$

This function is well defined: if M is connected, it is path connected, and so for any pair of points on the manifold there always exists a smooth curve connecting the two points. If there exists a length minimizing geodesic between the points, then the distance between equals the length of the corresponding geodesic segment. However, it is not necessary that such a geodesic exists. For example, consider a punctured open disk (i.e., an open disk with its center removed) in the plane and a pair of points symmetric with respect to the center. The Riemannian distance between these points is the Euclidean distance, although there is no

4.7. THE RIEMANNIAN DISTANCE

geodesic connecting them.

THEOREM 4.7.2 *(The Riemannian distance)*
The distance function defines a metric whose topology agrees with the topology of the manifold.

PROOF We first need to check the three distance properties from Example 1.2.2 (iii).

It is clear that the distance function is nonnegative. If $p \neq q$, we can find a geodesic ball $\bar{B}(p,\rho)$ at p, which does not contain q, for some $\rho > 0$. Hence any curve connecting p and q will have its length at least as large as the length of the curve segment from p to the geodesic sphere $S(p,\rho)$. Hence $d(p,q) > 0$. This proves (i).

The symmetry property (ii) is obvious.

For (iii), if c_1 is curve segment from p to r and c_2 is a curve segment from q to r, then patching the parametrizations of c_1 and c_2 together yields a curve c from p to q, which is not necessarily smooth at r, but is still piecewise smooth (this is the reason why we want to define the distance by the means of piecewise smooth curves instead of smooth ones). So $d(p,q) \leq L(c) = L(c_1) + L(c_2)$. By taking the infimum over all curves c_1 and c_2, we obtain (iii).

The topological space M together with the distance function defines a metric space. A geodesic ball $B(p,\rho)$ is a ball with respect to the Riemannian distance. Since it is the diffeomorphic image of some open set in the tangent space (which is just a copy of $\mathbb{R}^m$), it is open with respect to the topology of the smooth structure. On the other hand, for any open neighborhood U of a point p in M, there exists $\rho > 0$ such that $B(p,\rho) \subseteq U$, so the open sets with respect to the smooth structure are still open with respect to the Riemannian distance. Thus, the two corresponding topologies coincide. ∎

COROLLARY 4.7.3
The distance function $d: M \times M \to \mathbb{R}$ is continuous.

From the topological point of view, it is important to know whether the above metric space is complete. A metric space is said to be complete if every Cauchy sequence is convergent.

DEFINITION 4.7.4

A *Riemannian manifold M is geodesically complete if for each $p \in M$, every geodesic $\gamma_{p,v} : I \to M$ through p can be extended to a geodesic from $\mathbb{R}$ to M.*

The homogeneity property of geodesics means $\gamma_{p,v}(t) = \gamma_{p,tv}(1) = \exp_p(tv)$ whenever $\gamma_{p,v}(t)$ is defined. The condition of being geodesically complete is equivalent to the exponential map $\exp_p$ on the whole of $T_p M$ for every $p \in M$.

For example, every geodesic on the sphere S^n or on the cylinder C can be extended indefinitely, so these two surfaces are geodesically complete. On the other hand, an open disk in the $\mathbb{R}^n$ is not geodesically complete, since its geodesics cannot be defined for all $t \in \mathbb{R}$.

THEOREM 4.7.5 (Hopf-Rinow Theorem)

A Riemannian manifold is geodesically complete if and only if it is complete with respect to the Riemannian distance.

PROOF *Direct part: suppose that the Riemannian distance is complete; we want to prove that every geodesic can be extended to all of $\mathbb{R}$.*

Let $\gamma : I \to M$ be a unit speed geodesic defined on its maximal domain of definition. From the theory of ordinary differential equations we know that the maximal interval I must be open. Suppose that $I = (a, b)$ and $b < +\infty$. Let $\{t_n\}$ be an increasing sequence of numbers with $t_n \to b$. Since $\gamma|_{[t_n, t_m]}$ is a smooth curve connecting $\gamma(t_n)$ and $\gamma(t_m)$, we have $d(\gamma(t_n), \gamma(t_m)) \leq L_{[t_n, t_m]}(\gamma) = |t_n - t_m|$ for all $t_n < t_m$ (we do not necessarily have equality since we do not know whether $\gamma|_{[t_n, t_m]}$ is a minimizing geodesic). It results that the sequence $\{\gamma(t_n)\}$ is Cauchy, so, due to completeness, it converges to a point $p \in M$. Note that the point p depends only on γ and not on the particular sequence $\{t_n\}$.

Let W be a uniformly geodesic neighborhood of p and $\eta > 0$ such that $W \subseteq B(q, \eta)$, for each $q \in W$. There exists $t_j \in \mathbb{R}$ with $\gamma([t_j, b)) \subseteq W$. Then the geodesic ball $B(\gamma(t_j), \eta)$ contains the point p and every unit speed geodesic emanating from $\gamma(t_j)$ is defined at least for a time η. In particular, the domain of $\gamma|_{[t_j, b)}$ can be extended to $[t_j, t_j + \eta)$, hence γ is defined past b. See Figure 4.7.1. This contradicts the maximality of the interval I, so $b = \infty$. A similar argument shows that $a = \infty$.

Converse part: suppose that M is a geodesically complete manifold;

4.7. THE RIEMANNIAN DISTANCE

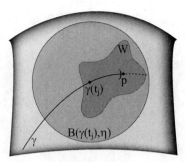

FIGURE 4.7.1
Proof of the direct part of Hopf-Rinow theorem.

we want to prove that the metric is complete.

Step 1. We prove that any points of M are joined by a length minimizing geodesic. That is, for any p and q in M, there is a vector $v \in T_p M$ with $\exp_p(v) = q$ and $\|v\| = d(p,q)$.

Let $d = d(p,q)$. Our claim is obvious if $p = q$ (take $v = 0$), so we assume that $d > 0$. Choose $\delta > 0$ small enough so that $\delta < d$ and the geodesic ball $B(p, \delta)$ is a geodesic neighborhood of p, as defined in Proposition 4.5.2. The function $s \in S(p, \delta) \to d(s, q)$ is continuous and it is defined on a compact set, so it attains its minimum at some point $r \in S(p, \delta)$. Let v be the unit vector in $T_p M$, such that $\exp_p(\delta v) = r$ and let $\gamma = \gamma_{p,v}$, so that γ is the unit speed geodesic with $\gamma(0) = p$ and $\gamma(\delta) = r$. See Figure 4.7.2. Since M is geodesically complete, $\gamma(t)$ is defined at all $t \in \mathbb{R}$.

Our task now is to prove that $\gamma(d) = q$. We prove this by showing that $d(p,q) = t + d(\gamma(t), q)$ for $\delta \leq t \leq d$. When $t = d$, this equation reduces to $0 = d(\gamma(d), q)$, which implies that $\gamma(d) = q$.

Claim 1. $d(p, q) = \delta + d(\gamma(\delta), q)$.

The triangle inequality gives us

$$d(p,q) \leq d(p, \gamma(\delta)) + d(\gamma(\delta), q) = \delta + d(r, q).$$

To prove the reverse inequality, note that any piecewise smooth curve $c : [a, b] \to M$ with $c(a) = p$ and $c(b) = q$ crosses $S(p, \delta)$ at some $t = \tau$. The point $c(\tau)$ cannot be closer to q than $r = \gamma(\delta)$. Thus

$$\begin{aligned} L_{[a,b]}(c) &\geq L_{[a,\tau]}(c) + L_{[\tau,b]}(c) \\ &\geq \delta + d(c(\tau), q) \\ &\geq \delta + d(\gamma(\delta), q). \end{aligned}$$

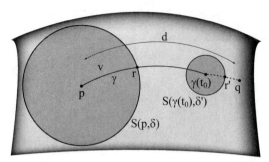

FIGURE 4.7.2
Proof of the converse part of Hopf-Rinow theorem.

Taking the infimum over all curves c gives us $d(p,q) \geq \delta + d(r,q)$.

Claim 2. $d(p,q) = t + d(\gamma(t), q)$ for $\delta \leq t \leq d$.

Let t_0 be the supremum of those $t \in [\delta, d]$ for which the equation holds. Step 1 ensures that t_0 is well defined, and it follows from the continuity of the distance function that $d(p,q) = t_0 + d(\gamma(t_0), q)$.

We now show by contradiction that $t_0 < d$ is impossible. If $t_0 < d$, we $d(\gamma(t_0), p) \leq t_0 < d = d(p,q)$, so $\gamma(t_0) \neq q$. Choose $\delta' > 0$ small enough so that $\delta' < d(\gamma(t_0), q)$ and $B(\gamma(t_0), \delta')$ is a geodesic neighborhood of $\gamma(t_0)$. Choose $r' \in S(\gamma(t_0), \delta')$ with $d(r', q)$ as small as possible. See Figure 4.7.2. The argument used to establish Claim 1 now gives us

$$d - t_0 = d(\gamma(t_0), q) = d(\gamma(t_0), r') + d(r', q)$$
$$= \delta' + d(r', q),$$

and so $d(r', q) = d - t_0 - \delta'$. It now follows from the triangle inequality that

$$d(p, r') \geq d(p,q) - d(r', q)$$
$$= d - (d - t_0 - \delta') = t_0 + \delta'.$$

Now consider the piecewise smooth curve $c : [0, t_0 + \delta'] \to M$ such that $c(t) = \gamma(t)$ for $0 \leq t \leq t_0$ and $c_{[t_0, t_0 + \delta']}$ is the unit speed geodesic of length δ' from $\gamma(t_0)$ to r'. Then c joins p to r' and $L(c) = t_0 + \delta' \leq d(p, r')$. By Theorem 4.6.2, this is possible only if $d(p, r') = t_0 + \delta'$ and c is a geodesic. It follows from the uniqueness of geodesics with respect to initial conditions that $\gamma(t) = c(t)$ for $0 \leq t \leq t_0$. In particular $\gamma(t_0 + \delta') = c(t_0 + \delta') = r'$. Since $d(r', q) = d - t_0 - \delta'$, we obtain

$d(p,q) = t_0 + \delta' + d(\gamma(t_0 + \delta'), q)$. This contradicts the definition of t_0, and ends Step 1.

Step 2. We show that bounded subsets of M have compact closure.

Step 1 tells us that, for any $p \in M$ and any $r \geq 0$, the set $\{q \in M \mid d(p,q) \leq r\}$ lies in $\exp_p(\{v \in T_pM \mid \|v\| \leq r\})$, which is compact, since it is the continuous image of a compact set.

Step 3. We show that every Cauchy sequence in M is convergent.

Let q_n be a Cauchy sequence in M. Then $\sup_{n\geq 1} d(q_1, q_n)$ is finite; call this number r. All the q_n for $n \geq 1$ lie in the set $\{q \in M \mid d(p,q) \leq r\}$, which is compact by Step 2. Now the sequence q_n must converge, because Cauchy sequences in compact metric spaces converge. ∎

COROLLARY 4.7.6
In a geodesically complete manifold any two points can be joined by a geodesic of minimal length.

REMARK 4.7.7 The converse of the above corollary is not true. For example, any pair of points of an open disk in the plane can be connected by a geodesic of minimal length, although the open disk is not geodesically complete.

Note also that in a geodesically complete manifold, a minimizing geodesic between a given pair of points may not be unique. For example, a sphere is geodesically complete, but there exist infinitely many geodesics of minimal length between each pair of antipodal points. ∎

4.8 Exercises

4.8.1 Consider $\mathbb{R}$ with the connection $\nabla_{(\partial/\partial x)}(\partial/\partial x) = \lambda$, for some $\lambda \in \mathbb{R}$. Let $c : [0,1] \to \mathbb{R}$ be a curve with $dc/dt(0) = \partial/\partial x$. Show that the parallel transport along c

$$P_{c(t),c(0)} : T_{c(0)}\mathbb{R} \to T_{c(t)}\mathbb{R}$$

is given by

$$P_{c(t),c(0)}(v\partial/\partial x) = ve^{-\lambda t}(\partial/\partial x),$$

for $v \in T_{c(t)}\mathbb{R}$. Note that $\lambda = 0$ gives the usual connection, and every $\lambda \neq 0$ determines a non-Euclidean parallelism on $\mathbb{R}$.

4.8.2 Prove Proposition 4.2.7.

4.8.3 Let $M = \mathbb{R}^m$ and define $\nabla_X Y(p) = X_p(Y)$. Here $X_p(Y)$ is the directional derivative of Y in the direction of X_p. Show that $\nabla_X Y$ is the Riemannian connection on $\mathbb{R}^m$.

4.8.4 Prove the existence part of Theorem 4.3.6.

4.8.5 Let M be an embedded surface in $\mathbb{R}^3$. Prove that the tangential connection $\nabla^t_X Y$ defined by (4.2.8) is the Riemannian connection on M.

4.8.6 (Naturality of the Riemannian connection) Let M, N be a Riemannian manifold equipped with the Riemannian connections ∇ and $\tilde{\nabla}$, respectively. Assume that M and N are isometric, with $f : M \to N$ an isometry. Show:

(i) f maps the Riemannian connection ∇ into the Riemannian connection $\tilde{\nabla}$ in the sense
$$(df)(\nabla_X Y) = \tilde{\nabla}_{(df)(X)}((df)Y);$$

(ii) f maps the covariant derivative D/dt on M into the covariant derivative $\tilde{D}/dt$ on N, in the sense that if $c : I \to M$ is a curve in M, X is a vector field along c, $\tilde{c} = f \circ c : I \to N$ is the image of c through f, and $(df)(X)$ is the corresponding vector field along $\tilde{c}$, then
$$(df)\left(\frac{DX}{dt}\right) = \frac{\tilde{D}((df)(X))}{dt};$$

(iii) f maps geodesics in M into geodesics in N: if γ is a geodesic in M, then $f \circ \gamma$ is a geodesic in N.

4.8.7 Suppose that M and N are two Riemannian manifolds and $f : M \to N$ is a diffeomorphism. Prove that f is distance preserving, i.e.
$$d_M(p, q) = d_N(f(p), f(q)),$$
for all $p, q \in M$, if and only if it is an isometry.

4.8.8 Prove that a Riemannian manifold M is complete if and only if the image of any geodesic $\gamma : I \to M$ (with I finite or infinite interval) with finite total length is relatively compact.

4.8. EXERCISES

4.8.9 Let M be a Riemannian manifold, and N be a closed submanifold of M. Show by an example that the Riemannian distance on N defined by the Riemannian metric induced by M on N is not the same as the distance induced by M on N. Prove that, nevertheless, if M is complete, then so is N.

4.8.10 Show that for each point p of an m-dimensional Riemannian manifold M there exist a neighborhood U, and m smooth vector fields $E_1,\ldots,E_m$ on U that are orthonormal at each point in U and satisfy $\nabla_{E_i} E_j = 0$ at p for all i,j. Such a family of vector fields is called a geodesic frame.

4.8.11 A geodesic $\gamma : [0,\infty) \to M$ is called a ray if it minimizes the distance between $\gamma(0)$ and $\gamma(s)$ for all $s \in [0,\infty)$. Show that if M is complete and non-compact, then there is a ray leaving from every point in M.

4.8.12 Suppose that for every smooth Riemannian metric on a manifold M, M is complete. Show that M is compact.

4.8.13 Let M be a connected Riemannian manifold with the property that for every pair p and q in M, there exists an isometry of M taking p to q. Show that M is geodesically complete.

4.8.14 Consider the parametrized curve $(f(v), g(v))$ in the xy-plane, where f and g are smooth functions, with $f \neq 0$ and $(f')^2 + (g')^2 \neq 0$. The rotation of this curve around the z-axis generates a surface of revolution $\phi : U \subseteq \mathbb{R}^2 \to \mathbb{R}^3$ given by $\phi(u,v) = (f(v)\cos u, f(v)\sin u, g(v))$ where
$$U = \{(u,v) \in \mathbb{R}^2 \,|\, u_0 < u < u_1, \, v_0 < v < v_1\}.$$
The images by ϕ of the curves $u =$ constant are called meridians, and the images of the curves $v =$ constant are called parallels.

(i) Show that ϕ is an immersion;

(ii) Show that the induced metric in (u,v)-coordinates is given by
$$g_{11} = f^2, \quad g_{12} = g_{21} = 0, \quad g_{22} = (f')^2 + (g')^2;$$

(iii) Show that the geodesic equations are
$$\frac{d^2 u}{dt^2} + \frac{2ff'}{f^2} \frac{du}{dt}\frac{dv}{dt} = 0,$$
$$\frac{d^2 v}{dt^2} - \frac{ff'}{(f')^2 + (g')^2}\left(\frac{du}{dt}\right)^2 + \frac{f'f'' + g'g''}{(f')^2 + (g')^2}\left(\frac{dv}{dt}\right)^2 = 0;$$

FIGURE 4.8.1
Geodesics of a paraboloid.

(iv) Deduce the following interpretation for geodesics γ that are neither meridians nor parallels: the second equation says that the 'energy' $\|\gamma'(t)\|^2$ is constant along γ, and the first equation says that the angle $\beta(t) < \pi$ made by γ with a parallel through $\gamma(t)$ corresponding to r, satisfies the Clairaut equation $r \cos \beta(t) = $ constant;

(v) Consider a paraboloid obtained by rotating $(f(v), g(v)) = (v, v^2)$ about the z-axis, with $0 \leq u \leq 2\pi$ and $0 < v < \infty$. Use Clairaut's equation to show that any geodesic which is not a meridian intersects itself infinitely many times. See Figure 4.8.1;

(vi) Deduce the following interpretation of the Clairaut equation (Oprea (2004)). Consider a particle of mass 1 moving on a surface of revolution, subject to the constraint force to the surface. Let $c(t) = (f(v(t)) \cos u(t), f(v(t)) \sin u(t), g(v(t)))$ be the trajectory of the particle, and assume $\|c'(t)\| = 1$. The constraint force is in the direction of the unit normal vector to the surface, according to D'Alembert's principle. This implies that the tangential component of the acceleration $c''(t)$ vanishes, so the trajectory describes a geodesic. The angular momentum vector of the trajectory is, by definition, $L(t) = c(t) \times c'(t)$. Show that Clairaut's equation says that the z-component of the angular momentum is conserved.

4.8.15 Show that the geodesics of the Poincaré disk $\mathbb{B}^2$ are either diameters of the disk or arcs of circles orthogonal to the boundary circle.

4.8.16 Show that the geodesics of the hyperboloid model $\mathcal{H}^2$ of the hyperbolic plane are the 'great hyperbolas', that is, the hyperbolas obtained by slicing $\mathcal{H}^2$ with planes through the origin of $\mathbb{R}^3$.

Chapter 5

Curvature

5.1 Introduction

Loosely speaking, curvature measures the extent to which a manifold differs in the way it bends from flat space.

There are two kinds of curvature: extrinsic and intrinsic. Extrinsic curvature describes the manifold from the point of view of an external observer, who compares the bending of curves that lie on the manifold with the straight lines that go off it. Intrinsic curvature describes the manifold from the point of view of an observer confined to the manifold, who performs measurements only along paths that lie on the manifold.

We start by discussing curvature from the extrinsic point of view. In the case of a regular curve $c : I \to \mathbb{R}^3$ in the Euclidean space, the curvature at a point p can be obtained by taking the limiting position of the circles determined by p, p_1 and p_2, where p_1 and p_2 are two points of the curve that tend to p. See Figure 5.1.1 (a). The limit circle is called the osculating circle, and the curvature of the curve at p is, by definition, the reciprocal $1/R$ of the radius R of the osculating circle. This agrees with the intuitive perception of curvature: a bigger circle requires less bending than a smaller circle. The plane containing the osculating circle is called the osculating plane. See Figure 5.1.1 (b). For a curve c parametrized by arc length ($|c'(t)| = 1$ for all t), the curvature measures how rapidly the velocity vector is turning, and is given by the size of the acceleration vector $\kappa = |c''(s)|$. Thus, the radius of the osculating circle is $1/\kappa$. The radius vector at the contact point with the curve is $c''(s)$, and the osculating plane is spanned by the velocity vector $c'(s)$ and by the acceleration $c''(s)$.

In the case of an embedded surface S the bending generally varies

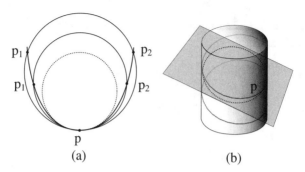

FIGURE 5.1.1
(a) An osculating circle (dashed) to a parabola, as a limit of circles. (b) An osculating circle (dashed) to a helix.

with the direction. Given a unit normal vector N_p at a point p on the surface, by slicing the surface by a plane Π through the normal vector one obtains a curve lying on the surface. For such a curve, one can compute the curvature by the method described above. One can define a signed curvature, by assigning a positive sign when the normal vector points outwards the osculating circle, and a negative sign when it points inwards. This signed curvature is called the normal curvature determined by Π. The maximum normal curvature κ_1 and the minimum normal curvature κ_2 are called the principal curvatures at p, and the directions in which they occur are called principal directions. See Figure 5.1.2. Of course, their sign depends on the choice of the unit normal vector at p. The directions of maximum and minimum curvature are orthogonal (at exceptional points called umbilics the curvature is the same in all directions), so one can always choose a pair of principal directions $\{v_1, v_2\}$ that constitute an orthonormal basis of the tangent plane at p. If v is another direction vector in the tangent plane, then the normal curvature in the direction of v is given by $\kappa_1 \cos^2 \theta + \kappa_2 \sin^2 \theta$, where θ is the angle between v and v_1. Certain averages of the principal curvatures are of special interest: the sum $H = \kappa_1 + \kappa_2$ is called the mean curvature, and the product $K = \kappa_1 \kappa_2$ is called the Gaussian curvature.

Another way in which one could measure the curvature at a point p is by the rate of change in the normal unit vector as its foot point approaches p. This is given by the derivative of the Gauss map $G : S \to S^2$, defined in Section 1.6. The Gauss map assigns a unit normal vector N_p to each point p of the surface. The derivative map $(dG)_p : T_p S \to T_{N_p} S^2$

5.1. INTRODUCTION

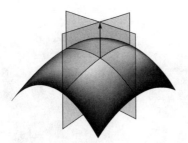

FIGURE 5.1.2
The principal curvatures of a surface.

can be viewed as a linear map into T_pS, since T_pS and $T_{N_p}S^2$ are parallel planes. The quadratic form $v \in T_pM \to -\langle (dG)_p(v), v \rangle \in \mathbb{R}$ is called the second fundamental form. The negative sign is put in front of dG_p by convention. The matrix associated to $-(dG)_p$ is called the coefficient matrix of the second fundamental form. If c is the curve obtained by slicing the surface with a plane through N_p and v, it follows that the signed curvature of c at p is exactly $-\langle (dG)_p(v), v \rangle$, provided c is arc length parametrized. It turns out that the principal curvature k_1 and k_2 are the maximum and minimum values of the second fundamental form over all unit tangent vectors $v \in T_pM$. Hence the principal curvatures are the eigenvalues of the matrix of $(dG)_p$, and the corresponding eigenvectors are principal directions. Therefore the mean curvature H is the trace and the Gaussian curvature K is the determinant of this matrix.

A crucial result, called the Theorema Egregium of Gauss, reveals that the Gaussian curvature is 'intrinsic'. This means that if one deforms a surface in a way that preserves the Riemannian distance between any pair of points (that is, with no stretching and shrinking), the Gaussian curvature K is preserved. In other words, K is invariant under local isometries, or, equivalently, K depends only on the coefficients of the first fundamental form $(g_{ij})_{i,j=1,2}$. In particular, K_p is independent of the choice of a unit normal vector N_p at p. The mean curvature H_p changes under local isometries, so it is 'extrinsic'. See Figure 5.1.3.

To define the curvature of a general Riemannian manifold, one approaches the idea of curvature differently. In Section 4.1, we noticed that parallel transport in a manifold is path dependent, while parallel transport in a Euclidean space is not. The extent to which parallel transport fails to be path independent can be used as a measure for curvature. Let X, Y and Z be three smooth vector fields on M. First, suppose that X and Y commute, that is $[X, Y] = 0$. Denote by ϕ^t the

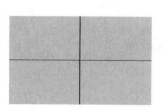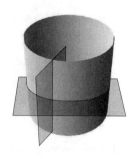

FIGURE 5.1.3
By rolling a rectangular sheet of paper one obtains a cylinder. This type of transformation preserves the distances between pairs of points. The principal curvatures of the rectangle are $\kappa_1 = \kappa_2 = 0$. The principal curvatures of the cylinder are $\kappa_1 = 1/R$ and $\kappa_2 = 0$, where R is the radius of the cylinder. The Gaussian curvature $K = 0$ is the same for both surfaces. However, the mean curvature of the rectangle is $H = 0$, while the mean curvature of the cylinder is $H = 1/R$.

local flow generated by X and by ψ^t the local flow generated by Y. We construct a 'small parallelogram' lying on M as follows. We start from p and travel along the flow ϕ^t between $p = \phi^0(p)$ and $q = \phi^h(p)$, for some small $h > 0$. Then we travel along ψ^t between $q = \psi^0(q)$ and $r = \psi^k(q)$, for some $k > 0$. Then we travel from r along ϕ^{-t} for $0 \le t \le h$ to a new point s, and finally we travel from s along ψ^{-t} for $0 \le t \le k$. Since $[X, Y] = 0$, from Proposition 2.4.7 it follows that the arrival point coincides with the departure point p. The resulting parallelogram is the image of the rectangle $[0, h] \times [0, k] \in \mathbb{R}^2$ through $s(x, y) = \psi^y \phi^x(p)$.

Now we parallel transport the vector field Z along the contour of the parallelogram. Let $Z = Z(x, y)$ be the restriction of Z restricted to S, where $(x, y) \in [0, h] \times [0, k]$. We first parallel transport $Z(h, k)$ from r to q along ψ^t obtaining the vector $P_{q,r} Z(h, k)$, then we parallel transport this vector from q to p along ϕ^t, obtaining $P_{p,q} P_{q,r} Z(h, k)$. We then parallel transport vector Z along the other pair of sides of the parallelogram, obtaining $P_{p,s} P_{s,r} Z(h, k)$. The two resulting tangent vectors at p are, in general, different. See Figure 5.1.4. We describe the curvature at p through the limit

$$\lim_{h,k \to 0} \frac{(P_{p,q} P_{q,r} - P_{p,s} P_{s,r}) Z(h, k)}{hk}. \tag{5.1.1}$$

5.1. INTRODUCTION

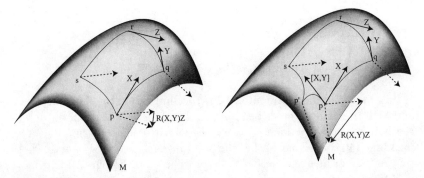

FIGURE 5.1.4
The curvature tensor as a measure of the path dependence of the parallel transport. On the left, the case when X and Y commute is considered. On the right, the case when X and Y do not commute is considered.

We first claim

$$\lim_{h,k \to 0} \frac{1}{hk} \left[P_{p,q} P_{q,r} Z(h,k) - P_{p,q} Z(h,0) \right.$$
$$\left. - P_{p,s} Z(0,k) + Z(0,0) \right] = \nabla_X \nabla_Y Z(0,0). \quad (5.1.2)$$

Using (4.2.9), we expand the left side of (5.1.2) as

$$\lim_{h \to 0} \frac{1}{h} \left[P_{p,q} \left(\lim_{k \to 0} \frac{P_{q,r} Z(h,k) - Z(h,0)}{k} \right) - \lim_{k \to 0} \frac{P_{p,s} Z(0,k) - Z(0,0)}{k} \right]$$
$$= \lim_{h \to 0} \frac{P_{p,q} \nabla_Y Z(h,0) - \nabla_Y Z(0,0)}{h} = \nabla_X \nabla_Y Z(0,0),$$

which proves our claim. Similarly, we obtain that

$$\lim_{h,k \to 0} \frac{1}{hk} \left(P_{p,s} P_{s,r} Z(h,k) - P_{p,q} Z(h,0) \right.$$
$$\left. - P_{p,s} Z(0,k) + Z(0,0) \right) = \nabla_Y \nabla_X Z(0,0). \quad (5.1.3)$$

Subtracting (5.1.3) from (5.1.2), we obtain that the limit in (5.1.1) is equal to

$$\nabla_X \nabla_Y Z(0,0) - \nabla_Y \nabla_X Z(0,0).$$

If we replace both k and h by $\sqrt{h}$ in (5.1.1), we obtain

$$\lim_{h \to 0} \frac{(P_{p,q} P_{q,r} - P_{p,s} P_{s,r}) Z}{h} = (\nabla_X \nabla_Y Z - \nabla_Y \nabla_X Z)(0,0). \quad (5.1.4)$$

The failure of parallel transport to be path independent is thus reflected by the quantity

$$R(X,Y)Z = \nabla_X \nabla_Y Z - \nabla_Y \nabla_X Z.$$

If X and Y do not commute, then travelling along the flow lines of X and Y in the manner described above will not close that path, but it will result in an arriving point $p' = p'(\sqrt{h})$ which tends to p as $h \to 0$. As mentioned in the Remark 2.4.8, the point $p'(\sqrt{h})$ describes a curve whose tangent vector at p is exactly $[X,Y](p)$. Thus, for the parallel transport of Z from r to p via s, one has to take into account the parallel transport of Z from p' to p. See Figure 5.1.4. The parallel transport from p' to p contributes to $\nabla_Y \nabla_X Z$ with an additional quantity of

$$\lim_{h \to 0} \frac{P_{p,p'}Z - Z}{h} = \nabla_{[X,Y]}Z(0,0).$$

This suggests a definition of curvature by the means of the quantity

$$R(X,Y)Z = \nabla_X \nabla_Y Z - \nabla_Y \nabla_X Z - \nabla_{[X,Y]}Z.$$

At this point, it not clear whether the two above approaches to the notion of curvature are compatible. This will be clarified in the following sections.

5.2 The curvature tensor

DEFINITION 5.2.1
The curvature tensor on a manifold M, endowed with a connection ∇, is a smooth mapping R that assigns to every triple X, Y, Z of smooth vector fields on M another vector field $R(X,Y)Z$ defined by

$$R(X,Y)Z = \nabla_X \nabla_Y Z - \nabla_Y \nabla_X Z - \nabla_{[X,Y]}Z. \qquad (5.2.1)$$

When the manifold is Riemannian and the connection is the Riemannian connection, the corresponding curvature tensor is called the Riemannian curvature tensor.

5.2. THE CURVATURE TENSOR

The smoothness of the mapping means that the resulting vector field $R(X,Y)Z$ is smooth. This is clear since all operations involved in (5.2.1) are smooth. The term 'tensor' will be justified in Section 6.4: for the moment, it is just a way to express the facts, which we will soon see, that R is linear in X, Y and Z, and its value at a point $p \in M$ depends only on the vectors X_p, Y_p and Z_p.

PROPOSITION 5.2.2
The curvature tensor R on a manifold M with a connection ∇ satisfies the following properties:

(i) *antisymmetry in variables X and Y:*
$$R(X,Y)Z = -R(Y,X)Z,$$

(ii) *bilinearity in variables X and Y over smooth functions:*
$$R(f_1 X_1 + f_2 X_2, Y)Z = f_1 R(X_1, Y)Z + f_2 R(X_2, Y)Z,$$
$$R(X, f_1 Y_1 + f_2 Y_2)Z = f_1 R(X, Y_1)Z + f_2 R(X, Y_2)Z,$$

for all smooth vector fields X_1, X_2, and all smooth functions f_1, f_2 on M,

(iii) *linearity in variable Z over smooth functions:*
$$R(X,Y)(f_1 Z_1 + f_2 Z_2) = f_1 R(X,Y) Z_1 + f_2 R(X,Y) Z_2,$$

for all smooth vector fields Z_1, Z_2, and all smooth functions f_1, f_2 on M.

PROOF Property (i) follows from the antisymmetry of the Lie bracket (Proposition 2.4.5).

For (ii), we use the properties of $\nabla_X Y$ from Definition 4.2.1, and the properties of $[X,Y]$ from Proposition 2.4.5, to obtain

$$\begin{aligned}
R(f_1 X_1 + f_2 X_2, Y)Z &= (f_1 \nabla_{X_1} + f_2 \nabla_{X_2})\nabla_Y Z - \nabla_Y (f_1 \nabla_{X_1} + f_2 \nabla_{X_2})Z \\
&\quad - \nabla_{f_1 [X_1, Y] + f_2 [X_2, Y]} Z \\
&= f_1 \nabla_{X_1} \nabla_Y Z - f_1 \nabla_Y \nabla_{X_1} Z - Y(f_1) \nabla_{X_1} Z \\
&\quad + f_2 \nabla_{X_2} \nabla_Y Z - f_2 \nabla_Y \nabla_{X_2} Z - Y(f_2) \nabla_{X_2} Z \\
&\quad - f_1 \nabla_{[X_1, Y]} Z - f_2 \nabla_{[X_2, Y]} Z \\
&\quad + Y(f_1) \nabla_{X_1} Z + Y(f_2) \nabla_{X_2} Z \\
&= f_1 R(X_1, Y)Z + f_2 R(X_2, Y)Z.
\end{aligned}$$

The analogue in Y follows from (i).

For (iii), we first notice that $R(X,Y)Z$ is additive in variable Z. Then we calculate

$$\nabla_X \nabla_Y (fZ) - \nabla_Y \nabla_X (fZ) = f(\nabla_X \nabla_Y Z - \nabla_Y \nabla_X Z) + [X,Y](f)Z,$$

therefore

$$\begin{aligned} R(X,Y)(fZ) &= f(\nabla_X \nabla_Y Z - \nabla_Y \nabla_X Z) + [X,Y](f)Z \\ &\quad - f\nabla_{[X,Y]}(Z) - [X,Y](f)Z \\ &= fR(X,Y)Z. \end{aligned}$$

We conclude

$$\begin{aligned} R(X,Y)(f_1 Z_1 + f_2 Z_2) &= R(X,Y)(f_1 Z_1) + R(X,Y)(f_2 Z_2) \\ &= f_1 R(X,Y) Z_1 + f_2 R(X,Y) Z_2. \end{aligned}$$

∎

It is useful to express R in local coordinates. For this purpose, let $(x_1,\ldots,x_m)$ be a local coordinate system near a point $p \in M$. We express the vector fields X, Y, Z in local coordinates

$$X = \sum_{i=1}^{m} u_i \frac{\partial}{\partial x_i},\ Y = \sum_{j=1}^{m} v_j \frac{\partial}{\partial x_j},\ Z = \sum_{k=1}^{m} w_k \frac{\partial}{\partial x_k}.$$

Due to the multi-linearity of R, we can write

$$R(X,Y)Z = \sum_{i,j,k=1}^{m} u_i v_j w_k R\left(\frac{\partial}{\partial x_i}, \frac{\partial}{\partial x_j}\right) \frac{\partial}{\partial x_k}. \tag{5.2.2}$$

We express $R(\partial/\partial x_i, \partial/\partial x_j)\partial/\partial x_k$ in local coordinates

$$R\left(\frac{\partial}{\partial x_i}, \frac{\partial}{\partial x_j}\right) \frac{\partial}{\partial x_k} = \sum_{l=1}^{m} R^l_{ijk} \frac{\partial}{\partial x_l}, \tag{5.2.3}$$

for some smooth functions R^l_{ijk}, hence

$$R(X,Y)Z = \sum_{l=1}^{m} \left[\sum_{i,j,k=1}^{m} u_i v_j w_k R^l_{ijk}\right] \frac{\partial}{\partial x_l}. \tag{5.2.4}$$

5.2. THE CURVATURE TENSOR

The above equation shows that the value of $R(X,Y)Z$ at a point $p \in M$ depends only on the vectors X_p, Y_p, Z_p, and not on the vector fields X, Y, Z themselves.

REMARK 5.2.3 The point dependence shown above reflects a more general principle. Suppose that $F(X,Y,\ldots)$ is a quantity that depends smoothly on the vector fields $X, Y, \ldots$, and is multi-linear over smooth functions. Then the value of F at a point depends only on the values of the vector fields at that point. In other words, F is a tensor. This can be easily seen by writing F in local coordinates. If $X = \sum_{i=1}^{m} u_i \partial/\partial x_i$, $Y = \sum_{i=1}^{m} v_i \partial/\partial x_i$, etc..., then

$$F(X,Y,\ldots) = \sum_{i,j=1}^{m} u_i v_j \ldots F(\partial/\partial x_i, \partial/\partial x_j, \ldots),$$

and so

$$F(X,Y,\ldots)_p = \sum_{i,j=1}^{m} u_i(p) v_j(p) \ldots F(\partial/\partial x_i, \partial/\partial x_i \partial/\partial x_j, \ldots)_p.$$

For example, Riemannian metrics are tensors, while affine connections are not tensors. ∎

We express the coefficients R^i_{jkl} in terms of the Christoffel coefficients. From (5.2.3), and using the fact that $[\partial/\partial x_i, \partial/\partial x_j] = 0$, we obtain

$$\begin{aligned}
R\left(\frac{\partial}{\partial x_i}, \frac{\partial}{\partial x_j}\right)\frac{\partial}{\partial x_k} &= \nabla_{\frac{\partial}{\partial x_i}} \nabla_{\frac{\partial}{\partial x_j}} \frac{\partial}{\partial x_k} - \nabla_{\frac{\partial}{\partial x_j}} \nabla_{\frac{\partial}{\partial x_i}} \frac{\partial}{\partial x_k} \\
&= \nabla_{\frac{\partial}{\partial x_i}} \left(\sum_{h=1}^{m} \Gamma^h_{jk} \frac{\partial}{\partial x_h}\right) - \nabla_{\frac{\partial}{\partial x_j}} \left(\sum_{h=1}^{m} \Gamma^h_{ik} \frac{\partial}{\partial x_h}\right) \\
&= \sum_{h=1}^{m} \Gamma^h_{jk} \nabla_{\frac{\partial}{\partial x_i}}\left(\frac{\partial}{\partial x_h}\right) + \sum_{h=1}^{m} \frac{\partial \Gamma^h_{jk}}{\partial x_i} \frac{\partial}{\partial x_h} \\
&\quad - \sum_{h=1}^{m} \Gamma^h_{ik} \nabla_{\frac{\partial}{\partial x_j}}\left(\frac{\partial}{\partial x_h}\right) - \sum_{h=1}^{m} \frac{\partial \Gamma^h_{ik}}{\partial x_j} \frac{\partial}{\partial x_h} \\
&= \sum_{l=1}^{m} \left[\sum_{h=1}^{m} (\Gamma^h_{jk}\Gamma^l_{ih} - \Gamma^h_{ik}\Gamma^l_{jh}) + \frac{\partial \Gamma^l_{jk}}{\partial x_i} - \frac{\partial \Gamma^l_{ik}}{\partial x_j}\right] \frac{\partial}{\partial x_l},
\end{aligned}$$

whence

$$R^l_{ijk} = \sum_{h=1}^{m} \left(\Gamma^h_{jk}\Gamma^l_{ih} - \Gamma^h_{ik}\Gamma^l_{jh} \right) + \frac{\partial \Gamma^l_{jk}}{\partial x_i} - \frac{\partial \Gamma^l_{ik}}{\partial x_j}. \qquad (5.2.5)$$

The above discussion applies to the curvature tensor for a general connection on a manifold. From now on we consider the special case when the manifold and the connection are Riemannian. The coefficient R^l_{ijk} represents the component of $R(\partial/\partial x_i, \partial/\partial x_j)\partial/\partial x_j$ in the $\partial/\partial x_l$ direction. The vectors $(\partial/\partial x_l)_{l=1,\ldots,m}$ do not form, in general, an orthonormal basis, so R^l_{ijk} is not equal to the orthogonal projection of $R(\partial/\partial x_i, \partial/\partial x_j)\partial/\partial x_k$ onto the direction of $\partial/\partial x_l$. Let

$$\left\langle R\left(\frac{\partial}{\partial x_i}, \frac{\partial}{\partial x_j}\right) \frac{\partial}{\partial x_k}, \frac{\partial}{\partial x_l} \right\rangle = R_{ijkl}. \qquad (5.2.6)$$

Then the orthogonal projection is given by $R_{ijkl}/\|\partial/\partial x_l\|$.

The functions R_{ijkl} present an interest of their own. Taking the inner product of (5.2.3) with $\partial/\partial x_l$, we derive the relationship

$$R_{ijkl} = \sum_{h=1}^{m} R^h_{ijk} g_{lh}. \qquad (5.2.7)$$

More generally, we can define the function

$$R(X, Y, Z, T) = \langle R(X,Y)Z, T \rangle,$$

where X, Y, Z, T are smooth vector fields on the manifold. It is clear that $R(X, Y, Z, T)$ is also a tensor, since it depends smoothly on X, Y, Z, and T, and is multi-linear over smooth functions. If $X = (u_i)_{i=1,\ldots,m}$, $Y = (v_j)_{j=1,\ldots,m}$, $Z = (w_i)_{i=1,\ldots,m}$, and $T = (s_i)_{i=1,\ldots,m}$, in local coordinates, then we have

$$R(X, Y, Z, T) = \sum_{i,j,k,l=1}^{m} u_i v_j w_k s_l R_{ijkl}.$$

This tensor and its coefficients R_{ijkl} satisfy certain symmetry relations which are important when we compute curvature in local coordinates.

5.2. THE CURVATURE TENSOR

PROPOSITION 5.2.4
The following identities hold true

(i) $\langle R(X,Y)Z, T \rangle = -\langle R(Y,X)Z, T \rangle$, and so $R_{ijkl} = -R_{jikl}$.

(ii) $\langle R(X,Y)Z, T \rangle = -\langle R(X,Y)T, Z \rangle$, and so $R_{ijkl} = -R_{ijlk}$.

(iii) $\langle R(X,Y)Z, T \rangle = \langle R(Z,T)X, Y \rangle$, and so $R_{ijkl} = R_{klij}$.

(iv) Bianchi identity: $R(X,Y)Z + R(Y,Z)X + R(Z,X)Y = 0$, and so $R_{ijkl} + R_{jkil} + R_{kijl} = 0$.

PROOF Assertion (i) follows immediately from the antisymmetry of the curvature tensor.

Assertion (ii) is a consequence of the compatibility of the affine connection with the Riemannian metric (Corollary 4.3.3). We first compute

$$\langle \nabla_X \nabla_Y Z, T \rangle = X \langle \nabla_Y Z, T \rangle - \langle \nabla_Y Z, \nabla_X T \rangle$$
$$= XY \langle Z, T \rangle - X \langle Z, \nabla_Y T \rangle - Y \langle Z, \nabla_X T \rangle + \langle Z, \nabla_Y \nabla_X T \rangle,$$

and, similarly,

$$\langle \nabla_Y \nabla_X Z, T \rangle = YX \langle Z, T \rangle - Y \langle Z, \nabla_X T \rangle - X \langle Z, \nabla_Y T \rangle + \langle Z, \nabla_X \nabla_Y T \rangle.$$

We then compute

$$\langle \nabla_{[X,Y]} Z, T \rangle = [X, Y] \langle Z, T \rangle - \langle Z, \nabla_{[X,Y]} T \rangle.$$

Combining the above equations, we conclude

$$\langle R(X,Y)Z, T \rangle = \langle Z, \nabla_Y \nabla_X T - \nabla_X \nabla_Y T + \nabla_{[X,Y]} T \rangle$$
$$= -\langle Z, R(X,Y)T \rangle.$$

We postpone the proof of (iii) for a moment, and prove (iv). We use the symmetry of the Riemannian connection (Definition 4.3.4) to obtain:

$$R(X,Y)Z + R(Y,Z)X + R(Z,X)Y = \nabla_X \nabla_Y Z - \nabla_Y \nabla_X Z - \nabla_{[X,Y]} Z$$
$$+ \nabla_Y \nabla_Z X - \nabla_Z \nabla_Y X - \nabla_{[Y,Z]} X + \nabla_Z \nabla_X Y - \nabla_X \nabla_Z Y - \nabla_{[Z,X]} Y$$
$$= \nabla_X [Y, Z] - \nabla_{[Y,Z]} X + \nabla_Y [Z, X] - \nabla_{[Z,X]} Y + \nabla_Z [X, Y] - \nabla_{[X,Y]} Z$$
$$= [X, [Y, Z]] + [Y, [Z, X]] + [Z, [X, Y]] = 0,$$

with the last identity representing Jacobi's identity from Proposition 2.4.5.

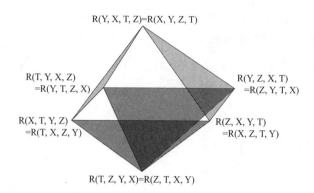

FIGURE 5.2.1
The use of Bianchi's identity.

We now prove (iii) using (i), (ii), (iv), and the diagram in Figure 5.2.1 (due to Milnor (1963)).

For each of the four shaded triangles, the sum of quantities at the vertices equals to zero, by Bianchi's identity. Thus, the sum of the quantities for the upper two shaded triangles equals twice the quantity at the top plus the sum of quantities at the base, and equals zero. Similarly, the sum of the quantities for the lower two shaded triangles equals twice the quantity at the top plus the sum of quantities at the base, and equals zero. When we subtract the quantities for the lower shaded triangles from the quantities for the upper shaded triangles, we obtain twice the quantity at the top vertex minus twice the quantity at the bottom vertex, and this must equal zero. Hence

$$\langle R(X,Y)Z,T \rangle = \langle R(Z,T)X,Y \rangle.$$

∎

Example 5.2.5
If $M = \mathbb{R}^n$, then $[\partial/\partial x_j, \partial/\partial x_k] = 0$, for all j,k, hence the Christoffel symbols $\Gamma^i_{jk} = 0$ for all i,j,k. Thus $R^l_{ijk} = 0$ for all i,j,k,l, due to (5.2.5). Thus, the curvature tensor R of the Euclidean space is identically equal to 0. □

We conclude this section with a first geometric interpretation of the curvature tensor. For a vector field along an embedded surface in a mani-

5.2. THE CURVATURE TENSOR

fold, the curvature tensor measures the non-commutativity of the second order covariant derivatives computed in two independent directions.

PROPOSITION 5.2.6
Let U be some open set in $\mathbb{R}^2$, $(t,s) \in U \to f(t,s) \in M$ be an embedded surface in M and V be a smooth vector field along the surface. Then

$$\frac{D}{\partial t}\frac{D}{\partial s}V - \frac{D}{\partial s}\frac{D}{\partial t}V = R\left(\frac{\partial f}{\partial t}, \frac{\partial f}{\partial s}\right)V. \quad (5.2.8)$$

PROOF In the above formula, the covariant derivatives are taken along the appropriate choices of coordinate curves $t \to f(t,s)$ and $s \to f(t,s)$. We will omit the subscript specifying which curve is considered in each case.

We choose a local coordinate system $(x_1, \ldots, x_m)$ near some point p on the embedded surface, and we express V in local coordinates as

$$V(t,s) = \sum_{i=1}^{m} v_i(t,s)\frac{\partial}{\partial x_i}.$$

We compute the covariant derivatives in the order

$$\frac{D}{\partial s}V = \sum_{i=1}^{m}\left(\frac{\partial v_i}{\partial s}\frac{\partial}{\partial x_i} + v_i\frac{D}{\partial s}\frac{\partial}{\partial x_i}\right),$$

and then

$$\frac{D}{\partial t}\frac{D}{\partial s}V = \sum_{i=1}^{m}\left(\frac{\partial^2 v_i}{\partial t \partial s}\frac{\partial}{\partial x_i} + \frac{\partial v_i}{\partial s}\frac{D}{\partial t}\frac{\partial}{\partial x_i} + \frac{\partial v_i}{\partial t}\frac{D}{\partial s}\frac{\partial}{\partial x_i} + v_i\frac{D}{\partial t}\frac{D}{\partial s}\frac{\partial}{\partial x_i}\right).$$

Taking the covariant derivatives in the reverse order, we obtain

$$\frac{D}{\partial s}\frac{D}{\partial t}V = \sum_{i=1}^{m}\left(\frac{\partial^2 v_i}{\partial s \partial t}\frac{\partial}{\partial x_i} + \frac{\partial v_i}{\partial t}\frac{D}{\partial s}\frac{\partial}{\partial x_i} + \frac{\partial v_i}{\partial s}\frac{D}{\partial t}\frac{\partial}{\partial x_i} + v_i\frac{D}{\partial s}\frac{D}{\partial t}\frac{\partial}{\partial x_i}\right),$$

hence

$$\frac{D}{\partial t}\frac{D}{\partial s}V - \frac{D}{\partial s}\frac{D}{\partial t}V = \sum_{i=1}^{m}v_i\left(\frac{D}{\partial t}\frac{D}{\partial s}\frac{\partial}{\partial x_i} - \frac{D}{\partial s}\frac{D}{\partial t}\frac{\partial}{\partial x_i}\right). \quad (5.2.9)$$

Now we express the right side of (5.2.9) in terms of the Riemannian connection. The components of $f(t,s)$ are given in local coordinates by $x_1(t,s), \ldots, x_m(t,s)$. We have

$$\frac{D}{\partial s}\frac{\partial}{\partial x_i} = \nabla_{\left(\sum_{j=1}^m \frac{\partial x_j}{\partial s}\frac{\partial}{\partial x_j}\right)}\frac{\partial}{\partial x_i} = \sum_{j=1}^m \frac{\partial x_j}{\partial s}\nabla_{\frac{\partial}{\partial x_j}}\frac{\partial}{\partial x_i},$$

and

$$\frac{D}{\partial t}\frac{D}{\partial s}\frac{\partial}{\partial x_i} = \sum_{j=1}^m \left(\frac{\partial^2 x_j}{\partial t \partial s}\nabla_{\frac{\partial}{\partial x_j}}\frac{\partial}{\partial x_i} + \frac{\partial x_j}{\partial s}\nabla_{\left(\sum_{k=1}^m \frac{\partial x_k}{\partial t}\frac{\partial}{\partial x_k}\right)}\nabla_{\frac{\partial}{\partial x_j}}\frac{\partial}{\partial x_i}\right)$$

$$= \sum_{j=1}^m \frac{\partial^2 x_j}{\partial t \partial s}\nabla_{\frac{\partial}{\partial x_j}}\frac{\partial}{\partial x_i} + \sum_{j,k=1}^m \frac{\partial x_k}{\partial t}\frac{\partial x_j}{\partial s}\nabla_{\frac{\partial}{\partial x_k}}\nabla_{\frac{\partial}{\partial x_j}}\frac{\partial}{\partial x_i}.$$

Switching the order of t and s, and the order of j and k, we have

$$\frac{D}{\partial s}\frac{D}{\partial t}\frac{\partial}{\partial x_i} = \sum_{j=1}^m \frac{\partial^2 x_j}{\partial s \partial t}\nabla_{\frac{\partial}{\partial x_j}}\frac{\partial}{\partial x_i} + \sum_{j,k=1}^m \frac{\partial x_k}{\partial t}\frac{\partial x_j}{\partial s}\nabla_{\frac{\partial}{\partial x_j}}\nabla_{\frac{\partial}{\partial x_k}}\frac{\partial}{\partial x_i}.$$

Combining the two expressions from above in (5.2.9), and using the symmetry of the connection, we obtain

$$\frac{D}{\partial t}\frac{D}{\partial s}V - \frac{D}{\partial s}\frac{D}{\partial t}V$$

$$= \sum_{i=1}^m v_i \left[\sum_{j,k=1}^m \frac{\partial x_k}{\partial t}\frac{\partial x_j}{\partial s}\left(\nabla_{\frac{\partial}{\partial x_k}}\nabla_{\frac{\partial}{\partial x_j}} - \nabla_{\frac{\partial}{\partial x_j}}\nabla_{\frac{\partial}{\partial x_k}}\right)\frac{\partial}{\partial x_i}\right]$$

$$= \sum_{i,j,k=1}^m v_i \frac{\partial x_k}{\partial t}\frac{\partial x_j}{\partial s} R\left(\frac{\partial}{\partial x_k}, \frac{\partial}{\partial x_j}\right)\frac{\partial}{\partial x_i} = R\left(\frac{\partial f}{\partial t}, \frac{\partial f}{\partial s}\right)V,$$

where the last equality is due to (5.2.2). ∎

5.3 The second fundamental form

We give a second geometric interpretation of the curvature tensor. We show that the principal curvatures, the mean curvature, and the

5.3. THE SECOND FUNDAMENTAL FORM

Gaussian curvature described in Section 5.1 can be recovered from the curvature tensor induced on a surface. We first consider a more general situation, and then we derive the desired formulae for surfaces.

Let M and M' be two smooth manifolds and $f : M \to M'$ be an immersion. If M' is a Riemannian manifold, then M inherits an induced Riemannian metric (the first fundamental form) from M' by

$$I(v, w) = \langle (df)(v), (df)(w) \rangle.$$

The mapping f becomes an isometric immersion. By Corollary 1.9.6, for each $p \in M$ there exists an open neighborhood U of p such that $f|_U : U \subseteq M \to f(U) \subseteq M'$ is an embedding. Since the computations of curvature are all local, we can assume that $U = M$ and identify M with $f(M)$. Consequently, we can assume $T_pM \subseteq T_pM'$ and $(df)_p = \text{id}$, for each $p \in M$. We will also denote $I(\cdot, \cdot)$ by $\langle \cdot, \cdot \rangle$.

Let ∇' denote the Riemannian connection on M'. If X and Y are smooth vector fields on M, then $\nabla'_X Y$ is not necessarily a vector field on M. At each point $p \in M$, $(\nabla'_X Y)_p \in T_pM'$ can be written as the sum of its tangential component in T_pM, which we will denote by $(\nabla_X Y)_p$, and its normal component in the orthogonal complement $(T_pM)^\perp$ of T_pM in T_pM', which we will denote by $B_p(X, Y)$. So we have

$$(\nabla'_X Y)_p = (\nabla_X Y)_p + B_p(X, Y). \tag{5.3.1}$$

The notation $\nabla_X Y$ suggests that the tangential component actually represents an affine connection. The following lemma shows that this is indeed the case.

LEMMA 5.3.1
$\nabla_X Y$ is the Riemannian connection on M with the induced metric.

PROOF The mapping $p \to (\nabla_X Y)_p$ is the composition of the orthogonal projection of $(\nabla'_X Y)_p \in T_pM'$ onto T_pM, and is as smooth as $\nabla'_X Y$ is. It is easy to check that ∇ satisfies the properties in Definition 4.2.1 and is therefore a connection. Now we show that ∇ has the two properties that characterize the Riemannian connection.

Step 1. We claim that $\nabla_X Y$ is symmetric.

We locally extend X and Y to smooth vector fields X' and Y' defined in a neighborhood of p in M'. Although such extensions are not unique, they satisfy

$$[X', Y']_p = [X, Y]_p,$$

$$(\nabla'_{X'}Y')_p = (\nabla'_X Y)_p,$$
$$(\nabla'_{Y'}X')_p = (\nabla'_Y X)_p,$$

whenever p is in M. Since ∇' is symmetric, we have

$$\begin{aligned} 0 &= \nabla'_{X'}Y' - \nabla'_{Y'}X' - [X',Y'] \\ &= \nabla_X Y - \nabla_Y X - [X,Y] \\ &\quad + B(X,Y) - B(Y,X) \end{aligned} \tag{5.3.2}$$

on M. Since $\nabla_X Y - \nabla_Y X - [X,Y]$ and $B(X,Y) - B(Y,X)$ represent the tangential component and, respectively, the normal component of the zero vector, it results

$$\nabla_X Y - \nabla_Y X - [X,Y] = 0, \tag{5.3.3}$$
$$B(X,Y) - B(Y,X) = 0. \tag{5.3.4}$$

Step 2. We claim that $\nabla_X Y$ is compatible with the metric on M.

For X, Y, and Z smooth vector fields on M, we consider X', Y' and Z' local extensions near p to smooth vector fields on M'. Since ∇' is compatible to the metric, we have

$$X'\langle Y', Z'\rangle = \langle \nabla'_{X'} Y', Z'\rangle + \langle Y', \nabla_{X'} Z'\rangle,$$

which translates into the following identity on M

$$\begin{aligned} X\langle Y, Z\rangle &= \langle \nabla_X Y, Z\rangle + \langle B(X,Y), Z\rangle \\ &\quad + \langle Y, \nabla_X Z\rangle + \langle Y, B(X,Z)\rangle. \end{aligned}$$

Since $B(X,Y), B(X,Z) \in TM^\perp$, and $Z, Y \in TM$, $\langle B(X,Y), Z\rangle = 0$ and $\langle Y, B(X,Z)\rangle = 0$, which proves the claim. ∎

LEMMA 5.3.2
$B(X,Y)$ *is symmetric and bilinear over smooth functions, and hence $B_p(X,Y)$ depends only on X_p and Y_p.*

PROOF We have already shown in (5.3.4) that B is symmetric. Since $\nabla_X Y$ and $\nabla'_X Y$ are both linear over smooth functions in the X variable, so is $B(X,Y)$. The linearity of $B(X,Y)$ over smooth functions in the Y variable follows from symmetry. The point dependence of $B(X,Y)$ follows from the general principle discussed in Remark 5.2.3. ∎

5.3. THE SECOND FUNDAMENTAL FORM

DEFINITION 5.3.3
Let p be a point in M and $N_p \in (T_pM)^\perp$ a normal vector to M. The second fundamental form at p along the normal vector N_p is the symmetric bilinear form on T_pM defined by

$$II_{N_p}(X_p, Y_p) = \langle N_p, B_p(X_p, Y_p)\rangle.$$

The symmetry and bilinearity of II follow from Lemma 5.3.2. The second fundamental form is a measure of the difference between the ambient Riemannian connection on M' and the induced Riemannian connection on M. The reason for considering the projection of $B(X, Y)$ onto some normal direction will be explained in Example 5.3.5. Historically, the term 'second fundamental form' was used for the quadratic form $X \to II_N(X, X)$.

The following theorem will provide us with a convenient way to compute the second fundamental form in the case of surfaces.

THEOREM 5.3.4 *(The Weingarten equation)*
Let $p \in M$, $X_p, Y_p \in T_pM$, and $N_p \in (T_pM)^\perp$. Let X, Y be local extensions of X_p, Y_p, respectively, to smooth vector fields on M, and let N be a local extension of N_p to a smooth vector field normal to M in M'. Then

$$II_N(X, Y) = \langle N, B(X, Y)\rangle = -\langle \nabla'_X N, Y\rangle. \tag{5.3.5}$$

PROOF Since $\nabla_X Y$ is orthogonal to N on M, we have

$$\langle N, B(X,Y)\rangle = \langle N, \nabla'_X Y\rangle - \langle N, \nabla_X Y\rangle = \langle N, \nabla'_X Y_p\rangle.$$

By the compatibility of ∇' with the metric, we have

$$\langle N, \nabla'_X Y\rangle = X\langle N, Y\rangle - \langle \nabla'_X N, Y\rangle = -\langle \nabla'_X N, y\rangle,$$

since N is orthogonal to Y on M. ∎

Example 5.3.5
Let $M \subseteq \mathbb{R}^3$ be an embedded surface. The above constructions show that the connection induced on M by the standard connection on $\mathbb{R}^3$ is the tangential connection defined by (4.2.8).

The Gauss map (Example 1.7.3) $q \in M \to G(q) = N_q \in S^2$, where N_q represents the unit normal vector to M at p, can be defined as a smooth map in some neighborhood of any given point $p \in M$, whether M is orientable or not. Let X_p be a tangent vector at p and $c : I \to M$ a curve through p with $dc/dt(0) = X_p$. The derivative $(dG)_p : T_pM \to T_pS^2 \simeq T_pM$ of the Gauss map is given by

$$(dG)_p(X_p) = \frac{d}{dt}(N \circ c(t))|_{t=0} = \frac{D_c(N_{c(t)})}{dt}|_{t=0} = \nabla^{\mathbb{R}^3}_{X_p} N,$$

where $\nabla^{\mathbb{R}^3}$ represents the standard Riemannian connection on $\mathbb{R}^3$. Using the Weingarten equation and the symmetry of the second fundamental form we obtain that

$$\langle (dG)_p(X_p), Y_p \rangle = \langle \nabla^{\mathbb{R}^3}_{X_p} N, Y_p \rangle = -II_{N_p}(X_p, Y_p) = -II_{N_p}(Y_p, X_p)$$
$$= \langle X_p, (dG)_p(Y_p) \rangle.$$

That is, $(dG)_p$ is symmetric with respect to $\langle \cdot, \cdot \rangle$. If we let $X_p = Y_p$, we obtain $II_{N_p}(X_p, X_p) = -\langle (dG)_p(X_p), X_p \rangle$, which is exactly the way we introduced the second fundamental form in Section 5.1. From linear algebra, it follows that $-(dG)_p$ always has two real eigenvalues κ_1 and κ_2, and the corresponding eigenvectors v_1 and v_2 are orthogonal.

If we slice the surface by a normal plane at p containing a unit tangent vector v, the intersection is some curve $c : I \to M$ with $c(0) = p$ and $c'(0) = v$ (we assume c is parametrized by arc length). Since $c''(t)$ lies in the normal plane and is orthogonal to $c'(t)$ (because $c'(t) \cdot c'(t) = 1$ so $2c''(t) \cdot c'(t) = 0$), $c''(t)$ is collinear to the unit normal vector $N_{c(t)}$. Thus

$$\langle c'(t), G(c(t)) \rangle = 0.$$

Taking the derivative with respect to t and then letting $t = 0$, we obtain

$$\langle c''(t), G(c(t)) \rangle = -\langle c'(t), (dG)_{c(t)} c'(t) \rangle,$$
$$\langle c''(0), N_p \rangle = -\langle v, (dG)_p(v) \rangle = II_{N_p}(v, v). \quad (5.3.6)$$

The left hand side of the above equation is exactly the signed curvature of the curve c at p. Hence the extremal values of the curvature of a normal section curve correspond to the extremal values of the quadratic form $II_{N_p}(v, v)$, with v unit tangent vector. From linear algebra, we know that these extrema occur in the directions of the eigenvectors of the matrix associated with II_{N_p}. This shows that the numbers κ_1 and κ_2 are the principal curvatures and the vectors v_1 and v_2 are the principal directions. The Gaussian curvature at p is, by definition, the geometric mean squared $\kappa_1 \kappa_2$ of the principal curvatures.

5.3. THE SECOND FUNDAMENTAL FORM

Now we compute the second fundamental form in terms of a local parametrization $\phi : U \subseteq \mathbb{R}^2 \to \mathbb{R}^3$ near some point $p \in M$. The vectors $\partial\phi/\partial x_1$ and $\partial\phi/\partial x_2$ form a basis (not necessarily orthonormal) of the tangent plane $T_p M$. A unit normal vector at p is given by

$$N_p = \frac{\partial\phi/\partial x_1 \times \partial\phi/\partial x_2}{\|\partial\phi/\partial x_1 \times \partial\phi/\partial x_2\|}.$$

We have

$$B(\partial\phi/\partial x_i, \partial\phi/\partial x_j) = \nabla^{\mathbb{R}^3}_{\partial\phi/\partial x_i} \partial\phi/\partial x_j - \nabla^t_{\partial\phi/\partial x_i} \partial\phi/\partial x_j,$$

where ∇^t is the induced (tangential) connection. By (4.2.8), it follows that

$$B(\partial\phi/\partial x_i, \partial\phi/\partial x_j) = \langle \nabla^{\mathbb{R}^3}_{\partial\phi/\partial x_i} \partial\phi/\partial x_j, N \rangle N,$$

thus

$$II_N(\partial\phi/\partial x_i, \partial\phi/\partial x_j) = \langle \nabla^{\mathbb{R}^3}_{\partial\phi/\partial x_i} \partial\phi/\partial x_j, N \rangle.$$

□

Example 5.3.6
Using (5.3.6) from above, we can provide a natural interpretation of the second fundamental form. Recall that the covariant derivative of a vector field on a surface is the tangential component of the ordinary derivative of the vector field. If c is an arc length parametrized curve on an embedded surface M, we have

$$c'' = \langle c'', N \rangle N + \frac{Dc'}{dt} = II_N(c', c')N + \frac{Dc'}{dt}.$$

The quantity $\kappa_g = \|Dc'/dt\|$ is called the geodesic curvature of c. An arc length parametrized curve c is a geodesic if and only if $\kappa_g = 0$. For a geodesic c, the above equation becomes

$$c'' = II_N(c', c')N.$$

This says that $II_N(c', c')$ represents the (scalar) acceleration of the geodesic. □

THEOREM 5.3.7 (The Gauss equation)
Let R' be the curvature tensor on M', and R be the induced curvature tensor on M. For any smooth vector fields X, Y, Z, T on M, the

following identity holds:

$$\langle R'(X,Y)Z,T\rangle = \langle R(X,Y)Z,T\rangle - \langle B(X,T),B(Y,Z)\rangle$$
$$+ \langle B(X,Z),B(Y,T)\rangle. \quad (5.3.7)$$

PROOF We compute $\langle R'(X,Y)Z,T\rangle$ by using (5.3.1) twice:

$$\nabla'_X \nabla'_Y Z = \nabla_X \nabla_Y Z + B(X,\nabla_Y Z) + \nabla'_X B(Y,Z).$$

Since $\langle B(\cdot,\cdot),T\rangle = 0$, we obtain

$$\langle \nabla'_X \nabla'_Y Z, T\rangle = \langle \nabla_X \nabla_Y Z, T\rangle + \langle \nabla'_X B(Y,Z), T\rangle.$$

Similarly, we have

$$\langle \nabla'_Y \nabla'_X Z, T\rangle = \langle \nabla_Y \nabla_X Z, T\rangle + \langle \nabla'_Y B(X,Z), T\rangle,$$
$$\langle \nabla'_{[X,Y]} Z, T\rangle = \langle \nabla_{[X,Y]} Z, T\rangle.$$

Combining these equations, we obtain

$$\langle R'(X,Y)Z,T\rangle = \langle R(X,Y)Z,T\rangle + \langle \nabla'_X B(Y,Z), T\rangle$$
$$- \langle \nabla'_Y B(X,Z), T\rangle.$$

Finally, we use the Weingarten equation (with $B(Y,Z)$ and $B(X,Z)$ as normal vectors) and obtain

$$\langle R'(X,Y)Z,T\rangle = \langle R(X,Y)Z,T\rangle - \langle B(X,T),B(Y,Z)\rangle$$
$$+ \langle B(X,Z),B(Y,T)\rangle.$$

∎

COROLLARY 5.3.8
If $\dim(M) = \dim(M')-1$, *and* X, Y, Z, *and* T *are smooth vector fields on* M, *then*

$$\langle R'(X,Y)Z,T\rangle = \langle R(X,Y)Z,T\rangle - \langle II_N(X,T), II_N(Y,Z)\rangle$$
$$+ \langle II_N(X,Z), II_N(Y,T)\rangle.$$

PROOF Since $\dim(M) = \dim(M') - 1$, the tangent space $T_p M'$ is spanned by $T_p M$ and N_p. Thus

$$B(X_p, Y_p) = II_p(X_p, Y_p) N_p.$$

5.3. THE SECOND FUNDAMENTAL FORM

Since N is a unit normal vector, we obviously have $\langle B(X,T), B(Y,Z)\rangle = II_N(X,T)II_N(Y,Z)$ and $\langle B(X,Z), B(Y,T)\rangle = II_N(X,Z)II_N(Y,T)$. ∎

By letting $Z = Y$ and $T = X$ in the Gauss equation, we obtain the following formula, which is a key ingredient in the extrinsic computation of the Gaussian curvature of a surface.

COROLLARY 5.3.9
Assume $\dim(M) = \dim(M') - 1$. *For any smooth vector fields X, Y on M, the following identity holds:*

$$\langle R'(X,Y)Y, X\rangle = \langle R(X,Y)Y, X\rangle + II_N(X,Y)^2 - II_N(X,X)II_N(Y,Y).$$

Example 5.3.10
We continue the investigation of the curvature of surfaces started in Example 5.3.5. We will prove the following formula for the Gaussian curvature

$$K = \frac{\langle R(X,Y)Y, X\rangle}{\|X\|^2\|Y\|^2 - \langle X,Y\rangle^2}. \tag{5.3.8}$$

This represents a form of the Theorema Egregium of Gauss. The numerator depends only on the curvature tensor, which is uniquely defined by the choice of a Riemannian metric. The denominator represents the area of the parallelogram determined by X_p and Y_p in T_pM, so it also depends only on the Riemannian metric. It follows that the Gaussian curvature is invariant under local isometries.

To check (5.3.8), we need to compute the matrix of the second fundamental form with respect to some orthonormal basis of the tangent plane, and then take its determinant. Suppose that $\{X,Y\} = \{e_1, e_2\}$ is already an orthonormal basis of T_pM. In this case, $K = \det(II_N(e_i, e_j)) = II_N(e_1, e_1)II_N(e_2, e_2) - II_N(e_1, e_2)^2$. By using Corollary 5.3.9 and the fact that $R' = 0$, we obtain $\langle R(e_1, e_2)e_2, e_1\rangle = II_N(e_1, e_1)II_N(e_2, e_2) - II_N(e_1, e_2)^2$ and $\|e_1\|^2\|e_2\|^2 - \langle e_1, e_2\rangle^2 = 1$, so (5.3.8) holds.

If $\{X,Y\}$ is not an orthonormal basis, we apply the Gram-Schmidt procedure to obtain an equivalent basis which is orthonormal. Set

$$e_1 = \frac{X}{\|X\|}, \text{ and } e_2 = \frac{Y - \langle Y, e_1\rangle e_1}{\|Y - \langle Y, e_1\rangle e_1\|} = \frac{Y - \langle Y, \frac{X}{\|X\|}\rangle \frac{X}{\|X\|}}{\left\|Y - \langle Y, \frac{X}{\|X\|}\rangle \frac{X}{\|X\|}\right\|}.$$

By the previous step, and using the symmetries of the curvature tensor, we have

$$K = \langle R(e_1, e_2)e_2, e_1 \rangle = \frac{\left\langle R\left(X, Y - \frac{\langle Y, X \rangle}{\|X\|^2} X\right) \left(Y - \frac{\langle Y, X \rangle}{\|X\|^2} X\right), X \right\rangle}{\|X\|^2 \left\| Y - \frac{\langle Y, X \rangle}{\|X\|^2} X \right\|^2}$$

$$= \frac{\langle R(X, Y)Y, X \rangle}{\|X\|^2 \left(\|Y\|^2 - 2\frac{\langle Y, X \rangle^2}{\|X\|^2} + \frac{\langle Y, X \rangle^2}{\|X\|^4} \|X\|^2 \right)}$$

$$= \frac{\langle R(X, Y)Y, X \rangle}{\|X\|^2 \|Y\|^2 - \langle X, Y \rangle^2}.$$

This ends the proof of (5.3.8). In classical notation, the coefficients $II_N(e_1, e_1)$, $II_N(e_1, e_2)$, $II_N(e_2, e_2)$ of the second fundamental form are denoted by e, f, g, respectively, while the coefficients of the first fundamental form (3.1.2) are denoted by E, F, G. Then (5.3.8) becomes

$$K = \frac{eg - f^2}{EG - F^2}.$$

This above computation relies heavily on the fact that a surface is a 2-dimensional object. For a manifold of dimension greater than two, one can similarly compute the Gaussian curvatures of the 2-dimensional manifolds cut off the original manifold along pairs of independent directions. This will be discussed in the next section.

The formula (5.3.8) can be taken as the definition of the Gaussian curvature for any abstract Riemannian manifold of dimension two. □

Example 5.3.11

We compute the Gaussian curvatures of the standard surfaces.

(i) We showed that the curvature tensor of $\mathbb{R}^n$ is $R = 0$. Thus, the Gaussian curvature K of the Euclidean plane is zero.

(ii) For the sphere S_r^2 of radius r, the Gauss map $G : S_r^2 \to S^2$ is given by $G(x) = x/r$, where $x \in S_r^2$, and the orientation is given by outward pointing unit normal vectors. Its derivative is a multiple of the identity mapping $(dG)_x = (1/r)\text{id}$. The eigenvalues of the matrix associated with $-(dG)_x$ are $\kappa_1 = \kappa_2 = -1/r$, and so $K = 1/r^2$. In particular, for the unit sphere S^2 we have $K = 1$.

(iii) We compute the Gaussian curvature of the Poincaré plane using (5.3.8). We already computed the Christoffel symbols in Example 4.4.3:

5.3. THE SECOND FUNDAMENTAL FORM

$\Gamma^1_{12} = \Gamma^1_{21} = \Gamma^2_{22} = -\Gamma^2_{11} = -1/y$, the rest being equal to zero. For $X = \partial/\partial x$ and $Y = \partial/\partial y$, we have

$$\langle R(X,Y)Y, X \rangle = \left\langle R^1_{122} \frac{\partial}{\partial x} + R^2_{122} \frac{\partial}{\partial y}, \frac{\partial}{\partial x} \right\rangle = \frac{1}{y^2} R^1_{122}.$$

Using (5.2.5), we calculate

$$\begin{aligned} R^2_{122} &= (\Gamma^1_{22}\Gamma^1_{11} - \Gamma^1_{12}\Gamma^1_{21}) + (\Gamma^2_{22}\Gamma^1_{12} - \Gamma^2_{12}\Gamma^1_{22}) + \frac{\partial \Gamma^1_{22}}{\partial x} - \frac{\partial \Gamma^1_{12}}{\partial y} \\ &= \left(0 - \left(-\frac{1}{y}\right)\left(-\frac{1}{y}\right)\right) + \left(\left(-\frac{1}{y}\right)\left(-\frac{1}{y}\right) - 0\right) + 0 - \frac{1}{y^2} \\ &= -\frac{1}{y^2}. \end{aligned}$$

Since $\|X\|^2\|Y\|^2 - \langle X, Y \rangle^2 = 1/y^4$, we conclude $K = -1$. □

REMARK 5.3.12 The hyperbolic plane cannot be smoothly isometrically embedded (or even immersed) in $\mathbb{R}^3$. Hilbert proved that no complete, C^4-differentiable surface of curvature $K = -1$ can be immersed in $\mathbb{R}^3$. Such a surface has positive curvature at a point where the distance from the origin is maximized, since the surface is tangent to the sphere centered at the origin that passes through the point, and lies inside this sphere. ∎

The curvature of each of the standard surfaces is the same at all points. It turns out that every surface can be endowed with a metric of constant curvature.

THEOREM 5.3.13 (Uniformization theorem)
Every connected surface is diffeomorphic to the quotient of one of the three standard surfaces by the free and properly discontinuous action of a subgroup of isometries. Consequently, every connected surface can be endowed with a Riemannian metric that is complete and has constant curvature.

The way in which a metric on a standard surface induces a metric on a quotient was described in Section 3.3. The canonical projection becomes a local isometry from the standard surface to the manifold, thus the Gaussian curvature of the manifold is equal to the Gaussian curvature of the corresponding standard surface.

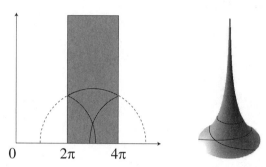

FIGURE 5.3.1
The pseudosphere.

The standard proof of this theorem relies on complex analysis and can be found in Rudin (1966).

Example 5.3.14

The orientation preserving isometries of the Poincaré half plane are described by the group $PSL_2(\mathbb{R})$ (see Section 3.3). Consider the subgroup generated by the translation $z \to z + 2\pi$. The quotient space of $\mathbb{H}^2$ by the action of this subgroup is called the pseudosphere, and is diffeomorphic to an infinite cylinder. The level sets of the pseudosphere are circles of radius approaching infinity as $y \to 0$, and of radius approaching zero 0 as $y \to \infty$. The equation of the pseudosphere is

$$z = -\sqrt{1 - x^2 - y^2} - \ln(\sqrt{x^2 + y^2}) + \ln(1 + \sqrt{1 - x^2 - y^2}),$$

and its curvature is $K = -1$. There are geodesics of the pseudosphere that self-intersect transversally for a finite number of times. See Figure 5.3.1. According to Remark 5.3.12, there is no immersion or embedding of the entire hyperbolic plane in the pseudosphere. However, it is possible to embed portions of the hyperbolic plane isometrically into the pseudosphere. □

5.4 Sectional and Ricci curvatures

We now provide a third geometric interpretation of the curvature tensor. Let p be a point of a Riemannian manifold M, X_p, Y_p be two linearly independent tangent vectors at p, and $\Pi = \Pi_{(X_p,Y_p)}$ be the plane spanned by these two vectors in T_pM. The area of the parallelogram determined by X_p and Y_p is given by $A(X_p, Y_p) = \|X_p\|^2 \|Y_p\|^2 - \langle X_p, Y_p \rangle^2$.

DEFINITION 5.4.1
The sectional curvature of M at p, determined by Π, is

$$K(\Pi) = \frac{\langle R(X_p, Y_p)Y_p, X_p \rangle}{A(X_p, Y_p)}. \tag{5.4.1}$$

We need to check that this definition makes sense, i.e., it does not depend on the basis $\{X_p, Y_p\}$ of the plane Π. Indeed, if $\{X'_p, Y'_p\}$ is another basis of Π, and $X'_p = a_{11}X_p + a_{21}Y_p$ and $Y'_p = a_{12}X_p + a_{22}Y_p$, then Proposition 5.2.4 implies that

$$\langle R(X'_p, Y'_p)Y'_p, X'_p \rangle = \det(a_{ij})^2 \langle R(X_p, Y_p)Y_p, X_p \rangle.$$

Since the area of the parallelogram changes by the same factor, i.e., $A(X'_p, Y'_p) = \det(a_{ij})^2 A(X_p, Y_p)$, it results that

$$K(\Pi_{(X'_p, Y'_p)}) = K(\Pi_{(X_p, Y_p)}).$$

On a surface, there is only one sectional curvature at every point, which is the Gaussian curvature. On a Riemannian manifold of dimension greater than 2, the sectional curvature at p along Π is the Gaussian curvature of the surface formed by all 'small' geodesics departing from p with initial velocities in Π. See Figure 5.4.1.

PROPOSITION 5.4.2
Let p be a point in M, Π be a plane in T_pM, and U be a neighborhood of 0 in T_pM on which $\exp_p$ is well defined. The sectional curvature $K(\Pi)$ equals the Gaussian curvature K_p of the surface $S_\Pi = \exp_p(\Pi \cap U)$.

PROOF First, notice that the canonical inclusion $i : S_\Pi \to M$ is an immersion, provided that the neighborhood U is chosen sufficiently

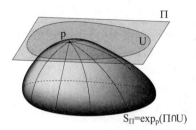

FIGURE 5.4.1
A plane of section and the surface it generates.

small. Denote by R the curvature tensor on M and by R_S the curvature tensor induced on S_Π. Since the Gaussian curvature of S_Π is given by

$$K(\Pi) = \frac{\langle R_S(X_p, Y_p)Y_p, X_p \rangle}{A(X_p, Y_p)},$$

where X_p, Y_p form a basis $T_p(S_\Pi)$, it is sufficient to prove that $R_S = R$ along S_Π.

Step 1. We prove that the second fundamental form is 0 at p along every normal vector to S_Π.

We pick a normal vector N_p to S_Π at p, and we locally extend it to a normal vector field to S_Π. We pick a vector X_p in U, and consider the geodesic $\gamma(t) = \exp_p(tX_p)$ from p in the direction of X_p. We locally extend X_p to a smooth tangent vector field X to M. By the compatibility of the Riemannian connection ∇ on M with the Riemannian metric, we have

$$X \langle N, X \rangle = \langle \nabla_X N, X \rangle + \langle N, \nabla_X X \rangle.$$

At p, we have $\langle N_p, X_p \rangle = 0$, and $(\nabla_X X)_p = 0$, since γ is a geodesic. Using the Weingarten equation (5.3.5), we obtain

$$II_{N_p}(X_p, X_p) = -\langle \nabla_{X_p} N, X_p \rangle = 0.$$

Using the bilinearity and symmetry of the second fundamental form, easy algebra shows that $II_{N_p}(X_p, Y_p) = 0$ for all tangent vectors $X_p, Y_p \in T_p(S_\Pi)$.

Step 2. We prove $R_S = R$ on S_Π.

Since N_p is an arbitrary vector in $(T_p S_\Pi)^\perp$, from Step 1 we infer that $B_p(X, Y) = 0$. We use the Gauss equation (5.3.7) to conclude $\langle R(X,Y)Z, T \rangle = \langle R_S(X,Y)Z, T \rangle$ for all vector fields X, Y, Z, T on S_Π.

5.4. SECTIONAL AND RICCI CURVATURES

The curvature tensor determines all sectional curvatures. The converse is also true: the sectional curvatures uniquely determine the curvature tensor. This follows from the following:

PROPOSITION 5.4.3
Assume that R_1, R_2 are two tensors on M satisfying the properties (i) - (iv) from Proposition 5.2.4. If $\langle R_1(X,Y)Y, X \rangle = \langle R_2(X,Y)Y, X \rangle$ for all smooth vector fields X, Y on M, then $R_1 = R_2$.

PROOF Let $R = R_1 - R_2$. Then R satisfies the properties (i)-(iv) from Proposition 5.2.4. We have $\langle R(X,Y)Y, X \rangle = 0$ for all X and Y, and we need to prove $R = 0$.

We claim that $\langle R(X,Y)Z, X \rangle = 0$. Indeed, we have

$$0 = \langle R(X, Y+Z)(Y+Z), X \rangle = \langle R(X,Y)Y, X \rangle + \langle R(X,Y)Z, X \rangle$$
$$+ \langle R(X,Z)Y, X \rangle + \langle R(X,Z)Z, X \rangle = \langle R(X,Y)Z, X \rangle + \langle R(X,Z)Y, X \rangle$$
$$= 2\langle R(X,Y)Z, X \rangle.$$

We use the previous identity for X replaced by $X + T$:

$$0 = \langle R(X+T, Y)Z, X+T \rangle = \langle R(X,Y)Z, X \rangle + \langle R(X,Y)Z, T \rangle$$
$$+ \langle R(T,Y)Z, X \rangle + \langle R(T,Y)Z, T \rangle = \langle R(X,Y)Z, T \rangle + \langle R(T,Y)Z, X \rangle$$
$$= \langle R(X,Y)Z, T \rangle + \langle R(X,Z)Y, T \rangle, \tag{5.4.2}$$

hence $\langle R(X,Y)Z, T \rangle$ is antisymmetric in any pair of adjacent arguments. By Bianchi's identity we have

$$0 = \langle R(X,Y)Z, T \rangle + \langle R(Y,Z)X, T \rangle + \langle R(Z,X)Y, T \rangle \tag{5.4.3}$$
$$= \langle R(X,Y)Z, T \rangle - \langle R(Y,X)Z, T \rangle - \langle R(X,Z)Y, T \rangle, \tag{5.4.4}$$

which yields
$$3\langle R(X,Y)Z, T \rangle = 0.$$

∎

We immediately obtain:

COROLLARY 5.4.4
Let R_1 and R_2 be two curvature tensors on a Riemannian manifold M, i.e., two tensors satisfying the properties (i)-(iv) from Proposition 5.2.4.

Assume that for every $p \in M$ and every plane $\Pi = \Pi_{(X_p, Y_p)} \subseteq T_p M$ we have

$$K_1(\Pi) = \frac{\langle R_1(X_p, Y_p)Y_p, X_p \rangle}{A(X_p, Y_p)} = K_2(\Pi) = \frac{\langle R_2(X_p, Y_p)Y_p, X_p \rangle}{A(X_p, Y_p)}.$$

Then $R = R'$.

REMARK 5.4.5 We know that for a fixed Riemannian metric there is a unique Riemannian connection, so, implicitly, a unique curvature tensor. The previous statement says that the curvature tensor is uniquely determined by the sectional curvatures, hence by the knowledge of $n(n-1)/2$ coefficients R_{ijij}. This means that the coefficients R_{ijij}, determine, in some sense, the metric. More precisely, if the curvature tensors associated to two Riemannian metrics have equal coefficients R_{ijij}, then the second derivatives of the two metrics coincide. See Spivak (1999) for details. ∎

A manifold is said to have constant sectional curvature if the sectional curvatures are the same for all planes of section at all points. A complete manifold of constant curvature is called a space form. The standard surfaces are examples of space forms. For manifolds of constant sectional curvature, the curvature tensor is essentially given by the sectional curvature:

LEMMA 5.4.6
If M has constant sectional curvature equal to K, then

$$R(X, Y)Z = K\left(\langle Y, Z \rangle X - \langle X, Z \rangle Y\right).$$

PROOF By Definition 5.4.1, we have

$$\langle R(X, Y)Y, X \rangle = K\left(\|X\|^2 \|Y\|^2 - \langle X, Y \rangle^2\right).$$

Define a tensor $\bar{R}(X, Y)Z$ by

$$\bar{R}(X, Y)Z = K\left(\langle Y, Z \rangle X - \langle X, Z \rangle Y\right).$$

It is easy to verify that $\bar{R}$ satisfies properties (i)-(iv) from Proposition 5.2.4. Clearly $\langle \bar{R}(X, Y)Y, X \rangle = \langle R(X, Y)Y, X \rangle$ for all smooth vector fields X, Y on M. Proposition 5.4.3 implies that $R = \bar{R}$. ∎

5.4. SECTIONAL AND RICCI CURVATURES

Another tool for measuring the curvature is the Ricci curvature tensor. We will give only a brief presentation, and recommend Lee (1997) for details. Consider the mapping $X \to R(X,Y)Z$ where Y and Z are fixed smooth vector fields. This induced a linear mapping $X_p \to R(X_p, Y_p)Z_p$ at each $p \in M$. One can prove that the trace of this mapping is independent of the choice of a basis of the tangent space T_pM.

DEFINITION 5.4.7
The Ricci curvature tensor is the smooth mapping Ric that assigns to every pair of smooth vector fields Y, Z on M the smooth function $Ric(Y, Z)$ on M defined by

$$Ric(Y,Z) = \text{trace}(X \to R(X,Y)Z).$$

In local coordinates, if

$$Y = \sum_{i=1}^{m} v^i \frac{\partial}{\partial x_i}, \text{ and } Z = \sum_{i=1}^{m} w^i \frac{\partial}{\partial x_i},$$

one can write

$$Ric(Y,Z) = \sum_{i,j=1}^{m} v_i w_j R_{ij},$$

where $R_{ij} = Ric(\partial/\partial x_i, \partial/\partial x_j)$. The coefficients R_{ij} can be computed as

$$R_{ij} = \sum_{k=1}^{m} R_{kij}^{k} = \sum_{k,l=1}^{m} g^{kl} R_{kijl},$$

where (g^{kl}) is the inverse of the matrix (g_{ij}) associated to the Riemannian metric. The last identity follows directly from (5.2.7).

We obtain a geometrical interpretation of the Ricci curvature in relationship to the sectional curvature. Let X be a smooth vector field and $\{X_1, \ldots, X_m\}$ an orthonormal frame defined on some coordinate neighborhood, such that $X_1 = X$. Then

$$Ric(X,X) = \sum_{j=2}^{m} K(\Pi_{(X,X_j)}).$$

That is, for every unit tangent vector X, $Ric(X,X)$ is the sum of all sectional curvatures determined by the planes in T_pM spanned by X and the other vectors of an orthonormal basis.

REMARK 5.4.8   The Ricci curvature can also be interpreted as a measure of the growth rate of the volume of metric balls in a manifold. If the Ricci curvature of a complete m-dimensional Riemannian manifold is non-negative at all points, then the volume of a metric ball $B_d(p, R) = \{q \mid d(q,p) < R\}$, measured with respect to the Riemannian metric, is smaller than or equal to the volume of a ball $B^m(0, R)$ of the same radius in $\mathbb{R}^m$, measured with respect to the Euclidean metric. Moreover, the ratio of these volumes is a non-increasing function in R. These results are due to Bishop and Crittenden (1964) and Gromov (1981). ∎

REMARK 5.4.9   The Ricci curvature gives rise to a remarkable flow, called the Ricci flow, which spreads out the Ricci curvature more evenly across a Riemannian manifold. This flow, introduced by Hamilton (1982), is a flow on the space of Riemannian metrics on a manifold. In local coordinates, a metric is given by a matrix $(g_{ij})_{i,j=1,\ldots,m}$, which is symmetric and positive definite. To each Riemannian metric corresponds a unique curvature tensor, and hence a Ricci tensor. The differential equation

$$\frac{dg_{ij}}{dt} = -2R_{ij},$$

defines a flow on the space of the Riemannian metrics. Each solution curve $(g_{ij}(t))$ represents a Riemannian metric that varies smoothly in time. Due to the negative sign in front of the Ricci tensor, the metric 'shrinks' in the directions corresponding to negative Ricci tensor coefficients, and it 'expands' in the directions corresponding to positive coefficients. One can rescale $(g_{ij}(t))$ so that the shrinking and expanding in various directions balance each other out, and so $(g_{ij}(t))$ preserves the Riemannian volume of the manifold in time. The existence of time-dependent, volume-preserving metrics is important in physics, especially in general relativity (see Schutz (1985)). An example of a metric with the above property is the Einstein metric, for which the Ricci tensor is a scalar multiple of the metric at each point. In general, finding metrics for which some type of curvature is constant at all points is one of the fundamental problems in differential geometry. The existence of such metrics on a manifold has important topological implications.

In general, the Ricci flow, even after rescaling, is not defined for all time. Singularities of the Ricci flow are created when the Ricci curvature becomes unbounded. The existence of singular metrics has important topological applications. It is possible that the recent work of Perelman (2002, 2003a,b) provides a complete classification of all compact

FIGURE 5.5.1
Variations through geodesics on a surface of positive curvature and on a surface of negative curvature.

3-dimensional manifolds in terms of the singularities of the Ricci flow. This would imply the geometrization conjecture, discussed in Remark 3.3.1. See Anderson (2004) for a report on this work. ∎

5.5 Jacobi fields

We now provide a fourth geometric interpretation of the curvature tensor. When a geodesic is slid across a manifold so that all intermediate positions are still geodesics, the velocity at which the nearby geodesics are driven apart defines a vector field along the original geodesic. This vector field contains all of the information on the curvature tensor along the geodesic.

One can visualize this by stretching an elastic band across a round ball. By slightly sliding the band to the left or right, one generates a smooth family of geodesics of the sphere. One notices that these geodesics eventually come together. When one repeats the experiment using a surface of a saddle instead of a ball, one notices that the elastic band stretches out geodesics that eventually spread out. See Figure 5.5.1. These experiments suggest that positive curvature causes nearby geodesics to converge, and negative curvature causes nearby geodesics to diverge.

DEFINITION 5.5.1

Let $\gamma : [a,b] \to M$ be a geodesic on M and $\epsilon > 0$. A variation of γ through geodesics is a smooth mapping

$$(s,t) \in (-\epsilon, \epsilon) \times [a,b] \to \Gamma(s,t) \in M,$$

such that

(i) $\Gamma(0, \cdot) = \gamma$,

(ii) $\Gamma(s, \cdot)$ is a geodesic for each $s \in (-\epsilon, \epsilon)$.

Example 5.5.2

Consider a geodesic γ on M given by an exponential map $\gamma(t) = \exp_p(tv)$, for $t \in [0,1]$, where $v \in T_pM$. Let $w \in T_pM$, and ϵ be small enough such that $v + sw$ is in the domain of $\exp_p$ for all $s \in (-\epsilon, \epsilon)$. The mapping $t \to \exp_p(t(v+sw))$ represents a geodesic segment from p to $\exp_p(v+sw)$, for each $s \in (-\epsilon, \epsilon)$, and coincides with γ for $s=0$. The mapping

$$(s,t) \to \exp_p(t(v+sw))$$

is smooth, so it represents a variation of γ through geodesics. □

In order to measure the velocity of each point of the original geodesic as it varies through geodesics, we define

$$J(t) = \frac{\partial \Gamma(s,t)}{\partial s}\Big|_{s=0}. \tag{5.5.1}$$

Each $J(t)$ is a tangent vector to the curve $s \to \Gamma(s,t)$, hence $t \to J(t)$ defines a vector field along γ. See Figure 5.5.2. Since the family of curves Γ is smooth, the vector field J is also smooth. The next theorem shows this vector field is related to the curvature tensor.

THEOREM 5.5.3

Let γ be a geodesic and Γ a variation of γ through geodesics. The vector field J along γ obtained as in (5.5.1), satisfies the equation

$$\frac{D^2 J}{dt^2}(t) + R(J(t), \gamma'(t))\gamma'(t) = 0. \tag{5.5.2}$$

5.5. JACOBI FIELDS

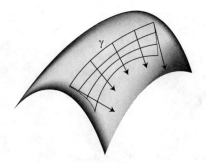

FIGURE 5.5.2
A Jacobi field.

PROOF We extend $(s,t) \to \Gamma(s,t)$ in some open neighborhood of $(-\epsilon, \epsilon) \times [a, b]$ in $\mathbb{R}^2$, obtaining an embedded surface in M. We have

$$\frac{\partial \Gamma}{\partial s}(0, t) = J(t), \text{ and } \frac{D}{dt}\frac{\partial \Gamma}{\partial t}(s, t) = 0,$$

since each $t \to \Gamma(s, t)$ is a geodesic. Using the symmetry of the Riemannian connection as in Proposition 4.3.5, and then using Proposition 5.2.6, we obtain

$$\frac{D^2 J}{dt^2}(t) = \frac{D}{dt}\frac{D}{dt}\frac{\partial \Gamma}{\partial s}(0, t) = \frac{D}{dt}\frac{D}{ds}\frac{\partial \Gamma}{\partial t}(0, t)$$
$$= \left(\frac{D}{ds}\frac{D}{dt}\frac{\partial \Gamma}{\partial t} + R\left(\frac{\partial \Gamma}{\partial t}, \frac{\partial \Gamma}{\partial s}\right)\frac{\partial \Gamma}{\partial t}\right)(0, t)$$
$$= R(\gamma'(t), J(t))\gamma'(t) = -R(J(t), \gamma'(t))\gamma'(t).$$

∎

This leads to the following:

DEFINITION 5.5.4
A vector field J along a geodesic $\gamma : [a, b] \to M$ is called a Jacobi field if it satisfies the second order differential equation

$$\frac{D^2 J}{dt^2}(t) + R(J(t), \gamma'(t))\gamma'(t) = 0. \tag{5.5.3}$$

Example 5.5.5
Along each geodesic γ, there always exist two trivial Jacobi fields. One is $J(t) = \gamma'(t)$. Since $D\gamma'/dt = 0$ and $R(\gamma', \gamma')\gamma' = 0$, the velocity vector

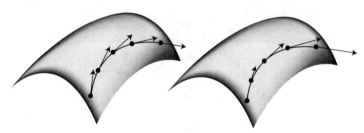

FIGURE 5.5.3
Trivial Jacobi fields along a geodesic.

field $J(t) = \gamma'(t)$ satisfies the Jacobi equation (5.5.3), so it is a Jacobi field. The other one is $J(t) = t\gamma'(t)$. Since $D^2 J/dt^2 = D\gamma'/dt = 0$ and $R(t\gamma'(t), \gamma'(t))\gamma'(t) = tR(\gamma'(t), \gamma'(t))\gamma'(t) = 0$, the vector field $J(t) = t\gamma'(t)$ satisfies the Jacobi equation (5.5.3), so is also a Jacobi field. See Figure 5.5.3.

It is clear that these two Jacobi fields are linearly independent. Moreover, any other Jacobi field whose vectors $J(t)$ are scalar multiples of the velocity vectors $\gamma'(t)$ is a linear combination of the trivial Jacobi fields. Indeed, if $J(t) = f(t)\gamma'(t)$ is a Jacobi field, then we must have that $f''(t) = 0$, so $f(t) = at + b$, for some $a, b \in \mathbb{R}$.

The trivial Jacobi fields do not provide any information on the behavior of geodesics in the proximity of γ. One is usually interested only in Jacobi fields normal to γ. □

The components of a Jacobi field satisfy a system of second order differential equations. We will express the differential equations with respect to an orthonormal frame along γ. Let p be a point on γ, and $E_1, \ldots, E_m$ be an orthonormal basis of $T_p M$. We extend $E_1, \ldots, E_m$ to parallel, orthonormal vector fields $E_1(t), \ldots, E_m(t)$ along the geodesic γ. We express

$$J(t) = \sum_{i=1}^{m} h_i(t) E_i(t), \quad \text{and}$$

$$\gamma'(t) = \sum_{i=1}^{m} \gamma'_i(t) E_i(t).$$

We obtain

$$R(J, \gamma')\gamma' = \sum_{j,k,l=1}^{m} h_j \gamma'_k \gamma'_l R(E_j, E_k) E_l$$

5.5. JACOBI FIELDS

$$= \sum_{i,j,k,l=1}^{m} h_j \gamma'_k \gamma'_l R^i_{jkl} E_i.$$

The components h_i of J are the solution of the linear system of second order differential equations

$$h''_i + \sum_{j,k,l=1}^{m} h_j \gamma'_k \gamma'_l R^i_{jkl} = 0, \qquad (5.5.4)$$

with $i = 1, \ldots, m$. The solution of this system is uniquely determined by the initial values of J and DJ/dt at a single point.

PROPOSITION 5.5.6
Let $\gamma : [a,b] \to M$ be a geodesic and $p = \gamma(t_0)$ be a point on γ. For any pair of tangent vectors v and w at p, there exists a unique Jacobi field J along γ such that $J(t_0) = v$ and $DJ/dt(t_0) = w$.

PROOF If $E_1(t), \ldots, E_m(t)$ are parallel, orthonormal vector fields along γ near $t = t_0$, then we obtain the linear system of m second order differential equations (5.5.4). By letting $g'_i = h_i$, we convert it into a linear system of $2m$ first order differential equations:

$$h'_i = g_i, \qquad (5.5.5)$$

$$g'_i = -\sum_{j,k,l=1}^{m} h_j \gamma'_k \gamma'_l R^i_{jkl},$$

$i = 1, \ldots, m$. By Theorem 2.3.1, for any given initial conditions $J(t_0)$ and $DJ/dt(t_0)$, there exists a unique solution of the system (5.5.5), defined in some neighborhood of t_0. Such a solution extends uniquely to the whole of $[a, b]$. ∎

We obtain the following easy consequence.

COROLLARY 5.5.7
The Jacobi fields along a geodesic γ on an m-dimensional manifold M form a $2m$-dimensional vector space.

Example 5.5.8
For a geodesic $\gamma(t) = \exp_p(tv)$, where $t \in [0,1]$, $v \in T_pM$, we can explicitly compute the Jacobi field J with $J(0) = 0$ and $DJ/dt = w$.

Consider the variation $\Gamma(s,t) = \exp_p(t(v+sw))$ of γ through geodesics, where s takes values in some open interval about 0. Then

$$\frac{\partial \Gamma}{\partial s}(s,t) = (d\exp_p)_{t(v+sw)}(tw),$$

hence

$$J(t) = \frac{\partial \Gamma}{\partial s}(0,t) = (d\exp_p)_{(tv)}(tw).$$

We have

$$\frac{DJ}{\partial t} = \frac{D}{dt}(d\exp_p)_{(tv)}(tw) = \frac{D}{\partial t}\left(t(d\exp_p)_{(tv)}(w)\right)$$

$$= (d\exp_p)_{(tv)}(w) + t\frac{D}{\partial t}(d\exp_p)_{(tv)}(w),$$

so $DJ/dt(0) = (d\exp_p)_0(w) = w$. □

We defined a Jacobi field as the solution of a certain second order differential equation. We shown that a variation through geodesics gives rise to a Jacobi field. Now we prove the converse statement.

THEOREM 5.5.9

Every Jacobi field along a geodesic $\gamma : [0,1] \to M$ can be obtained by a variation of γ through geodesics.

PROOF Let $v = \gamma'(0)$ be the initial velocity of the geodesic, and $J(t)$ be a Jacobi field along the geodesic. The exponential map $\exp_p(tv)$ is defined for all $t \in [0,1]$. The mapping $(q,w) \in TM \to (q, \exp_q w) \in M \times M$ is well defined and smooth in some neighborhood $\mathcal{U}$ of (p,v) in TM. We construct a curve $s \in (-\epsilon, \epsilon) \to \alpha(s) \in M$, for some $\epsilon > 0$, such that $d\alpha/ds(0) = J(0)$, and a vector field $V(s)$ along $\alpha(s)$ such that $V(0) = v$, and $DV/ds(0) = DJ/dt(0)$. See Figure 5.5.4. Such a vector field always exists, due to the theorem on the existence and uniqueness of solutions of differential equations. By choosing $\epsilon > 0$ sufficiently small, we can ensure that $(\alpha(s), V(s)) \in \mathcal{U}$, hence $\exp_{\alpha(s)}(tV(s))$ is defined for all $s \in (-\epsilon, \epsilon)$, and all $t \in [0,1]$.

We claim that

$$\Gamma(s,t) = \exp_{\alpha(s)}(tV(s)), \quad (s,t) \in (-\epsilon, \epsilon) \times [0,1],$$

is a variation of γ through geodesics that generates precisely the Jacobi field J. It is easy to see that for each fixed $s \in [0,1]$, the curve $t \to \Gamma(s,t)$

5.5. JACOBI FIELDS

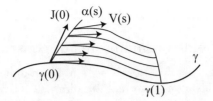

FIGURE 5.5.4
Construction of a variation through geodesics that generates a given Jacobi field.

is a geodesic, and for $s = 0$, $t \to \Gamma(0, t)$ is exactly γ. We have

$$\frac{\partial \Gamma}{\partial s}(s, 0)|_{s=0} = \frac{d\alpha}{ds}(s)|_{s=0} = J(0).$$

By the symmetry of the connection, we have

$$\frac{D}{dt}\frac{\partial \Gamma}{\partial s}(s, t) = \frac{D}{ds}\frac{\partial \Gamma}{\partial t}(s, t) = \frac{D}{ds}(d\exp_{\alpha(s)})_{(tV(s))}(V(s)).$$

By the properties of the exponential map

$$\frac{D}{dt}\frac{\partial \Gamma}{\partial s}(s, 0) = \frac{D}{ds}(d\exp_{\alpha(s)})_0(V(s)) = \frac{D}{ds}(V(s)),$$

thus

$$\frac{D}{dt}\frac{\partial \Gamma}{\partial s}(0, 0) = \frac{DJ}{dt}(0).$$

We know from Theorem 5.5.3 that $\partial\Gamma/\partial s(t, 0)$ represents a Jacobi field along γ. From the above, this Jacobi field satisfies the same initial conditions as J at $\gamma(0)$. Since a Jacobi field is uniquely determined by its initial conditions (Proposition 5.5.6), we conclude that $\partial\Gamma/\partial s(t, 0) = J(t)$ for all $t \in [0, 1]$. ∎

The previous theorem works as is only for finite geodesic segments. For geodesics $\gamma : \mathbb{R} \to M$ we need to assume that M is geodesically complete. If M is not complete, it may be impossible to produce any variation of γ through geodesics. For example, let M be $\mathbb{R}^2$ with closed disks of radius $1/2^n$ centered at $(\pm n, \pm 1/2^{n-1})$ removed for all $n \geq 1$, and γ be the x-axis. There exists no variation of γ through geodesics in M.

5.6 Jacobi fields on manifolds of constant curvature

For any given geodesic $\gamma : [0,1] \to M$ in an m-dimensional manifold M, the Jacobi fields along γ form a $2m$-dimensional vector space. Any Jacobi field whose vectors are scalar multiples of $\gamma'(t)$ is a linear combination of $t \to \gamma'(t)$ and $t \to t\gamma'(t)$ (see Example 5.5.5). It follows that the Jacobi fields normal to γ form a $(2m-2)$-dimensional vector subspace.

On a manifold with constant sectional curvature one can find explicitly all Jacobi fields. Consider a Riemannian manifold M of constant sectional curvature equal to K, and let γ be a unit speed geodesic on M. Let J be a Jacobi field along γ, normal to γ. By Lemma 5.4.6, we have that $R(J,\gamma')\gamma' = K(\langle \gamma',\gamma'\rangle J - \langle J,\gamma'\rangle \gamma') = KJ$. The Jacobi equation (5.5.3) now becomes

$$\frac{D^2 J}{dt^2} + KJ = 0. \qquad (5.6.1)$$

PROPOSITION 5.6.1
Every Jacobi field J along γ, normal to γ and vanishing at $t = 0$, is of the form

$$J(t) = h(t)E(t),$$

where $E(t)$ is some parallel, normal vector field along γ, and $h(t)$ is given by

$$h(t) = \begin{cases} t, & \text{if } K = 0, \\ \sin(t\sqrt{K}), & \text{if } K > 0, \\ \sinh(t\sqrt{-K}), & \text{if } K < 0. \end{cases} \qquad (5.6.2)$$

PROOF Suppose that $J(t)$ is of the form $h(t)E(t)$, where $h(t)$ is a smooth function, and $E(t)$ is a parallel, normal vector field along γ. Substituting $h(t)E(t)$ in (5.6.1) we obtain

$$h''(t) + Kh(t) = 0.$$

The general solution of this ordinary differential equation is given by

$$h(t) = \begin{cases} at + b, & \text{if } K = 0, \\ a\sin(t\sqrt{K}) + b\cos(t\sqrt{K}), & \text{if } K > 0, \\ a\sinh(t\sqrt{-K}) + b\cosh(t\sqrt{-K}), & \text{if } K < 0. \end{cases}$$

5.6. MANIFOLDS OF CONSTANT CURVATURE

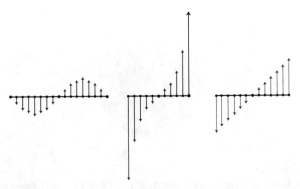

FIGURE 5.6.1
A Jacobi field on a manifold with constant positive curvature (left), constant negative curvature (center), and zero curvature (right).

A solution $h(t)$ that vanishes at $t = 0$ is a constant multiple of one of the functions in (5.6.2). On the one hand, all Jacobi fields normal to γ form a $(2m - 2)$-dimensional vector space, so all Jacobi fields normal to γ that vanish at $t = 0$ form a $(m - 1)$-vector space. On the other hand, all Jacobi fields of the form $J(t) = h(t)E(t)$, with $E(t)$ a parallel, normal vector field to γ, also form a $(m - 1)$-dimensional vector space. It must be the case that these are all of the normal Jacobi fields that vanish at $t = 0$. ∎

REMARK 5.6.2 This result agrees with the heuristics at the beginning of this section. A typical variation through geodesics makes nearby geodesics get together on a manifold with constant positive curvature, makes them spread out in an exponential fashion on a manifold with constant negative curvature, and makes them evolve in a linear fashion on a manifold with zero curvature. See Figure 5.6.1. ∎

Example 5.6.3
Consider a 2-sphere S_r^2 of radius r in $\mathbb{R}^3$, a geodesic on the sphere given by a meridian great circle $t \to (r\sin t, 0, r\cos t)$, for $t \in [0, 2\pi]$. By a parameter change, we obtain a unit speed geodesic with the same geometric image

$$\gamma(t) = \left(r\sin\left(\frac{t}{r}\right), 0, r\cos\left(\frac{t}{r}\right)\right).$$

FIGURE 5.6.2
A Jacobi field along a meridian circle on the sphere.

Let Γ be variation of γ through meridian great circles given by

$$\Gamma(s,t) = \left(r\sin\left(\frac{t}{r}\right)\cos s, r\sin\left(\frac{t}{r}\right)\sin s, r\cos\left(\frac{t}{r}\right)\right),$$

where $s \in (-\epsilon, \epsilon)$, for some $0 < \epsilon < \pi$. We have $\Gamma(0,t) = \gamma(t)$, and

$$J(t) = \frac{\partial \Gamma}{\partial s}(0,t) = \left(0, r\sin\left(\frac{t}{r}\right), 0\right) = r\sin\left(\frac{t}{r}\right)(0,1,0),$$

in agreement with Proposition 5.6.1. See Figure 5.6.2. □

5.7 Conjugate points

As an application of Jacobi fields, we investigate the largest possible domain on which an exponential mapping is a local diffeomorphism. From Proposition 4.5.2, we know that any $p \in M$ has a geodesic neighborhood V which is the diffeomorphic image of some open set in $T_p M$ through the exponential mapping $\exp_p$. This means that for every point $q \in V$ there is a unique geodesic joining p to q. If the neighborhood V of p is chosen too large, this property may fail. Consider the case of a two-sphere of radius r. If p is an arbitrary point on S_r^2, and V is the sphere with the antipodal point $A(p)$ removed, every point $q \in V$ is reached by moving away from p along a geodesic in some uniquely determined direction $v \in T_p M$. If we enlarge the neighborhood V to the entire sphere, this is no longer the case. The antipodal point $A(p)$

5.7. CONJUGATE POINTS

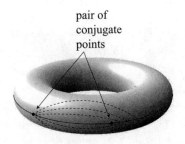

FIGURE 5.7.1
Pairs of conjugate points.

can be reached by moving from p along geodesics in any direction. This means that the exponential map $\exp_p$ ceases to be even a local diffeomorphism at the the points $v \in T_p S_r^2$ with $\|v\| = \pi r$. Example 5.6.3 shows that the antipodal point of p is precisely the place of occurrence of the first zero, other than p itself, of a Jacobi field J with $J = 0$ at p, along any geodesic emerging from p. This suggests that the largest domain to which an exponential mapping can be extended as a local diffeomorphism can be described in terms of Jacobi fields.

DEFINITION 5.7.1
Let γ be a geodesic in M. The points $\gamma(t_1)$ and $\gamma(t_2)$ on γ are said to be conjugate points provided that there exists a non-zero Jacobi field along γ which vanishes at $\gamma(t_1)$ and $\gamma(t_2)$. The largest number of linearly independent Jacobi fields that vanish at $\gamma(t_1)$ and $\gamma(t_2)$ is called the multiplicity of the conjugacy.

See Figure 5.7.1. Since tangential Jacobi fields vanish at most at one point, the dimension of the space of Jacobi fields that vanish at two points is at most $m - 1$. Thus, the maximum order of multiplicity of two conjugate points is $m - 1$.

Example 5.7.2
For a manifold with constant sectional curvature, the conjugate points along a given unit speed geodesic γ can be found by using Proposition 5.6.1. If $K \leq 0$, then there are no conjugate points along any geodesic. If

$K > 0$, the conjugate points of a Jacobi field as in (5.6.2) correspond to the values of $t = m\pi/\sqrt{K}$, with $m \in \mathbb{Z}$. That is, there are no conjugate points along a geodesic γ if the length of γ is less than $\pi/\sqrt{K}$, and there are conjugate points along γ otherwise. See Figure 5.6.1. □

PROPOSITION 5.7.3
Let $\gamma : [0,1] \to M$ be a geodesic given by $\gamma(t) = \exp_p(tv)$, with $v \in T_pM$. Let $q = \exp_p(v) = \gamma(1)$. Then $\exp_p$ is a local diffeomorphism in a neighborhood of v if and only if p and q are not conjugate points. If p and q are conjugate points, the multiplicity of the conjugacy equals the dimension of the kernel of the linear mapping $(d\exp_p)_v$.

PROOF By the inverse function theorem, $\exp_p$ is a local diffeomorphism in some neighborhood of v if and only if the linear mapping $(d\exp_p)_v$ is invertible.

The proof Theorem 5.5.9 shows that every non-zero Jacobi field along γ with $J(0) = 0$ can be obtained from a variation of geodesics $\Gamma(s,t) = \exp_p(tV(s))$ for some curve of vectors $V(s) \in T_pM$ satisfying $V(0) = v$. Then
$$J(t) = (d\exp_p)_{(tv)}(tw),$$
for $w = V'(0) = DJ/dt(0) \neq 0$. In particular, $J(1) = (d\exp_p)_v(w)$. Hence, there exists a non-zero Jacobi field J with $J(0) = J(1) = 0$ if and only if there exists $w \neq 0$ with $(d\exp_p)_v(w) = 0$, that is, if and only if the kernel of $(d\exp_p)_v$ is non-trivial. This proves the first claim of the proposition. The second claim follows from the fact that the number of linearly independent vectors w in the kernel of $(d\exp_p)_v$ equals the number of linearly independent Jacobi fields $J(t) = (d\exp_p)_{tv}(tw)$ with $J(1) = 0$. ∎

We know that a Jacobi field is uniquely defined by its initial value and its covariant derivative initial value. We now show that a Jacobi field is uniquely defined also by its values at the endpoints of a geodesic segment provided that the endpoints are not conjugate points.

PROPOSITION 5.7.4
Let $\gamma : [0,1] \to M$ be a geodesic. There exists a unique Jacobi field J along γ with $J(0) = V_0$ and $J(1) = V_1$ for any choice of V_0 and V_1, if and only if $\gamma(0)$ and $\gamma(1)$ are not conjugate points.

5.8. HORIZONTAL AND VERTICAL SUB-BUNDLES

PROOF Assume that $\gamma(0)$ and $\gamma(1)$ are not conjugate points. We define a linear mapping L from the vector space $\mathcal{J}_{V_0}$ of all Jacobi fields J along γ with $J(0) = V_0$ into the tangent space $T_{\gamma(1)}M$ by setting $L(J) = J(1)$. Both vector spaces $\mathcal{J}_{V_0}$ and $T_{\gamma(1)}M$ are m-dimensional.

We claim that the mapping L is one-to-one. Indeed, if $L(J_1) = L(J_2)$, then $J_1 - J_2$ is a Jacobi field along γ with $(J_1 - J_2)(0) = (J_1 - J_2)(1) = 0$. Since $\gamma(0)$ and $\gamma(1)$ are non-conjugate points, $J_1 - J_2$ must be the zero vector field, so $J_1 = J_2$.

Since L is an injective linear mapping between vector spaces of equal dimensions, it is also surjective. Hence, for any given $V_1 \in T_{\gamma(1)}M$, there exists a Jacobi field J in $\mathcal{J}_{V_0}$ with $J(1) = V_1$.

If $\gamma(0)$ and $\gamma(1)$ are not conjugate points, there exists a non-trivial Jacobi field J_0 that vanishes at both 0 and 1. Then J and $J + J_0$ are two distinct Jacobi fields with the same values at 0 and 1. ∎

5.8 Horizontal and vertical sub-bundles

This section is devoted to the tangent bundle of the tangent bundle of a manifold. If M is an m-dimensional manifold, TTM can decomposed as a sum of a 'horizontal' and a 'vertical' sub-bundle. These sub-bundles play a role similar to the coordinate axes in the cartesian plane. We will use the horizontal and vertical sub-bundles in the study of the geodesic flow in Section 5.9.

If M is m-dimensional, TM is locally diffeomorphic to $\mathbb{R}^m \times \mathbb{R}^m$. We will first define the horizontal and vertical sub-bundles of TTM in the special case when $M = \mathbb{R}^m$. Every vector ξ tangent to $TM = \mathbb{R}^m \times \mathbb{R}^m$ at a point $\theta = (x,v) \in TM$ is the tangent vector to some curve of tangent vectors $V(t)$ in TM with $V(0) = \theta$. As a curve in $\mathbb{R}^m \times \mathbb{R}^m$, $V(t)$ can be described in terms of its components $(h(t), v(t))$ satisfying $(h(0), v(0)) = (x, v)$, where $h(t)$ and $v(t)$ are curves in $\mathbb{R}^m$. We want to identify $h(t)$ and $v(t)$ with objects in TM, that is, with curves of tangent vectors. The curve $h(t)$ is the curve in M traced out by the vectors $V(t)$. We can identify $h(t)$ with the vector field along $h(t)$ obtained by 'attaching a copy' of the vector v at every point of $h(t)$, i.e., by parallel translating v along the curve $h(t)$. We can identify $v(t)$ with the curve $(x, v(t))$ in $\{x\} \times T_xM = \{x\} \times \mathbb{R}^m$. For the tangent vector ξ to $V(t)$, the horizontal component ξ_h is given by the derivative $dh/dt(0)$, and the vertical component ξ_v of ξ by the derivative $dv/dt(0)$. The horizontal component ξ_h is the tangent vector at $t=0$ to the curve of footpoints; the vertical component ξ_v measures how $V(t)$ moves inside the tangent space. We can define the horizontal direction in TTM as the collection of all vectors in TTM tangent to the curves of vectors in TM obtained by parallel translating a given tangent vector along a curve in M. The vertical direction consists of all vectors in TTM tangent to the curves of vectors in TM obtained by fixing the foot point of a tangent vector and moving the vector in the tangent space to M at that point. This process of defining the horizontal and vertical directions is illustrated in Figure 5.8.1 for $M = \mathbb{R}^2$.

In this particular case when $M = \mathbb{R}^m$, there is a global choice of horizontal direction since parallel translation of vectors is path-independent. In order to adapt this construction to general manifolds, we use the Riemannian connection to provide a notion of horizontal direction along a given curve.

We can still define a vertical subspace in a canonical way. It is the same for any Riemannian metric. We call a vector $\xi \in TTM$ a vertical vector at $\theta = (x, v) \in TM$ if it is tangent to a curve $\sigma(t) = (x, v(t))$ of vectors in T_xM with $v(0) = 0$. Let $\pi : TM \to M$ denote the canonical projection. Since $\pi(\sigma(t)) = x$, its derivative $(d\pi)_\theta : T_\theta(TM) \to T_{\pi(\theta)}(M) = T_xM$ takes every vertical vector ξ to 0. Conversely, if $(d\pi)_\theta(\xi) = 0$, then ξ is tangent to some curve $\sigma(t) = (x, v(t)) \in TM$. Thus the vertical subspace V_θ at θ can be defined as

$$V_\theta = \ker(d\pi_\theta) = \{\xi \in T_\theta(TM) \mid (d\pi)_\theta(\xi) = 0\}.$$

Since a curve $(x, \sigma(t))$ has the same direction at $t = 0$ as the curve $t \to (x, v + tw)$, where $w = d\sigma/dt(0)$, it follows that $\ker(d\pi_\theta)$ is linearly isomorphic to T_xM.

The horizontal subspace is constructed in a way that depends on the Riemannian metric, as a complementary subspace to V_θ in $T_\theta(TM)$. We consider all vector fields obtained as parallel translations of v along geodesics that start at the foot point x of v. These vector fields represent curves in TM, given by the solutions of the equations $DV/dt = 0$ along geodesics starting at x. We define H_θ as the collection of all tangent vectors to these curves at θ. Each such curve is completely determined by the initial velocity of the geodesic at x, since the parallel vector field along this geodesic is uniquely determined by its initial vector v, which is fixed. Thus H_θ is linearly isomorphic to T_xM.

An alternative description of the horizontal sub-bundle can be given in terms of the connection map $K : TTM \to TM$, defined as follows. Let

5.8. HORIZONTAL AND VERTICAL SUB-BUNDLES

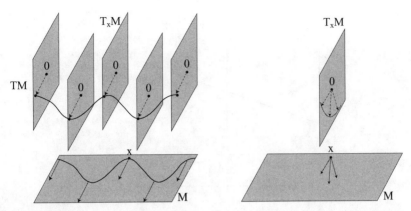

FIGURE 5.8.1
Curves of vectors generating horizontal and vertical directions, in the case $M = \mathbb{R}^2$.

$\xi \in T_\theta(TM)$, and $V(t)$ be a curve through θ in TM with $\xi = dV/dt(0)$. The curve $V(t)$ can be thought of as a vector field along the curve $h(t) = \pi \circ V(t)$ in M.

LEMMA 5.8.1
The covariant derivative $D_h V/dt(0)$ along $h(t) = \pi \circ V(t)$ is independent of the choice of a curve $V(t)$ in TM through θ with $dV/dt(0) = \xi$.

PROOF For $\theta = (x, v)$, let $(x_1, \ldots, x_m)$ be a local coordinate system near x, let $(v_1, \ldots, v_m)$ be the local coordinates of v, and let $(\xi_1, \ldots, \xi_m)$ the local coordinates of ξ. Let $(x_1(t), \ldots, x_m(t))$ be the curve $h(t)$ in local coordinates. We express V as a linear combination of $\partial/\partial x_1, \ldots, \partial/\partial x_m$, as

$$V(t) = \sum_{k=1}^{m} V_k(t) \frac{\partial}{\partial x_k}|_{h(t)}.$$

Then, by (4.2.6), we have

$$\frac{D_h V}{dt}(0) = \sum_{k=1}^{m} \left(\frac{dV_k}{dt}(0) + \sum_{i,j=1}^{m} V_j(0) \frac{dx_i}{dt}(0) \Gamma_{ij}^k \right) \frac{\partial}{\partial x_k}|_{h(0)}$$

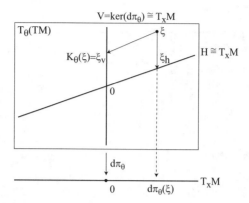

FIGURE 5.8.2
The horizontal and vertical subspaces of $T_\theta(TM)$.

$$= \sum_{k=1}^m \left(\xi_k + \sum_{i,j=1}^m v_j v_i \Gamma_{ij}^k \right) \frac{\partial}{\partial x_k}\bigg|_{h(0)}.$$

The fact that the right side of this equation depends only on ξ, θ and the Christoffel symbols Γ_{ij}^k yields the desired conclusion. ∎

Define the connection map $K : TTM \to TM$ by

$$K_\theta(\xi) = \frac{D_h V}{dt}(0),$$

where V is some curve in TM through θ with $dV/dt(0) = \xi$. The previous lemma shows that $K_\theta(\xi)$ is well defined. Then the horizontal subspace of $T_\theta(TM)$ at θ is precisely

$$H_\theta = \ker(K_\theta).$$

Indeed, every parallel vector field $V(t)$ along $\pi \circ V(t)$, as a curve in TM through θ has the same direction at $t = 0$ as the parallel vector field $W(t)$ along a geodesic through x, with $dW/dt(0) = dV/dt(0) = \xi$.

At this point, we have obtained the horizontal and vertical sub-bundles

$$H = \bigcup_{\theta \in TM} \{\theta\} \times H_\theta \quad \text{and} \quad V = \bigcup_{\theta \in TM} \{\theta\} \times V_\theta.$$

of TTM. Since the only vector $\xi \in T_\theta(TM)$ that is both horizontal and vertical is the zero vector, and since $\dim H_\theta + \dim V_\theta = \dim T_\theta(TM)$,

$$T_\theta(TM) = H_\theta + V_\theta.$$

Thus, every vector $\xi \in T_\theta(TM)$ can be written in terms of its horizontal and vertical components as
$$\xi = (\xi_h, \xi_v),$$
with $\xi_h \in H_\theta$ and $\xi_v \in V_\theta$. See Figure 5.8.2.

LEMMA 5.8.2
The map $i_\theta : T_\theta(TM) \to T_x M \times T_x M$ defined by
$$i_\theta(\xi) = (d\pi_\theta(\xi), K_\theta(\xi))$$
is a linear isomorphism.

PROOF The map i_θ is linear since both components are linear. If $i_\theta(\xi) = 0$ then $\xi = 0$, since it is the only vector which is both horizontal and vertical, thus i_θ is injective. Since the dimension of $T_\theta(TM)$ equals the dimension of $T_x M \times T_x M$, we conclude that i_θ is a linear isomorphism. ∎

LEMMA 5.8.3
The subspace $\ker(K_\theta)$ is linearly isomorphic to $\operatorname{im}(d\pi)_\theta$, and the subspace $\ker(d\pi)_\theta$ is linearly isomorphic to $\ker(d\pi)_\theta$.

PROOF From linear algebra, if $f : V \to W$ is a linear mapping, then $\operatorname{im}(f) \simeq V/\ker(f)$. In our case, for $(d\pi_p)_\theta : T_\theta(TM) \to T_x M$, $\ker((d\pi_p)_\theta) = V_\theta$, we obtain $\operatorname{im}((d\pi_p)_\theta) \simeq T_\theta(TM)/\ker((d\pi_p)_\theta) \simeq H_\theta$, and for $K_\theta : T_\theta(TM) \to T_x M$, from $\ker(K_\theta) = H_\theta$, we obtain $\operatorname{im}(K_\theta) \simeq T_\theta(TM)/\ker(K_\theta) \simeq V_\theta$, ∎

This result says that for any vector $\xi \in TTM$, we can identify the horizontal component ξ_h with the vector $d\pi_\theta(\xi) \in T_x M$, and the vertical component ξ_v with the vector $K_\theta(\xi) \in T_x M$.

5.9 The geodesic flow

The geodesic flow is a flow on the tangent bundle of a manifold. The main motivation of its study comes from mechanics (see Remark 5.9.4).

Let M be a geodesically complete m-dimensional manifold. Denote by $\gamma_{p,v}$ the unique geodesic with $\gamma_{p,v}(0) = p$ and $d\gamma_{p,v}/dt(0) = v$.

DEFINITION 5.9.1
The flow $\phi : \mathbb{R} \times TM \to TM$ on the tangent bundle TM of M defined by
$$\phi^t(p, v) = (\gamma_{p,v}(t), d\gamma_{p,v}/dt(t))$$
is called the geodesic flow.

The flow properties, $\phi^0(p,v) = (p,v)$, and $\phi^t(\phi^s(p,v)) = \phi^{t+s}(p,v)$, for all $t, s \in \mathbb{R}$, are satisfied due to the uniqueness of geodesics with respect to initial conditions. The fact that the geodesic flow acts on the tangent bundle TM and not on M implies that its flow lines are vector fields along geodesics, and not only geodesics. After all, geodesics may cross each other, but flow lines never do.

Since geodesics travel at constant speed, if the initial speed of a geodesic is $\|v\| = 1$, then $\|d\gamma_{p,v}/dt(t)\| = 1$ for all t. Let $T^1 M$ be the unit tangent bundle
$$T^1 M = \{(p, v) \in TM \mid \|v\| = 1\}.$$

The unit tangent bundle is invariant under the flow ϕ. We often consider the restriction of the geodesic flow to $T^1 M$. Such a restriction captures all important features of the geodesic flow, since tangent vectors of different length give the same flow with a constant speed reparametrization.

The geodesic flow is obtained by integrating the vector field G on TM whose local coordinates are given by the right-hand side of the system (4.4.4). Clearly G is a smooth map from the tangent bundle TM to the tangent bundle of the tangent bundle TTM. For each $\theta = (x, v) \in TM$, the vector $G(\theta)$ is tangent to the geodesic flow line through θ. The geodesic vector field is sometimes referred to as the geodesic spray (see Exercise 5.10.8).

We will show that this vector field has a simple form when TTM is represented as the sum of horizontal and vertical sub-bundles. Using the identification i_θ from Lemma 5.8.2, we obtain a simple expression of the vector field $G : TM \to TTM$ that generates the geodesic flow. For $(x, v) = \theta$, we denote by $\gamma(t)$ the geodesic with $\gamma(0) = x$, $\gamma'(0) = v$. We have
$$G(\theta) = \frac{d}{dt}(\gamma(t), \gamma'(t))|_{t=0}.$$

5.9. THE GEODESIC FLOW

Think of $(\gamma(t), \gamma'(t))$ as a curve of tangent vectors. The image of this curve under the projection $\pi : TM \to M$ is the geodesic γ, and hence $(d\pi)_\theta(G(\theta)) = (d\gamma/dt)(0) = v$. For the second component, note that $\gamma'(t)$ represents the parallel transport of v along γ, hence $K(G(\theta)) = 0$. Through the identification i_θ, the vector field $G : TM \to TTM$ that generates the geodesic flow takes the form

$$G(\theta) = (v, 0).$$

Using the same identification, we show that every Jacobi field can be obtained as a derivative of the geodesic flow.

PROPOSITION 5.9.2
Let $\theta = (x, v) \in TM$ and $\xi \in T_\theta(TM)$. The derivative of the diffeomorphism ϕ^t, expressed in terms of its horizontal and vertical components, is given by

$$(d(\phi^t))_\theta(\xi) = (J_\xi(t), DJ_\xi/dt(t)),$$

where $J_\xi(\theta)$ is the Jacobi field with initial conditions $J_\xi(0) = (d\pi)_\theta(\xi)$ and $DJ_\xi/dt(0) = K_\theta(\xi)$.

PROOF There exists a curve of tangent vectors $V(s)$ through θ in TM such that

$$\xi = \frac{dV}{ds}(s)|_{s=0}.$$

Each tangent vector $V(s)$ produces a geodesic $\gamma_{\pi(V(s)),V(s)}(t)$. Their assembly

$$\Gamma(s, t) = \gamma_{\pi(V(s)),V(s)}(t)$$

constitutes a variation through geodesics. Let J_ξ be the Jacobi field associated to the variation Γ. This means

$$J_\xi(t) = \frac{\partial}{\partial s}\Gamma(s, t)|_{s=0}, \text{ and } \frac{DJ_\xi}{dt} = \frac{D}{dt}\frac{\partial}{\partial s}\Gamma(s, t)|_{s=0}.$$

The geodesic flow can be expressed in terms of the variation as

$$\phi^t(V(s)) = \left(\Gamma(s, t), \frac{\partial}{\partial t}\Gamma(s, t)\right)|_{s=0}.$$

The derivative of the flow is given by

$$(d\phi^t)_\theta(\xi) = \frac{d}{ds}(\phi^t(V(s)))|_{s=0}.$$

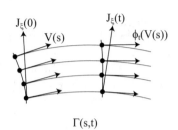

FIGURE 5.9.1
Jacobi field as derivative of the geodesic flow.

For t fixed, the projection π maps the curve of vectors $\phi^t(V(s))$ into the curve of foot points $s \to \Gamma(s,t)$. See Figure 5.9.1. Hence the horizontal component of $(d\phi^t)_\theta(\xi)$ is the image of $d(\phi^t(V(s)))/ds|_{s=0}$ through $(d\pi)_{\phi^t(\theta)}$, which is $\partial\Gamma(s,t)/\partial s|_{s=0}$, thus

$$((d\phi^t)_\theta(\xi))_h = J_\xi(t).$$

The vertical component of $(d\phi^t)_\theta(\xi)$ is the image of the vector field $\partial\Gamma(s,t)/\partial s|_{s=0}$ along $t \to \Gamma(s,t)|_{s=0}$ under $K_{\phi^t(\theta)}$, which is,

$$\frac{D}{dt}\frac{\partial}{\partial s}\Gamma(s,t)|_{s=0} = \frac{D}{ds}\frac{\partial}{\partial t}\Gamma(s,t)|_{s=0}$$

due to the symmetry of the connection. Thus

$$((d\phi^t)_\theta(\xi))_v = \frac{DJ_\xi}{dt}(t).$$

Making $t = 0$, we obtain $J_\xi(0) = (d\pi)_\theta(\xi)$ and $DJ_\xi/dt(0) = K_\theta(\xi)$. ∎

REMARK 5.9.3 The linear isomorphism i_θ described above can be used to induce a natural Riemannian metric on TM, by

$$\langle \xi, \eta \rangle_\theta = \langle d\pi_\theta(\xi), d\pi_\theta(\eta) \rangle_{\pi(\theta)} + \langle K_\theta(\xi), K_\theta(\eta) \rangle_{\pi(\theta)},$$

for $\xi, \eta \in T_\theta(TM)$. This is called the Sasaki metric. ∎

REMARK 5.9.4 The position of a particle subject to a force F is described by Newton's Second Law of Motion

$$m\frac{dx^2}{dt^2} = F(x,v),$$

5.9. THE GEODESIC FLOW

where v is the velocity of the particle. In many instances, the force depends only on the position, and can be represented as the gradient vector field $F(x) = -\text{grad}(U)$ of a potential function, so the equation becomes

$$\frac{d}{dt}(mv) = -\text{grad}(U).$$

The quantity $m\langle v, v\rangle/2$ is the kinetic energy, the quantity U is the potential energy, and the quantity

$$E = \frac{1}{2}m\langle v, v\rangle + U(x)$$

is the total energy. The total energy is preserved along the motion. The function $L : TM \to \mathbb{R}$ given by

$$L(x, v) = \frac{1}{2}m\langle v, v\rangle - U(x)$$

is called the Lagrangian. Newton's equation translates into the so-called Euler-Lagrange equation

$$\frac{d}{dt}\frac{\partial L}{\partial v} = \frac{\partial L}{\partial x}. \tag{5.9.1}$$

When a particle is subject to constraints, Newton's equation is difficult even to write, but the Euler-Lagrange equation is often simpler.

Suppose that the configuration space is represented by a manifold M. The kinetic energy is described by a positive definite quadratic form $K(v, v)$ smoothly depending on the position, described through a Riemannian metric $K(v, v) = g_x(v, v)/2$. We fix some total energy level h larger than the maximal value of the potential energy U on M. Define a new metric on M by $\rho_x = \sqrt{h - U(x)} g_x$. The condition that the total energy is sufficiently large is to avoid that the new metric ρ_x on M has singularities. On the boundary of the domain $h > U$, this metric becomes singular. That is, the length of a curve contained in $U = h$ is zero.

The action functional assigns to a smooth path $\gamma : [t_0, t_1] \to M$ the integral $A(\gamma) = \int_{t_0}^{t_1} L(\gamma, \gamma')dt$. A path γ is an extremal for A provided there is a constant $C > 0$ such that for all small enough $\epsilon > 0$ any path $\widehat{\gamma}[t_0, t_1] \to M$ with $d(\widehat{\gamma}(t), \gamma(t)) < \epsilon$ and $d(\widehat{\gamma}'(t), \gamma'(t)) < \epsilon$ for $t_0 < t < t_1$ satisfies $|A(\widehat{\gamma}) - A(\gamma)| < C\epsilon^2$. It can be shown that the extremals of the action functional are precisely the solutions of the Euler-Lagrange equation. This fact is known as the Maupertuis' least action principle.

If the configuration space is compact, these extremals give rise to a flow called the Lagrangian flow. In the case when the Lagrangian is of the form

$$L(x,v) = \frac{1}{2} g_x(v,v),$$

the Lagrangian flow is the geodesic flow (relative to the Riemannian metric g_x). In this way, the study of the evolution of certain mechanical systems can be reduced to the study of the geodesic flow on the corresponding phase space. See Arnold (1978) for details. ∎

5.10 Exercises

5.10.1 Show that all regular curves in $\mathbb{R}^3$ are locally isometric to $\mathbb{R}$. Conclude that the curvature of a regular curve is not 'intrinsic'.

5.10.2 Define

$$\nabla R(X,Y,Z,T,W) = W(\langle R(X,Y)Z,T\rangle)$$
$$- \langle R(\nabla_W X, Y)Z, T\rangle - \langle R(X, \nabla_W Y)Z, T\rangle$$
$$- \langle R(X,Y)\nabla_W Z, T\rangle - \langle R(X,Y)Z, \nabla_W T\rangle.$$

Show that it defines a tensor. Prove the differential Bianchi identity

$$\nabla R(X,Y,Z,T,W) + \nabla R(X,Y,T,W,Z) + \nabla R(X,Y,W,Z,T) = 0.$$

Show that the coefficients of the tensor $\nabla R(X,Y,Z,T,W)$ are $\partial R_{ijkl}/\partial x_m$, $i,j,k,l = 1,\ldots,m$, and they satisfy

$$\frac{\partial R_{ijkl}}{\partial x_m} + \frac{\partial R_{ijlm}}{\partial x_k} + \frac{\partial R_{ijmk}}{\partial x_l} = 0.$$

5.10.3 Let U be a normal neighborhood of a point p, and $\{E_1,\ldots,E_m\}$ an orthonormal basis of $T_p M$. Define a coordinate system on U by letting $(x_1,\ldots,x_m)$ be the coordinates of $q = \exp_p(x_1 E_1 + \ldots + x_m E_m)$. Such coordinates are called normal coordinates. Show:

(i) The components of the Riemannian metric are $g_{ij} = \delta_{ij}$ at p, where δ_{ij} is the Kronecker symbol; all $\partial g_{ij}/\partial x_k$ and all Γ_{ij} are equal to 0 at p.

5.10. EXERCISES

(ii) For any $v \in T_pM$ of normal coordinates $(v_1, \ldots, v_m)$, the geodesic $\gamma_{p,v}(t)$ starting at p with initial velocity v is represented in normal coordinates as a line segment $\gamma_{p,v}(t) = (tv_1, \ldots, tv_m)$.

(iii) The set in U given in normal coordinates by
$$\{x \mid x_1^2 + \cdots + x_m^2 < r^2\}$$
represents the geodesic ball $B(p, r)$.

5.10.4 Show that the components g_{ij} of the Riemannian metric have the following Taylor expansion in normal coordinates near a point
$$g_{ij}(x) = \delta_{ij} - \frac{1}{3} \sum_{k,l=1}^{m} R_{iklj} x_k x_l + R(x),$$
where the remainder $R(x)$ satisfies $\lim_{x \to 0} R(x)/\|x\|^3 = 0$.

5.10.5 Let p be a point in M, $v, w \in T_pM$ be a pair of orthonormal tangent vectors at p, and K be the sectional curvature determined by the plane spanned by v and w. Define a variation through geodesics $\Gamma(s,t) = \exp_p(t(v\cos s + w\sin s))$, for $0 \le t$ and $0 \le s \le 2\pi$. Let $J(t) = \partial \Gamma/\partial s(0,t)$ be the corresponding Jacobi field along the geodesic $\exp_p(tv)$. Show:

(i) The length of $J(t)$ has the following Taylor expansion about $t = 0$
$$\|J(t)\| = t - \frac{1}{6}Kt^3 + R(t),$$
where the remainder $R(t)$ satisfies $\lim_{t \to 0} R(t)/t^3 = 0$.

(ii) The arc length $L(r)$ of the geodesic 'circle' obtained by fixing $t = r$ in $\Gamma(s,t)$, has the following Taylor expansion about $t = 0$
$$L(r) = 2\pi\left(t - \frac{1}{6}Kr^3\right) + R(r),$$
where the remainder $R(r)$ satisfies $\lim_{t \to 0} R(r)/r^3 = 0$.

(iii) The area $A(r)$, defined as $\int_0^r L(t)dt$, of the geodesic 'disk' obtained by limiting the range of t in $\Gamma(s,t)$ to some closed interval $[0, r]$, has the following Taylor expansion about $t = 0$
$$A(r) = \pi\left(r^2 - \frac{1}{12}Kr^4\right) + R(r),$$
where the remainder $R(r)$ satisfies $\lim_{t \to 0} R(r)/r^4 = 0$.

5.10.6 Prove that any geodesic on a Riemannian manifold with non-positive sectional curvature has no conjugate points. Give an example of a manifold with constant positive curvature on which no geodesic has conjugate points.

5.10.7 Describe all Jacobi fields along geodesics on a cylinder in $\mathbb{R}^3$.

5.10.8 A second order differential equation on a manifold, is, by definition, a vector field ξ on the tangent bundle TM, with the property that every solution curve $X(t)$ (which is a curve of vectors in TM) is the velocity curve of $h(t) = \pi \circ X(t)$, where $\pi : TM \to M$ is the canonical projection. A curve $\gamma : (a,b) \to M$ is a solution to the second order differential equation provided that $d^2\gamma/dt^2(t) = \xi(d\gamma/dt(t))$. For $\theta = (x,v) \in TM$, denote by $\gamma_{x,v}$ the solution of initial conditions $\gamma_{x,v}(0) = x$ and $d\gamma_{x,v}/dt(0) = v$. Finally, ξ is a spray if for all $s, t \in \mathbb{R}$ and $\theta \in TM$, $\gamma_{x,sv}(t) = \gamma_{x,v}(st)$. This means that if one side is defined, so is the other (and they are then equal). Show:

(i) A vector field ξ on TM is a second order differential equation if and only if $(d\pi) \circ \xi = \mathrm{id}_{TM}$.

(ii) A second order differential equation ξ is a spray if and only if for all $s \in \mathbb{R}$ and all $\theta = (x,v) \in TM$, we have $\xi(sv) = (d\mu)(s\xi(\theta))$, where $\mu : TM \to TM$ represents the multiplication of tangent vectors by s.

(iii) Prove that on every manifold there exists a spray.

5.10.9 Let ξ be a spray on M. For any $x \in M$, show that there exists a neighborhood V of 0 in T_xM such that $\gamma_{x,v}(1)$ is well defined for all $v \in V$. Define $\exp_x : V \to M$ by $\exp_x(v) = \gamma_{(x,v)}(1)$. Show that $\exp_x$ takes lines through the origin in T_pM to solution curves of ξ, and $(d\exp_x)_0(v) = v$ for all $v \in T_xM$.

5.10.10 Assume $\theta = (x,v) \in TM$. Show that the mapping i_θ defined in Lemma 5.8.2 is an isomorphism from $T_\theta M$ to the space of all normal Jacobi fields to $\gamma_{x,v}$.

5.10.11 Let ϕ be the geodesic flow on TM. Show that $\phi^t : TM \to TM$ is an isometry for the Sasaki metric for all t if and only if M has constant sectional curvature equal to 1.

Chapter 6

Tensors and Differential Forms

6.1 Introduction

Tensor fields and differential forms are differentiable objects that are convenient for expressing the invariance of some physical laws and properties under smooth coordinate changes. Finding such invariants is a central problem in fluid mechanics, electromagnetism, general relativity, etc. In this section we describe some simple physical models that motivate the idea of differential forms.

A differential 1-form in $\mathbb{R}^3$ can be thought of as the infinitesimal work done by a force field as it moves a particle between two nearby points. If $F(x,y,z) = (F_1(x,y,z), F_2(x,y,z), F_3(x,y,z))$ is a force field on $\mathbb{R}^3$, the work done from $a = (x_1, y_1, z_1)$ to $b = (x_2, y_2, z_2)$ is $F_1(x,y,z)\Delta x + F_2(x,y,z)\Delta y + F_3(x,y,z)\Delta z$. Denote by dx, dy, dz the mappings that assign to a displacement vector its x-coordinate change, its y-coordinate change mapping, and its z-coordinate change, respectively. Then the work done between any two nearby points can be computed by applying the mapping

$$F_1(x,y,z)dx + F_2(x,y,z)dy + F_3(x,y,z)dz$$

to the corresponding displacement vector. This mapping takes infinitesimal vectors into real numbers, and defines a 1-form on $\mathbb{R}^3$.

A differential 2-form in $\mathbb{R}^3$ can be thought of as the infinitesimal volume of a spatial fluid flow crossing a small surface. Consider a fluid flowing at velocity $V(x,y,z) = (V_1(x,y,z), V_2(x,y,z), V_3(x,y,z))$ crossing a small triangle with vertices $a = (x_1, y_1, z_1)$, $b = (x_2, y_2, z_2)$, $c = (x_3, y_3, z_3)$ at some angle between $0°$ and $90°$ with the normal vector $N = (b-a) \times (c-a)$ to the triangle. The volume of fluid crossing the

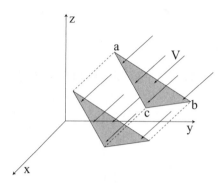

FIGURE 6.1.1
Fluid flow crossing a triangle.

triangle per unit of time is the volume of the prism determined by the vectors V, $b - a$ and $c - a$, which is half of

$$\begin{aligned}V \cdot ((b-a) \times (c-a)) &= V_1[(y_2 - y_1)(z_3 - z_1) - (y_3 - y_1)(z_3 - z_1)] \\&+ V_2[(z_2 - z_1)(x_3 - x_1) - (z_3 - z_1)(x_2 - x_1)] \\&+ V_3[(x_2 - x_1)(y_3 - y_1) - (y_2 - y_1)(x_3 - x_1)].\end{aligned}$$

See Figure 6.1.1. Denote by $dx \wedge dy$ the mapping that assigns to the pair of vectors $b - a$ and $c - a$ the area of the projection of the triangle into the xy-plane, which is $(x_2 - x_1)(y_3 - y_1) - (y_2 - y_1)(x_3 - x_1)$. Similarly, denote by $dy \wedge dz$ and $dz \wedge dx$ the mappings that assign to the pair of vectors $b - a$ and $c - a$ the areas of the projections of the triangle into the yz-plane and zx-plane, respectively. The order of the variables agrees with the positive orientation. The infinitesimal volume crossing the triangle is half of the result of applying the mapping

$$V_1(x,y,z)dy \wedge dz + V_2(x,y,z)dz \wedge dx + V_3(x,y,z)dx \wedge dy$$

to the ordered pair of vectors $(b - a, c - a)$. This mapping takes ordered pairs of vectors into real numbers, and defines a 2-form.

In general, differential n-forms are smooth mappings that take n-tuples of vectors into real numbers, and satisfy certain symmetry conditions. The properties of differential forms are independent of the choice of a coordinate system. Tensors are smooth mappings that take combinations of k-tuples of vectors and l-tuples of 1-forms into real numbers, and satisfy certain symmetry conditions. Tensor analysis was vital in Einstein's formulation of general relativity.

6.2. VECTOR BUNDLES

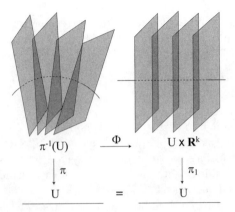

FIGURE 6.2.1
Vector bundle.

6.2 Vector bundles

The tangent bundle of a manifold is the smooth union of the tangent spaces associated to each point of the manifold. In many instances, a manifold can naturally be equipped with some other type of vector spaces smoothly attached to the points of the manifold.

DEFINITION 6.2.1
A smooth k-dimensional vector bundle is a pair of smooth manifolds E (the total space) and M (the base), together with a a map $\pi : E \to M$ (the projection) satisfying the following properties:

(i) *π is smooth and surjective;*

(ii) *for each $x \in M$, the fiber $\pi^{-1}(x)$ is a vector space;*

(iii) *for each $x \in M$, there exists a neighborhood U of x and a diffeomorphism $\Phi : U \times \mathbb{R}^k \to \pi^{-1}(U)$ which is a linear isomorphism from $\{y\} \times \mathbb{R}^k$ onto $\pi^{-1}(y)$, for every $y \in U$.*

The mapping Φ defined above is called a local trivialization: it expresses the fact that, locally, the manifold E is the product of a neighborhood in M by a vector space. See Figure 6.2.1. If the whole manifold E is a product of M by some vector space, or, equivalently, if the mapping Φ is globally defined, then the corresponding bundle is called trivial. The simplest examples of vector bundles are the trivial ones, of the form $M \times \mathbb{R}^k$, where $\pi : M \times \mathbb{R}^k \to M$ is the projection onto the first component.

Example 6.2.2
The tangent bundle of S^n with n even is non-trivial. Assume by contradiction that there exists a global trivialization $\Phi : S^n \times \mathbb{R}^n \to TS^n$. Then the image under Φ of a smooth graph $\{(p, v_p) \,|\, p \in S^n,\ v_p \neq 0\} \subset S^n \times \mathbb{R}^n$ is a nowhere vanishing vector field on S^n. This contradicts the 'hairy ball theorem' — Theorem 2.2.3.

In contrast, the tangent bundle of S^1 is trivial, so TS^1 is diffeomorphic to a cylinder. As it turns out, the tangent bundle TS^n is trivial only for $n = 1, 3$ and 7. □

Example 6.2.3
Consider the Möbius strip defined as the quotient set $[0,1] \times \mathbb{R}/\sim$ by the equivalence relation $(0, a) \sim (1, -a)$, with the induced smooth structure, and S^1 as the quotient of $[0, 1]$ to the equivalence relation $0 \sim 1$. The Möbius strip can be viewed as a vector bundle over S^1, with the projection π defined by $\pi(t, a) = t$ for $0 < t < 1$, and $\pi([(0, a)]) = \pi([(1, -a)]) = [0] = [1]$. See Figure 6.2.2. Such a bundle is obviously non-trivial, as the cylinder $S^1 \times \mathbb{R}$ is not diffeomorphic to the Möbius strip (the first is orientable while the second is not). □

Example 6.2.4 (Cotangent bundle)
Let M be an m-dimensional manifold. For each $x \in M$, the set of all linear functionals $L : T_xM \to \mathbb{R}$ forms a vector space T_x^*M under addition and scalar multiplication of functionals. The cotangent bundle is $T^*M = \bigcup_{x \in M} T_x^*M$, together with the projection $\pi : T^*M \to M$ that sends every functional $L \in T_x^*M$ to $x \in M$. Each linear functional on $\mathbb{R}^m$ can be uniquely represented in the form $(x_1, \ldots, x_m) \to a_1 x_1 + \ldots + a_m x_m$, so we can identify $\mathbb{R}^m$ with the space $(\mathbb{R}^m)^*$ of all linear functionals on $\mathbb{R}^m$.

Let $\phi : U_\alpha \to V_\alpha$ be a local parametrization on M. The mapping

$$\Phi_\alpha : U_\alpha \times \mathbb{R}^m \to \bigcup_{x \in U_\alpha} T_x^*M,$$

given by $\Phi_\alpha(x, L) = (\phi_\alpha(x), L \circ d(\phi_\alpha^{-1})_{\phi_\alpha(x)})$ is a local parametrization

6.2. VECTOR BUNDLES

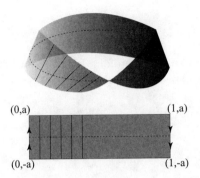

FIGURE 6.2.2
Non-trivial vector bundle.

on T^*M. It is clear that these Φ_α's defines a smooth structure on T^*M, since each is injective and the corresponding re-parametrizations

$$\Theta_{\alpha\beta} = \Phi_\alpha^{-1} \circ \Phi_\beta$$

are given by

$$\Theta_{\alpha\beta} = (\phi_\alpha^{-1} \circ \phi_\beta, d(\phi_\alpha^{-1} \circ \phi_\beta)^{-1}),$$

so they are all diffeomorphisms.

The projection π is obviously surjective and each fiber is a vector space of dimension m. The mapping Φ_α is a local trivialization since its restriction to $\{x\} \times \mathbb{R}^m$, taking $L \in (\mathbb{R}^m)^* \simeq \mathbb{R}^m$ to $L \circ (d\phi)^{-1}_{\phi_\alpha(x)} \in T_x^*M$, is a linear isomorphism.

The construction of the cotangent bundle has a physical motivation: it turns out that while velocity is best thought of as a tangent vector, linear momentum is best thought of as a cotangent vector. One reason is that there is a natural way to define certain operations on real valued functions (Poisson brackets) on the cotangent bundle, but not on the tangent bundle (see Robinson (1998)). □

DEFINITION 6.2.5
Two vector bundles $\pi_1 : E_1 \to M$ and $\pi_2 : E_2 \to M$ are said to be equivalent if there exists a diffeomorphism $h : E_1 \to E_2$ such that h maps each fiber $\pi_1^{-1}(x)$ isomorphically onto $\pi_2^{-1}(x)$.

As an example, the cotangent bundle is isomorphic to the tangent bundle (Exercise 6.11.1).

Example 6.2.6
Let M' be a Riemannian manifold and M be a submanifold of M'. We define the normal space at $p \in M$ by

$$(T_pM)^\perp = \{v \in T_pM' \mid \langle v, w \rangle = 0 \text{ for all } w \in T_pM\}.$$

Let

$$NM = \bigcup_{p \in M} (T_pM)^\perp,$$

and $\pi : NM \to M$ by $\pi(p, v) = p$. It is easy to see that this defines a vector bundle with total space TM, base space M and projection π, called the normal bundle of M in M'.

For example, the normal bundle of the sphere S^{n-1} in $\mathbb{R}^n$ is equivalent to the trivial bundle $S^{n-1} \times \mathbb{R}$. The normal bundle of S^1 in the Möbius strip is equivalent to the Möbius strip; thus it is non-trivial. □

The analogue of a smooth vector field in the case of a vector bundle is a smooth section. Given a vector bundle $\pi : E \to M$, a smooth section of E is a smooth map $\sigma : M \to E$ with $\sigma(x) \in \pi^{-1}(x)$ for all $x \in M$. The zero section σ_0 is the smooth section for which $\sigma_0(x)$ is the zero vector in $\pi^{-1}(x)$, for every $x \in M$. A handy criterion for triviality of a bundle is the following, whose proof is left as an exercise.

PROPOSITION 6.2.7
A k-dimensional vector bundle is trivial if and only if there exist k smooth sections $\sigma_1, \ldots, \sigma_k$ on M such that $\sigma_1(x), \ldots, \sigma_k(x)$ are linearly independent in $\pi^{-1}(x)$, for all $x \in M$.

In particular, if there is no non-vanishing section, it follows that the vector bundle is non-trivial.

REMARK 6.2.8 A (locally trivial) fiber bundle is a more general notion than a vector bundle. This structure consists of three topological spaces E (total space), B (base space), F (fiber space), and a continuous surjection $\pi : E \to B$, with the property that each fiber $\pi^{-1}(x)$ is a topological subspace of E homeomorphic to the fiber F, and for every

6.3. THE TUBULAR NEIGHBORHOOD THEOREM

point $x \in B$ there is a neighborhood U of x and a homeomorphism $\Phi : U \times F \to \pi^{-1}(U)$ which is a homeomorphism from each $\{x\} \times F$ onto $\pi^{-1}(x)$, for all $x \in U$. The unit tangent bundle $T^1 M$, introduced in Section 5.9, is an example of a fiber bundle. ∎

6.3 The tubular neighborhood theorem

Let M be a compact embedded surface in $\mathbb{R}^3$. At each point $p \in M$, we cut an open segment I_p of length 2ϵ off the normal line through p, for some $\epsilon > 0$. It seems reasonable (and is true) that if $\epsilon > 0$ is chosen sufficiently small, then any two line segments I_p, I_q with $p \neq q$ will be disjoint. This would define an open neighborhood $U = \bigcup_{p \in M} I_p$ of M in $\mathbb{R}^3$. See Figure 6.3.1. The manifold U together with the projection $\pi : U \to M$ that maps each point $x \in U$ to the unique point $p \in M$ with $x \in I_p$, defines a fiber bundle on M (see Remark 6.2.8). This fiber bundle is equivalent to the normal bundle, and its zero section represents the canonical embedding of M in U.

DEFINITION 6.3.1
Let M' be a Riemannian manifold and M be a submanifold of M'. A tubular neighborhood of M in M' is an open neighborhood of M in M' that is diffeomorphic to a neighborhood of the zero section $\sigma_0 : M \to E$ in some vector bundle $\pi : E \to M$ over M.

THEOREM 6.3.2 (*Tubular neighborhood theorem*)
Let M' be a Riemannian manifold, and M be a compact submanifold of M'. Then there exists a tubular neighborhood M_0 of M that is diffeomorphic to a neighborhood of the zero section in the normal bundle NM of M in M'.

PROOF Define the set M_ϵ of all points in M' at a distance less than ϵ from M, and the set U_ϵ of all $(p, v) \in NM$ with $|v| < \epsilon$, for $\epsilon > 0$. By Proposition 4.5.2, for each point $p \in M$ there exist a uniformly geodesic neighborhood W of p in M' and a number $\epsilon_p > 0$ such that the exponential map $\exp_q$ is defined for each $q \in W$ and is a dif-

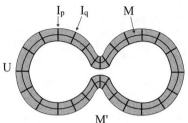

FIGURE 6.3.1
A tubular neighborhood of a compact surface.

feomorphism of the disk $B(0, \epsilon_p) \subseteq T_q M'$ onto its image. Moreover, as in the proof of Proposition 4.5.2, the map $E(q, v) = (q, \exp_q(v))$ takes the set $\bigcup_{q \in W}(\{q\} \times B(0, \epsilon_p)) \subseteq TM'$ diffeomorphically onto its image in $M' \times M'$. Using the compactness of M, we can cover M with finitely many such neighborhoods W, and obtain a value of $\epsilon > 0$, such that $\exp_q$ is defined at each $q \in M$, and the map $\exp$ given by $(q, v) \to \exp_q(v)$ is a diffeomorphism from each $\bigcup_{q \in W}(\{q\} \times B(0, \epsilon))$ onto its image in M'.

We claim that the mapping $\exp : U_\epsilon \to M'$, is a diffeomorphism onto M_ϵ, provided ϵ is chosen small enough.

First, we notice that the above mapping is one-to-one for sufficiently small ϵ. If this were not the case, there would exist sequences $(q_n, v_n) \neq (q'_n, v'_n) \in U_\epsilon$ with $\|v_n\|, \|v'_n\| < 1/n$ and $\exp_{q_n}(v_n) = \exp_{q'_n}(v'_n)$. By the compactness of M, we can assume that these sequences converge to $(q, 0)$ and $(q', 0)$, respectively. Since $\exp_q(0) = \exp_{q'}(0)$, we would have $q = q'$, and it would follow that q_n, q'_n are in the same uniformly geodesic neighborhood W, for all n large enough. This contradicts the fact that the map $(q, v) \to (q, \exp_q(v))$ is a local diffeomorphism.

Second, we notice that $\exp$ maps $M \times B(0, \epsilon)$ into M_ϵ; we actually need to prove that it does it surjectively. For an arbitrary $q \in M_\epsilon$, choose the closest point p from q in M. It is clear that $q = \exp_p(v)$ for some $v \in T_p M'$ with $\|v\| < \epsilon$. Let us say that the distance from q to p is $\rho < \epsilon$. The geodesic sphere $S(q, \rho)$ is tangent to M at p. By Corollary 4.5.5, the geodesic γ joining p to q is orthogonal to the geodesic sphere $S(q, \rho)$, hence γ is orthogonal to M and so $v \in (T_p M)^\perp$.

In conclusion, we obtained that M_ϵ is a neighborhood of M in M', and is the diffeomorphic image of U_ϵ which is the obvious tubular neighborhood of the zero section in NM. ∎

The result is true even if M is non-compact, as follows from a partition of unity-based argument. See Lang (1972) for details.

6.4 Tensor bundles

A real-valued function T on the cartesian product $V_1 \times \cdots \times V_k$ of vector spaces is called multilinear if it is separately linear in each variable, that is, for each $i \in \{1, \ldots, k\}$,

$$T(v_1, \ldots, av_i+bv'_i, \ldots, v_k) = aT(v_1, \ldots, v_i, \ldots, v_k) + bT(v_1, \ldots, v'_i, \ldots, v_k),$$

for all $a, b \in \mathbb{R}$, and $v_i, v'_i \in V_i$.

If V is a finite-dimensional vector space and V^* denotes the vector space of all linear functionals on V, then the vector space $V^{**} = (V^*)^*$ can be identified with V, through the mapping $v \in V \to v^{**} \in V^{**}$ given by

$$v^{**}(L) = L(v),$$

for every $L \in V^*$. Note that such an identification is independent of the choice of a basis on V. The vector spaces V and V^* are also isometric, but a natural identification is not possible: any isomorphism that can be constructed depends on the choice of a basis of V.

DEFINITION 6.4.1
Let M be a manifold. A multilinear map

$$T : \underbrace{T_p^*M \times \cdots \times T_p^*M}_{l \text{ times}} \times \underbrace{T_pM \times \cdots \times T_pM}_{k \text{ times}} \to \mathbb{R}$$

is called a $\binom{k}{l}$-tensor (k-covariant tensor, l-contravariant).

Due to the above identifications, the $\binom{0}{1}$-tensors are tangent vectors, which we will usually denote by the letters $X, Y, \ldots$. The $\binom{1}{0}$-tensors are called 1-forms, which we will usually denote by the letters $\omega, \eta, \ldots$. Because the sums and the scalar multiples of multilinear maps are still multilinear, the $\binom{k}{l}$-tensors form a vector space. We will denote this space by $\mathcal{T}_{\binom{k}{l}}(T_pM)$. Besides addition and scalar multiplication, one can multiply together tensors of different types, by an algebraic operation called tensor product. If $T \in \mathcal{T}_{\binom{k}{l}}(T_pM)$, and $S \in \mathcal{T}_{\binom{k'}{l'}}(T_pM)$ then the tensor product

$$T \otimes S : \underbrace{T_p^*M \times \cdots \times T_p^*M}_{l+l' \text{ times}} \times \underbrace{T_pM \times \cdots \times T_pM}_{k+k' \text{ times}} \to \mathbb{R}$$

is the multilinear map defined by

$$T \otimes S(\omega_1, \ldots, \omega_{l+l'}, X_1, \ldots, X_{k+k'})$$
$$= T(\omega_1, \ldots, \omega_l, X_1, \ldots, X_k) S(\omega_{l+1}, \ldots, \omega_{l+l'}, X_{k+1}, \ldots, X_{k+k'}).$$

This operation is associative but not commutative. The map $(T, S) \to T \otimes S$ is bilinear.

DEFINITION 6.4.2
The $\binom{k}{l}$-tensor bundle on a manifold M is the vector bundle of total space

$$\mathcal{T}_{\binom{k}{l}}(M) = \bigcup_{p \in M} \mathcal{T}_{\binom{k}{l}}(T_pM),$$

over the base space M, with the canonical projection. Local trivializations are given by charts in an obvious way. A $\binom{k}{l}$-tensor field is a section of this bundle.

While tensors at one point are multilinear over the reals, tensor fields are multilinear over the smooth functions, in the sense that

$$T(\ldots, f\alpha + g\beta, \ldots,) = fT(\ldots, \alpha, \ldots) + gT(\ldots, \beta, \ldots),$$

where f, g are smooth functions on M, and α, β are either vector fields or 1-forms, depending on their slot position.

Example 6.4.3
A Riemannian metric is a tensor field of type $\binom{2}{0}$.
Another example is the curvature tensor. This is a multilinear mapping that takes each triple of vector fields to another vector field. Denote by $\text{Lin}(T_pM \times \ldots \times T_pM, T_pM)$ the space of all multilinear mappings of k-tuples of vector fields into vector fields. In general, we have

$$\text{Lin}(\underbrace{T_pM \times \cdots \times T_pM}_{k \text{ times}}, T_pM) \simeq \mathcal{T}_{\binom{k}{1}}(T_pM),$$

6.4. TENSOR BUNDLES

through the canonical linear isomorphism that assigns to each L in $\text{Lin}(T_pM \times \cdots \times T_pM, T_pM)$ the tensor $T \in \mathcal{T}_{\binom{k}{1}}(T_pM)$ defined by

$$T(\omega, X_1, \ldots, X_k) = \omega(L(X_1, \ldots, X_k)).$$

This implies that the curvature tensor can be regarded as a tensor field of type $\binom{3}{1}$. □

Let $(x_1, \ldots, x_m)$ be a local coordinate system near a point p on M. This defines a basis $(\partial/\partial x_1, \ldots, \partial/\partial x_m)$ for the tangent space T_pM. A basis for the 1-forms is defined by the linear functionals $dx_i : T_pM \to \mathbb{R}$, given by $dx_i(\partial/\partial x_i) = 1$ and $dx_i(\partial/\partial x_j) = 0$, for $j \neq i$. Then

$$T = \sum_{\substack{(s_1, \ldots, s_l) \\ (t_1, \ldots, t_k)}} T^{s_1 \ldots s_l}_{t_1 \ldots t_k} \frac{\partial}{\partial x_{s_1}} \otimes \cdots \otimes \frac{\partial}{\partial x_{s_l}} \otimes dx_{t_1} \otimes \cdots \otimes dx_{t_k}$$

is obviously a $\binom{k}{l}$-tensor. Conversely, any $\binom{k}{l}$-tensor T can be uniquely written in this form, with

$$T^{s_1 \ldots s_l}_{t_1 \ldots t_k} = T\left(dx_{s_1}, \ldots, dx_{s_l}, \frac{\partial}{\partial x_{t_1}}, \ldots, \frac{\partial}{\partial x_{t_k}}\right).$$

Indeed, if T' is the tensor determined by these choices of $T^{s_1 \ldots s_l}_{t_1 \ldots t_k}$, then

$$T\left(dx_{\sigma_1}, \ldots, dx_{\sigma_l}, \frac{\partial}{\partial x_{\tau_1}}, \ldots, \frac{\partial}{\partial x_{\tau_k}}\right) = T'\left(dx_{\sigma_1}, \ldots, dx_{\sigma_l}, \frac{\partial}{\partial x_{\tau_1}}, \ldots, \frac{\partial}{\partial x_{\tau_k}}\right)$$

for any choices of $\sigma_1, \ldots, \sigma_l$ and $\tau_1, \ldots, \tau_k$. It follows easily using the multilinearity of T and T' that $T = T'$.

If T is a $\binom{k}{l}$-tensor, $\omega_1, \ldots, \omega_l$ are 1-forms and $X_1, \ldots, X_k$ are vector fields, the value at a point p of

$$T(\omega_1, \ldots, \omega_l, X_1, \ldots, X_k)$$

depends only on T and the values of $\omega_1, \ldots, \omega_l, X_1, \ldots, X_k$ at p.

The tensor components obey certain laws of transformations when the reference system is changed. If a tensor T is written as

$$T = \sum_{\substack{(s_1, \ldots, s_l) \\ (t_1, \ldots, t_k)}} T^{s_1 \ldots s_l}_{t_1 \ldots t_k} \frac{\partial}{\partial x_{s_1}} \otimes \cdots \otimes \frac{\partial}{\partial x_{s_l}} \otimes dx_{t_1} \otimes \cdots \otimes dx_{t_k},$$

236   6. TENSORS AND DIFFERENTIAL FORMS

with respect to a local coordinate system $(x_1, \ldots, x_m)$, and as

$$T = \sum_{\substack{(s_1,\ldots,s_l) \\ (t_1,\ldots,t_k)}} T'^{s_1\ldots s_l}_{t_1\ldots t_k} \frac{\partial}{\partial x'_{s_1}} \otimes \ldots \otimes \frac{\partial}{\partial x'_{s_l}} \otimes dx'_{t_1} \otimes \ldots \otimes dx'_{t_k},$$

with respect to another local coordinate system $(x'_1, \ldots, x'_m)$, then

$$T'^{s_1\ldots s_l}_{t_1\ldots t_k} = \sum_{\substack{(i_1,\ldots,i_k) \\ (j_1,\ldots,j_l)}} T^{j_1\ldots j_l}_{i_1\ldots i_k} \frac{\partial x_{i_1}}{\partial x'_{t_1}} \cdots \frac{\partial x_{i_k}}{\partial x'_{t_k}} \frac{\partial x'_{s_1}}{\partial x_{j_1}} \cdots \frac{\partial x'_{s_l}}{\partial x_{j_l}}. \tag{6.4.1}$$

A family of components that transform correctly under coordinate changes defines a unique tensor.

As a motivation for tensor algebra, we informally discuss an example from solid mechanics in which tensors emerge naturally.

Example 6.4.4
The moment of inertia of a solid is the rotational analogue of mass for linear motion: just as mass measures how easily an object accelerates due to a given force, the moment of inertia measures how easily an object rotates about a particular point of rotation due to a given torque. For a single particle of mass m with position vector $r = (x_1, x_2, x_3)$ rotating about the z-axis, the moment of inertia I is given by the quantity $m(x_1^2 + x_2^2)$, where $x_1^2 + x_2^2$ is the perpendicular distance from the point to the axis. In the case of a solid body, we will see that the moment of inertia is best described by a tensor. Consider a solid body consisting of a finite collection of mass elements, with the center of mass at the origin of some orthonormal frame $\{e_1, e_2, e_3\}$. We denote a generic mass element by m and its position vector by r, of coordinates (x_1, x_2, x_3). Assume that the solid rotates about some axis through the origin. Each element of mass has an instantaneous angular velocity $\omega = (\omega_1, \omega_2, \omega_3)$, so the velocity of the mass element m at r is given by the vector product $v = \omega \times r$, and the angular momentum about the origin by $mr \times (\omega \times r)$. See Arnold (1978) and Figure 6.4.1. The total angular momentum is given by

$$H = \sum mr \times (\omega \times r) = \sum m(\|r\|^2 \omega - (r \cdot \omega)r)$$
$$= \sum m \sum_{j=1}^{3} (mx_j^2 \omega - x_j \omega_j r) = \sum_{i=1}^{3} \left(\sum m \sum_{j=1}^{3} (mx_j^2 \omega_i - x_i x_j \omega_j) \right) e_i.$$

6.4. TENSOR BUNDLES

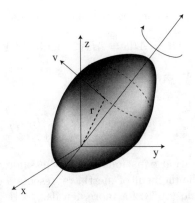

FIGURE 6.4.1
Angular momentum of a solid.

The summation is performed over all mass elements; for a continuous solid, this summation would be replaced by an integral. The components of H_1, H_2, H_3 of H are given by

$$H_i = I_{i1}\omega_1 + I_{i2}\omega_2 + I_{i3}\omega_3,$$

where

$$I_{ij} = \sum m\left(x_1^2 + x_2^2 + x_3^2\right)\delta_{ij} - \sum m x_i x_j,$$

where δ_{ij} is the Kronecker symbol, for all i and j in $\{1,2,3\}$. Hence the angular momentum H is given by

$$H = I\omega,$$

where I is the matrix of entries $(I_{ij})_{i,j=1,\ldots,3}$. The matrix I is called the moment of inertia tensor. The diagonal entries are referred to as the moments of inertia with respect to the coordinate axis, while the rest of the entries are referred to as products of inertia. Going back to the case of a single mass rotating about the z-axis, we have $\omega_1 = \omega_2 = 0$, and there is only one non-zero moment of inertia $I_{33} = m(x_1^2 + x_2^2)$.

To see that I behaves like a tensor, let us change the reference frame to $\{e_1', e_2', e_3'\}$, and assume that

$$H = I'\omega',$$

with respect to this new frame. As

$$e_i = \sum_{j=1}^{3} a_{ij} e_j', \text{ for all } i \in \{1,2,3\},$$

with $A = (a_{ij})_{i,j=1,2,3}$ an orthogonal matrix, we have

$$H' = AH = AI\omega = AIA^{-1}A\omega = AIA^t\omega',$$

and hence $I' = AIA^t$, or, in coordinates,

$$I'_{t_1 t_2} = \sum_{i_1,i_2=1}^{3} I_{i_1,i_2} a_{t_1 i_1} a_{t_2 i_2} = \sum_{i_1,i_2=1}^{3} I_{i_1,i_2} \frac{\partial x_{i_1}}{\partial x'_{t_1}} \frac{\partial x_{i_2}}{\partial x'_{t_2}}.$$

As these transformation equations are of the same type as in (6.4.1), we conclude that the moment of inertia represents a covariant tensor of type $\binom{2}{0}$. See Arnold (1978) for more details. □

6.5 Differential forms

A $\binom{k}{0}$-tensor

$$T: \underbrace{V \times \cdots \times V}_{k \text{ times}} \to \mathbb{R}$$

is said to be alternating if

$$T(\ldots, v_i, \ldots, v_j, \ldots) = -T(\ldots, v_j, \ldots, v_i, \ldots),$$

for each $i \neq j$. Equivalently, T is alternating if $T(v_1, \ldots, v_k) = 0$ whenever $v_i = v_j$, for some $i \neq j$. A $\binom{1}{0}$-tensor is alternating by default. An alternating $\binom{k}{0}$-tensor with $k > \dim V$ is always 0 since $T(v_1, \ldots, v_k) = 0$ provided $v_1, \ldots, v_k$ are dependent. The verification of this fact is left as an exercise.

One can always transform a k-covariant tensor T into an alternating one, by an operation called the 'alternation' of T

$$\text{Alt}(T)(v_1, \ldots, v_k) = \frac{1}{k!} \sum_{\substack{\sigma \text{ permutatation} \\ \text{of } k \text{ elements}}} \text{sign}(\sigma) T(v_{\sigma(1)}, \ldots, v_{\sigma(k)}),$$

where $\text{sign}(\sigma)$ is the sign of the permutation. The role of the scaling factor $1/k!$ is to ensure that $\text{Alt}(\text{Alt}(T)) = \text{Alt}(T)$.

6.5. DIFFERENTIAL FORMS

DEFINITION 6.5.1
Let M be a manifold and p be a point in M. An alternating $\binom{k}{0}$-tensor

$$\omega : \underbrace{T_pM \times \cdots \times T_pM}_{k \text{ times}} \to \mathbb{R}$$

is called a *k-form at p*. The set of all *k*-forms at p is denoted by $\Lambda^k(T_pM)$. The tensor bundle

$$\Lambda^k(TM) = \bigcup_{p \in M} \Lambda^k(T_pM)$$

is called the *bundle of k-forms*, and a smooth section of this bundle is called a *differential k-form*. Zero forms are just smooth functions on M.

In the sequel, we will discuss various operations with differential forms and their properties. The proofs will be often left as exercises. All forms will be assumed to be differential forms.

Addition and scalar multiplication of tensors extend naturally to forms. The tensor product of two forms however is not necessarily a form, since it may not be alternating. In order to get a form, one needs to apply the alternation to the tensor product. This defines a new operation, called the *exterior product*

$$\wedge : \Lambda^k(T_pM) \times \Lambda^l(T_pM) \to \Lambda^{k+l}(T_pM),$$

given by

$$\omega \wedge \eta = \frac{(k+l)!}{k!\, l!} \text{Alt}(\omega \otimes \eta).$$

The role of the scaling factor $(k+l)!/k!\,l!$ is to ensure that in $\mathbb{R}^m$, the exterior product of $dx_1 \wedge \ldots \wedge dx_m$ coincides with the determinant operator (see Example 6.5.3 below). Moreover, if $\omega, \ldots, \omega_k$ are 1-forms and $v_1, \ldots, v_k$ are vectors, then

$$(\omega_1 \wedge \ldots \wedge \omega_k)(v_1, \ldots, v_k) = \begin{pmatrix} \omega_1(v_1) & 0 & \cdots & 0 \\ 0 & \omega_2(v_2) & \cdots & 0 \\ \vdots & \vdots & \ddots & \vdots \\ 0 & 0 & \cdots & \omega_k(v_k) \end{pmatrix}.$$

Given a local coordinate system $(x_1, \ldots, x_m)$ about a point $p \in M$, a basis for $\Lambda^k(T_pM)$ is given by all combinations

$$dx_{i_1} \wedge dx_{i_2} \wedge \ldots \wedge dx_{i_k}, \text{ with } i_1 < i_2 < \cdots < i_k.$$

Thus, a k-form on M can be expressed locally as

$$\omega = \sum_{i_1 < i_2 < \ldots < i_k} f_{i_1 i_2 \ldots i_k} dx_{i_1} \wedge dx_{i_2} \wedge \ldots \wedge dx_{i_k}, \tag{6.5.1}$$

where $f_{i_1 i_2 \ldots i_k}$ are smooth functions.

The following summarizes the properties of the exterior product.

PROPOSITION 6.5.2

(i) $\omega \wedge (\eta_1 + \eta_2) = \omega \wedge \eta_1 + \omega \wedge \eta_2$;

(ii) $\omega \wedge \eta = (-1)^{kl} \eta \wedge \omega$, provided $\omega \in \Lambda^k(TM)$ and $\eta \in \Lambda^l(TM)$;

(iii) $\omega \wedge (\eta \wedge \delta) = (\omega \wedge \eta) \wedge \delta$.

Property (ii) results from the fact that differential forms are alternating tensors. Note, for example, that $dx \wedge dy = -dy \wedge dx$, and $dx \wedge (dy \wedge dz) = (dy \wedge dz) \wedge dx$.

Example 6.5.3

A 1-form $a dx + b dy + c dz$ in $\mathbb{R}^3$ applied to a vector $v = (v_1, v_2, v_3)$ gives $av_1 + bv_2 + cv_3$, which represents the length of the projection of v into the line of direction (a, b, c) through the origin, re-scaled by a factor of $\|(a, b, c)\|$. A 2-form $a dx_1 \wedge dx_2$ in $\mathbb{R}^3$ applied to a pair of vectors $v = (v_1, v_2, v_3), w = (w_1, w_2, w_3)$ in $\mathbb{R}^3$ represents the oriented area of the projection of the parallelogram P determined by v, w into the (x_1, x_2)-coordinate plane, re-scaled by a factor of a. A 2-form $a dx_1 \wedge dx_2 + b dx_1 \wedge dx_3 + c dx_2 \wedge dx_3$ applied to v, w is the combination of the oriented areas of the projections of the parallelogram P into the coordinate planes, re-scaled by factors of a, b, c, respectively. A 3-form $a dx_1 \wedge dx_2 \wedge dx_3$ in $\mathbb{R}^3$ applied to a triple of vectors $v = (v_1, v_2, v_3)$, $w = (w_1, w_2, w_3), u = (u_1, u_2, u_3)$ represents the oriented volume of the parallelepiped

$$\det \begin{bmatrix} v_1 & v_2 & v_3 \\ w_1 & w_2 & w_3 \\ u_1 & u_2 & u_3 \end{bmatrix}$$

determined by v, w, u, re-scaled by a factor of a. See Figure 6.5.1. In general, the n-form $dx_1 \wedge \ldots \wedge dx_n$ on $\mathbb{R}^n$ measures the oriented n-volume of parallelepipeds formed by n-vectors in $\mathbb{R}^n$, and it is called the canonical volume form. □

6.5. DIFFERENTIAL FORMS

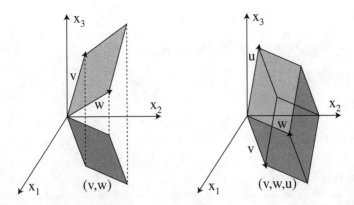

FIGURE 6.5.1
A 2-form and a 3-form representing an oriented area and an oriented volume, respectively.

Example 6.5.4
In the case of an orientable m-dimensional manifold M, there is no canonical volume form. On each coordinate neighborhood V_α, we can define an m-form $dx_1^\alpha \wedge \ldots \wedge dx_m^\alpha$, where $(x_1^\alpha, \ldots, x_m^\alpha)$ is the local coordinate system. We can choose a countable and locally finite family of charts $\{V_\alpha\}_\alpha$ on M, and consider a partition of unity $\{f_\alpha\}_\alpha$ subordinate to this covering. We define the m-form

$$\omega = \sum_\alpha f_\alpha dx_1^\alpha \wedge \ldots \wedge dx_m^\alpha.$$

Since $0 \leq f_\alpha \leq 1$ and $\sum_\alpha f_\alpha = 1$, $\omega \neq 0$ at each point $p \in M$.

By definition, any nowhere zero m-form on M is called a volume form. Note that if ω_1 and ω_2 are two volume forms, then there exists a nonvanishing smooth function f on M such that $\omega_1 = f\omega_2$. This simple fact yields the following orientability criterion: an m-dimensional manifold is orientable if and only if it supports a nowhere zero differential m-form.

The actual use of volume forms in computing the volume of a manifold will be discussed in the next section. □

Forms can be transported between manifolds by the pull-back operation. If $f : M \to N$ is a smooth mapping between two manifolds, and ω is a k-form on N, then

$$(f^*\omega)_p(v_1, \ldots, v_k) = \omega_{f(p)}((df)_p(v_1), \ldots, (df)_p(v_k)),$$

for $v_1, \ldots, v_k \in T_p M$, defines a k-form on M.

In local coordinates, if $(y_1, \ldots, y_m)$ is a local coordinate system on N and $\omega = \sum_{i_1 < i_2 < \ldots < i_k} f_{i_1 i_2 \ldots i_k} dy_{i_1} \wedge dy_{i_2} \wedge \ldots \wedge dy_{i_k}$, then

$$f^* \omega = \sum_{i_1 < i_2 < \cdots < i_k} (f_{i_1 i_2 \ldots i_k} \circ f) d(y_{i_1} \circ f) \wedge d(y_{i_2} \circ f) \wedge \ldots \wedge d(y_{i_k} \circ f).$$

Here $y_{i_j} \circ f$ represents the i_j-th component of f (in local coordinates on N), and $d(y_{i_j} \circ f)$ is the derived 1-form.

Example 6.5.5

If $f : M \to N$ is a smooth mapping between two orientable manifolds of equal dimension m, and $dy_1 \wedge \ldots \wedge dy_m$ is the local expression of a volume form on N, then its pull back to M is given by

$$f^*(dy_1 \wedge \ldots \wedge dy_m) = d(y_1 \circ f) \wedge \ldots \wedge d(y_m \circ f)$$
$$= \det(df)\, dx_1 \wedge \ldots \wedge dx_m,$$

where $\det(df)$ denotes the function $x \to \det(df_x)$ in local coordinates. If we think of $dy_1 \wedge \ldots \wedge dy_m$ as an 'integrand', the above equation represents the change of variable formula from integral calculus. □

PROPOSITION 6.5.6

(i) $f^*(\omega_1 + \omega_2) = f^* \omega_1 + f^* \omega_2$;

(ii) $f^*(\omega \wedge \eta) = f^* \omega \wedge f^* \eta$;

(iii) $(g \circ f)^*(\omega) = f^* g^* \omega$, provided $f : M \to N$ and $g : N \to P$ are smooth mappings.

The next operation, called the exterior derivative, is an analogue to the derivation of functions. It is more convenient to define this operation through local coordinates. If $\omega = \sum_{i_1 < \cdots < i_k} f_{i_1 \ldots i_k} dx_{i_1} \wedge \ldots \wedge dx_{i_k}$ is a k-form, then $d\omega$ is the $(k+1)$-form given by

$$d\omega = \sum_{i_1 < \cdots < i_k} \left(\sum_{j=1}^n \frac{\partial f_{i_1 \ldots i_k}}{\partial dx_j} dx_j \wedge dx_{i_1} \wedge \ldots \wedge dx_{i_k} \right).$$

It turns out that this definition does not depend on the coordinate system, as can be inferred from the following result (the proof is left as an exercise).

6.5. DIFFERENTIAL FORMS

PROPOSITION 6.5.7
If ω is a k-form on M, and $V_1, \ldots, V_{k+1}$ are smooth vector fields on M, then $d\omega$ is given by

$$d\omega(V_1, \ldots, V_{k+1}) = \sum_{i=1}^{k+1}(-1)^{i+1}V_i(\omega(V_1, \ldots, \widehat{V_i}, \ldots, V_{k+1}))$$
$$+ \sum_{i<j}(-1)^{i+j}\omega([V_i, V_j], V_1 \ldots, \widehat{V_i}, \ldots, \widehat{V_j}, \ldots, V_{k+1}),$$

where $\widehat{}$ means that the corresponding entry is omitted, and $[\cdot, \cdot]$ denotes the Lie bracket.

The above formula can be used as an alternative definition of the exterior derivative. In order to define $(d\omega)_p$ on a collection of $k+1$ vectors, one locally extends these vectors to vector fields and plugs them in the above formula. The outcome turns out to be independent of extensions, due to the fact that differential forms are tensors. The Lie bracket in the above formula ensures that the dependence on vector fields is linear. The properties (i), (ii), and (iii) from below can be easily verified. They uniquely define the exterior derivative operator.

PROPOSITION 6.5.8

(i) *If f is a 0-form on M, then df is the derivative of f;*

(ii) $d(\omega \wedge \eta) = d\omega \wedge \eta + (-1)^k \omega \wedge d\eta$, *provided ω is a k-form;*

(iii) $d(d\omega) = 0$;

(iv) $d(f^*\omega) = f^*(d\omega)$, *provided $f : M \to N$ is a smooth function and ω is a form on N.*

A form is called exact if is the exterior derivative of some other form. A form ω whose exterior derivative is zero is called a closed form. Any exact form is closed, by (iii). Any m-form on an m-dimensional manifold is automatically closed. The relationship between closed and exact forms is very important, as it can be used to describe the topology of the supporting manifold. The following theorem is basic to this relationship. A manifold M is said to be smoothly contractible to a point if there exists a smooth homotopy
$$H : M \times [0,1] \to M$$

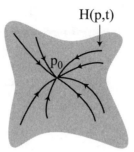

FIGURE 6.5.2
A manifold smoothly contractible to a point.

such that

$$H(\cdot, 1) = \mathrm{id}_M, \qquad (6.5.2)$$
$$H(\cdot, 0) = p_0,$$

for some point $p_0 \in M$. A manifold smoothly contractible to a point is shown in Figure 6.5.2.

THEOREM 6.5.9 (Poincaré lemma)

Let M be a manifold smoothly contractible to a point and $\omega \in \Lambda^k(TM)$ be a k-form on M, with $k \geq 1$. Then ω is exact if and only if it is closed.

PROOF If ω is exact, it is obviously closed, since $d \circ d = 0$. For the converse, let us first define a family of embeddings i_t of the manifold M into the manifold with boundary $M \times [0,1]$ by $i_t(x) = (x, t)$, where $t \in [0,1]$.

Part 1. We claim that there exists an operator $P : \Lambda^k(T(M \times [0,1])) \to \Lambda^{k-1}(TM)$ such that

$$P \circ d + d \circ P = i_1^* - i_0^*. \qquad (6.5.3)$$

This part of the proof is concerned with the relationship between forms on M and forms on $M \times [0,1]$, and does not need M to be contractible.

Step 1. We compute the operator on the right side of (6.5.3) applied to a form expressed in local coordinates.

6.5. DIFFERENTIAL FORMS

Any k-form Ω on $M \times [0,1]$, can be expressed, with respect to a local coordinate system $(x,t) = (x_1, \ldots, x_n, t)$ on $M \times [0,1]$, as

$$\Omega = \sum_{i_1 < \cdots < i_k} f_{i_1 \ldots i_k}(x,t) dx_{i_1} \wedge \ldots \wedge dx_{i_k} \qquad (6.5.4)$$

$$+ \sum_{j_1 < \cdots < j_{k-1}} g_{j_1 \ldots j_{k-1}}(x,t) dt \wedge dx_{j_1} \wedge \ldots \wedge dx_{j_{k-1}}.$$

Thus Ω is a sum of terms of the type

$$f(x,t) dx_{i_1} \wedge \ldots \wedge dx_{i_k} \text{ and } g(x,t) dt \wedge dx_{j_1} \wedge \ldots \wedge dx_{j_{k-1}}.$$

Applying $i_0^* - i_1^*$ to these terms we obtain

$$(i_1^* - i_0^*)(f(x,t) dx_{i_1} \wedge \ldots \wedge dx_{i_k}) \qquad (6.5.5)$$
$$= f(x,1) dx_{i_1} \wedge \ldots \wedge dx_{i_k} - f(x,0) dx_{i_1} \wedge \ldots \wedge dx_{i_k},$$

and

$$(i_1^* - i_0^*) g(x,t) dt \wedge dx_{j_1} \wedge \ldots \wedge dx_{j_{k-1}} = 0. \qquad (6.5.6)$$

Step 2. We define the linear operator P and apply $P \circ d + d \circ P$ to Ω. We define

$$P\Omega = \sum_{j_1 < \cdots < j_{k-1}} \left[\int_0^1 g_{j_1 \ldots j_{k-1}}(x,t) dt \right] dx_{j_1} \wedge \ldots \wedge dx_{j_{k-1}}.$$

Note that P vanishes on forms of the type $f(x,t) dx_{i_1} \wedge \ldots \wedge dx_{i_k}$.
Applying $P \circ d + d \circ P$ to each type of term in Ω, we obtain

$$(P \circ d + d \circ P)(f(x,t) dx_{i_1} \wedge \ldots \wedge dx_{i_k}) \qquad (6.5.7)$$
$$= (P \circ d)(f(x,t) dx_{i_1} \wedge \ldots \wedge dx_{i_k})$$
$$= P\left(\sum_j \frac{\partial f}{\partial x_j}(x,t) dx_j \wedge dx_{i_1} \wedge \ldots \wedge dx_{i_k} \right)$$
$$+ P\left(\frac{\partial f}{\partial t}(x,t) dt \wedge dx_j \wedge dx_{i_1} \wedge \ldots \wedge dx_{i_k} \right)$$
$$= \left[\int_0^1 \frac{\partial f}{\partial t}(x,t) dt \right] dx_j \wedge dx_{i_1} \wedge \ldots \wedge dx_{i_k}$$
$$= f(x,1) dx_{i_1} \wedge \ldots \wedge dx_{i_k} - f(x,0) dx_{i_1} \wedge \ldots \wedge dx_{i_k},$$

and

$$(P \circ d + d \circ P)\big(g(x,t)dt \wedge dx_{j_1} \wedge \ldots \wedge dx_{j_{k-1}}\big) \qquad (6.5.8)$$

$$= P\left(-\sum_j \frac{\partial g}{\partial x_j}(x,t)dt \wedge dx_j \wedge dx_{j_1} \ldots \wedge dx_{i_k}\right)$$

$$+ d\left(\left[\int_0^1 g(x,t)dt\right] dx_{j_1} \wedge \ldots \wedge dx_{j_{k-1}}\right)$$

$$= -\sum_j \left[\int_0^1 \frac{\partial g}{\partial x_j}(x,t)dt\right] dx_j \wedge dx_{j_1} \wedge \ldots \wedge dx_{j_{k-1}}$$

$$+ \sum_j \left[\int_0^1 \frac{\partial g}{\partial x_j}(x,t)dt\right] dx_j \wedge dx_{j_1} \wedge \ldots \wedge dx_{j_{k-1}}$$

$$= 0.$$

Comparing (6.5.5) and (6.5.6) with (6.5.7) and (6.5.8) concludes the proof of the claim made in Part 1.

Part 2. *We now prove that if ω is closed, then it is exact, provided M is contractible to a point.*

For a homotopy H with $H \circ i_1 = $ id and $H \circ i_0 = p_0$, we have $i_1^* \circ H^*(\omega) = \omega$ and $i_0^* \circ H^*(\omega) = 0$. We apply (6.5.3) from Part 1 to the form $\Omega = H^*\omega$ on $M \times [0,1]$. Using the facts that $d \circ H^* = H^* \circ d$, and $d\omega = 0$, we obtain

$$\omega = (d \circ P + P \circ d)(H^*\omega) = d(PH^*\omega).$$

This shows that ω is exact. ∎

See Exercise 6.11.3 for another way to look at Part 1 of the above proof. The condition on the manifold to be contractible to a point is important for this theorem. On the punctured plane $\mathbb{R}^2 \setminus \{(0,0)\}$ (which is not contractible to any point), there exist closed forms that are not exact (see Exercise 6.11.5). On the other hand, the contractibility condition is sufficient, but not necessary: on the sphere S^2 (which is not contractible to any point), any closed 1-form is exact, as we will see in Section 6.8.

6.6 Integration of differential forms

Integration of differential forms on manifolds naturally extends integration of functions on Euclidean spaces.

Suppose that M is an oriented m-dimensional manifold and ω is a differential m-form whose support $\mathrm{supp}(\omega) = \mathrm{cl}\{p \in M \mid \omega_p \neq 0\}$ is compact. We can choose a countable and locally finite family of charts $\{V_\alpha\}_\alpha$ of M, compatible with the orientation of M. The integral of ω on M is first defined on each chart V_α, and then extended to M through a partition of unity.

Suppose first that $\mathrm{supp}(\omega)$ is contained in some coordinate neighborhood V_α. Then ω can be uniquely written as

$$\omega = \sum_{i_1 < \cdots < i_m} f^\alpha_{i_1 \ldots i_n} dx^\alpha_1 \wedge \ldots \wedge dx^\alpha_{i_m},$$

with respect to the corresponding coordinate system $(x^\alpha_1, \ldots, x^\alpha_m)$. We define

$$\int_M \omega = \sum_{i_1 < \ldots < i_m} \int_{V_\alpha} f^\alpha_{i_1 \ldots i_m} dx^\alpha_1 \ldots dx^\alpha_{i_m},$$

where the integral on the right-hand side is the standard Riemannian integral. For this definition to be consistent, we need to show that it is independent of the choice of a local coordinate system. Suppose that $\mathrm{supp}(\omega)$ is contained in two coordinate neighborhoods V_α and V_β, and the corresponding coordinate systems $(x^\alpha_1, \ldots, x^\alpha_m)$ and $(x^\beta_1, \ldots, x^\beta_m)$ are consistent with the orientation of M. Hence the derivative $d\theta_{\alpha\beta}$ of the reparametrization has a positive determinant. Let

$$\omega = \sum_{i_1 < \ldots < i_m} f^\beta_{i_1 \ldots i_n} dx^\beta_1 \wedge \ldots \wedge dx^\beta_{i_m}$$

be the expression of ω with respect to the coordinate system on V_β. We have $f^\beta_{i_1 \ldots i_n} = \det(d\theta_{\alpha\beta}) \cdot f^\alpha_{i_1 \ldots i_n} = |\det(d\theta_{\alpha\beta})| \cdot f^\alpha_{i_1 \ldots i_n}$. Then, by the change of variable formula from calculus, we conclude

$$\sum_{i_1 < \ldots < i_m} \int_{V_\alpha} f^\beta_{i_1 \ldots i_m} dx^\beta_1 \ldots dx^\beta_{i_m}$$
$$= \sum_{i_1 < \ldots < i_m} \int_{V_\beta} |\det(d\theta_{\alpha\beta})| f^\alpha_{i_1 \ldots i_m} dx^\beta_1 \ldots dx^\beta_{i_m}$$

$$= \sum_{i_1<\ldots<i_m} \int_{V_\alpha} f^\alpha_{i_1\ldots i_m} dx^\alpha_1 \ldots dx^\alpha_{i_m}.$$

If supp(ω) is not contained in a single coordinate neighborhood, ω can be broken up using a partition of unity $\{g_\alpha\}_\alpha$ subordinate to $\{V_\alpha\}_\alpha$ as

$$\omega = \sum_\alpha (g_\alpha \omega).$$

By the local finiteness property of the partition of unity, the sum in the above equation has finitely many non-zero terms when evaluated at any point of M.

Finally we define

$$\int_M \omega = \sum_\alpha \int_M (g_\alpha \omega). \qquad (6.6.1)$$

We need to show that this definition does not depend on the choice of a partition of unity. Suppose that $\{h_\beta\}_\beta$ is another partition of unity, subordinate to a coordinate neighborhood covering $\{W_\beta\}_\beta$. Since $\sum_\alpha g_\alpha = \sum_\beta h_\beta = 1$, we have

$$\sum_\beta \int_M h_\beta \omega = \sum_\beta \int_M \left[\sum_\alpha h_\beta g_\alpha \omega\right] = \sum_{\beta,\alpha} \int_M h_\beta g_\alpha \omega$$

$$= \sum_\alpha \int_M \left[\sum_\beta h_\beta g_\alpha \omega\right] = \sum_\alpha \int_M g_\alpha \omega.$$

Integration of differential m-forms with compact support defines a linear functional, since

$$\int_M (\omega_1 + \omega_2) = \int_M \omega_1 + \int_M \omega_2, \quad \text{and} \quad \int_M c\omega = c \int_M \omega.$$

The following change of variable formula for forms follows in a manner similar to the above discussion.

PROPOSITION 6.6.1

If $f : M \to N$ is a diffeomorphism between two orientable m-dimensional manifolds M and N, and ω is a differential m-form with compact support on N, then

$$\int_N \omega = \pm \int_M f^* \omega, \qquad (6.6.2)$$

6.6. INTEGRATION OF DIFFERENTIAL FORMS

with positive sign if f is orientation preserving, and negative sign if f is orientation reversing.

Example 6.6.2
Volume forms were defined in Example 6.5.4. If ω is a volume form on a compact oriented manifold M, then the volume of M is

$$\int_M \omega.$$

In general, two different volume forms ω_1, ω_2 yield two different values of the volume. A sufficient condition to obtain the same value of the volume is that $\omega_1 = f^*\omega_2$ for some orientation preserving diffeomorphism f on M. We will see in Section 6.8 that this condition is also necessary.

A Riemannian manifold has a 'natural' volume form. Suppose that M is an oriented Riemannian m-dimensional manifold, (g_{ij}) is a Riemannian metric on M, $p \in M$, and $(v_1, \ldots, v_m)$ is a positively oriented orthonormal basis of T_pM. Define

$$\omega_p : \underbrace{T_pM \times \cdots \times T_pM}_{m \text{ times}} \to \mathbb{R},$$

by setting $\omega_p(v_1, \ldots, v_m) = 1$. If $(w_1, \ldots, w_m)$ is another positively oriented orthonormal basis of T_pM, we have $w_i = \sum_{i=1}^n a_{ij}v_j$, for some orthogonal matrix $(a_{ij})_{i,j}$, with $\det(a_{ij}) = 1$. Hence $\omega_p(w_1, \ldots, w_m) = \det(a_{ij})\omega_p(v_1, \ldots, v_m) = \omega_p(v_1, \ldots, v_m)$, so the definition of ω_p is independent of the orthonormal basis. Since it is possible to locally choose smoothly varying positively oriented orthonormal bases of the tangent spaces, we can extend the definition of ω to the entire manifold. If we want to express this form with respect to a local coordinate system $(x_1, \ldots, x_m)$, first we note that $\partial/\partial x_i = \sum_{j=1}^m b_{ij}v_j$, with the matrix $B = (b_{ij})$ satisfying $BB^t = (g_{ij})$. If $G = \det(g_{ij})$, then $\omega(\partial/\partial x_1, \ldots, \partial/\partial x_m) = \det(b_{ij})\omega(v_1, \ldots, v_n) = \sqrt{G}$. In conclusion, the canonical volume form on a compact orientable Riemannian manifold is

$$\omega = \sqrt{G}\, dx_1 \wedge \ldots \wedge dx_m.$$

□

Example 6.6.3
We define an $(n-1)$-form σ on $\mathbb{R}^n \setminus \{0\}$, called the solid angle form, whose restriction to the unit sphere S^{n-1} is an $(n-1)$-volume form.

In dimension $n = 2$, the solid angle form is defined by $\sigma_p = d\theta = (-y\,dx + x\,dy)/(x^2 + y^2)$, where $p = (x, y)$. It is only the usual notion of (signed) angle.

For higher dimensions, we start by defining an $(n-1)$-form σ' on S^{n-1} by
$$\sigma'_p(v_1, \ldots, v_{n-1}) = (dx_1 \wedge \ldots \wedge dx_n)(p, v_1, \ldots, v_{n-1}),$$
for any $v_1 \ldots, v_{n-1} \in T_p S^{n-1}$. Expanding the right-side determinant, we obtain
$$\sigma' = \sum_{j=1}^{n} (-1)^{j-1} x_j \, dx_1 \wedge \ldots \wedge \widehat{dx_j} \wedge \ldots \wedge dx_n.$$

Recall that a basis $(v_1, \ldots, v_{n-1})$ of $T_p S^{n-1}$ is positively oriented provided $(p, v_1, \ldots, v_{n-1})$ is positively oriented in $\mathbb{R}^n$. Since σ' is positive on any positively oriented basis of $T_p S^{n-1}$, σ' is a volume form on S^{n-1}.

We extend the definition of σ' to the whole of $\mathbb{R}^n \setminus \{0\}$ by pulling it back through the natural retraction $r : \mathbb{R}^n \setminus \{0\} \to S^{n-1}$ given by $r(p) = p/\|p\|$. We obtain the $(n-1)$-form $\sigma = r^* \sigma'$, called the solid angle form. Its expression in coordinates is given by
$$\sigma = \sum_{j=1}^{n} \frac{(-1)^{j-1} x_j}{(x_1^2 + \cdots + x_n^2)^{n/2}} \, dx_1 \wedge \ldots \wedge \widehat{dx_j} \wedge \ldots \wedge dx_n.$$

It is clear that this form is closed; however, it is not exact (Exercise 6.11.6). Its restriction on S^{n-1} is obviously the volume form σ', so $\int_{S^{n-1}} \sigma > 0$. One can actually obtain the 'surface area' (i.e., the $(n-1)$-dimensional volume) formula of the $(n-1)$-sphere
$$\int_{S^{n-1}} \sigma = \begin{cases} \dfrac{2\pi^{n/2}}{(n/2 - 1)!} & \text{if } n \text{ is even,} \\ \dfrac{2^{(n+1)/2} \pi^{(n-1)/2}}{(n-2)!!} & \text{if } n \text{ is odd,} \end{cases}$$
for $n \geq 2$. See Exercise 6.11.7. □

We can integrate k-forms ($k \leq m$) on k-dimensional submanifolds of M. Suppose that N is a k-dimensional oriented submanifold of M, $i : N \hookrightarrow M$ is the natural embedding, and ω is a k-form whose support intersects N in a compact subset. Then $i^*\omega$ is a k-form with compact support on N, so we can define
$$\int_{i(N)} \omega = \int_N i^*\omega.$$

6.7. STOKES' THEOREM

Example 6.6.4
Suppose that $c : \mathbb{R} \to \mathbb{R}^3$ is an embedded curve describing the trajectory of a particle of unit mass, subject to a force field of components $F_1(x_1, x_2, x_3)$, $F_2(x_1, x_2, x_3)$, $F_3(x_1, x_2, x_3)$. The work done by the particle along the path is given by

$$W = \int_{\mathbb{R}} (F_1(c(t))c'_1(t) + F_2(c(t))c'_2(t) + F_3(c(t))c'_3(t)) \, dt.$$

Since $c^*(dx_i) = dc_i = c'_i(t)dt$, for $i = 1, 2, 3$, the above integral can be interpreted as the path integral

$$W = \int_{c(\mathbb{R})} F_1 dx_1 + F_2 dx_2 + F_3 dx_3.$$

□

6.7 Stokes' theorem

The Fundamental Theorem of Calculus, translated in the language of forms, states that the integral of a 1-form $f(x)dx$ on the interval $[a, b] \subseteq \mathbb{R}$ is $F(b) - F(a)$, provided $dF(x) = f(x)dx$. The integration domain $M = [a, b]$ is a manifold with boundary $\{a, b\}$. The positive orientation of $[a, b]$ is given by the positive direction. At a, the outward direction is opposite to the positive direction, hence its orientation is negative, while at b, the outward direction agrees with the positive direction, hence its orientation is positive. If we define the oriented boundary $\partial[a, b] = \{b\} - \{a\}$, the Fundamental Theorem of Calculus can be summarized as

$$\int_M dF = \int_{\partial M} F.$$

Stokes' theorem extends the fundamental theorem of calculus to the higher dimensional case.

THEOREM 6.7.1 (Stokes' theorem)
Suppose that M is an m-dimensional oriented manifold with boundary, and ω is a differential $(m-1)$-form with compact support on M. Then

$$\int_M d\omega = \int_{\partial M} \omega.$$

Each component of ∂M has its orientation induced by M. If $\partial M = \emptyset$, then the right-hand side is considered to be 0.

PROOF By using a partition of unity, we can reduce the argument to the case when the support of ω is contained in some coordinate neighborhood V, with corresponding coordinate system $(x_1, \ldots, x_m)$. With respect to these coordinates, we can write

$$\omega = \sum_{j=1}^{m} (-1)^{j-1} f_j dx_1 \wedge \ldots \wedge \widehat{dx_j} \wedge \ldots \wedge dx_m,$$

where the factor $(-1)^{j-1}$ is added only for convenience, and $\widehat{}$ marks an omitted term. Through local coordinates, the computation will be actually carried out on $\mathbb{R}^m$, tacitly using the formula (6.6.2).

We have

$$d\omega = \sum_{j=1}^{m} (-1)^{j-1} \frac{\partial f_j}{\partial x_j} dx_j \wedge dx_1 \wedge \ldots \wedge \widehat{dx_j} \wedge \ldots \wedge dx_m$$

$$= \left(\sum_{j=1}^{m} \frac{\partial f_j}{\partial x_j} \right) dx_1 \wedge \ldots \wedge dx_j \wedge \ldots \wedge dx_m.$$

There are two cases: $V \cap \partial M = \emptyset$ and $V \cap \partial M \neq \emptyset$.

Case 1. $V \cap \partial M = \emptyset$.

Thus $\int_{\partial M} \omega = 0$ since $\mathrm{supp}(\omega) \subseteq V$ and so $\omega = 0$ on ∂M. On the other hand, using Fubini's theorem, we have

$$\int_M d\omega = \sum_{j=1}^{m} \int_{\mathbb{R}^m} \frac{\partial f_j}{\partial x_j} dx_1 \ldots dx_j \ldots dx_m$$

$$= \sum_{j=1}^{m} \int_{\mathbb{R}^{m-1}} \left[\int_{\mathbb{R}} \frac{\partial f_j}{\partial x_j} dx_j \right] dx_1 \ldots \widehat{dx_j} \ldots dx_m$$

$$= \sum_{j=1}^{m} \int_{\mathbb{R}^{m-1}} \left[f_j(x_1, \ldots, x_m) \Big|_{x_j=-\infty}^{x_j=+\infty} \right] dx_1 \ldots \widehat{dx_j} \ldots dx_m$$

$$= 0.$$

In the above computation we replaced the domain of integration M with $\mathbb{R}^m$ since f_j can be extended by zero outside M; the last identity holds since f_j vanishes for any sufficiently large values of $|x_j|$.

6.7. STOKES' THEOREM

Case 2. $V \cap \partial M \neq \emptyset$.

We choose a local coordinate system consistent with the orientation of M, such that

$$V \cap M \subseteq \{(x_1, \ldots, x_m) \mid x_m \geq 0\}, \text{ and } V \cap \partial M \subseteq \{(x_1, \ldots, x_m) \mid x_m = 0\}.$$

By a computation similar to the above, we have

$$\int_M d\omega = \sum_{j=1}^m \int_M \frac{\partial f_j}{\partial x_j} dx_1 \ldots dx_j \ldots dx_m$$

$$= \sum_{j=1}^{m-1} \int_{\mathbb{R}^{m-1}} \left[\int_{\mathbb{R}} \frac{\partial f_j}{\partial x_j} dx_j \right] dx_1 \ldots \widehat{dx_j} \ldots dx_m$$

$$+ \int_{\mathbb{R}^{m-1}} \left[\int_0^\infty \frac{\partial f_m}{\partial x_m} dx_m \right] dx_1 \ldots dx_{m-1}$$

$$= 0 + \int_{\mathbb{R}^{m-1}} \left[\int_0^\infty \frac{\partial f_m}{\partial x_m} dx_m \right] dx_1 \ldots dx_{m-1}$$

$$= \int_{\mathbb{R}^{m-1}} \left[f_m(x_1, \ldots, x_m) \Big|_{x_m=0}^{x_m=+\infty} \right] dx_1 \ldots dx_{m-1}$$

$$= -\int_{\mathbb{R}^{m-1}} f_m(x_1, \ldots, x_{m-1}, 0) dx_1 \ldots dx_{m-1}.$$

See Figure 6.7.1. On the other hand,

$$\int_{\partial M} \omega = \int_{V \cap \partial M} \omega$$

$$= \sum_{j=1}^m (-1)^{j-1} \int_{V \cap \partial M} f_j dx_1 \wedge \ldots \wedge \widehat{dx_j} \wedge \ldots \wedge dx_m$$

$$= (-1)^{m-1} \int_{V \cap \partial M} f_m dx_1 \wedge \ldots \wedge dx_{m-1}.$$

In the above computation we used the fact that f_j vanishes for any sufficiently large values of $|x_j|$, $j = 1, \ldots, m-1$. Notice that the natural embedding

$$(x_1, \ldots, x_{m-1}) \in \mathbb{R}^{m-1} \to (x_1, \ldots, x_{m-1}, 0) \in \partial\{(x_1, \ldots, x_m) \mid x_m \geq 0\}$$

may or may not be orientation preserving. The positive orientation on $\partial\{(x_1, \ldots, x_m) \mid x_m \geq 0\}$ is induced by $\{-\partial/\partial x_m, \partial/\partial x_1, \ldots, \partial/\partial x_{m-1}\}$,

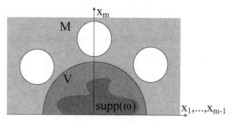

FIGURE 6.7.1
Proof of Stokes' theorem.

whose orientation is $(-1)^m$ times that induced by $\{\partial/\partial x_1, \ldots, \partial/\partial x_m\}$. Hence, according to (6.6.2), $\int_{V \cap \partial D} f_m dx_1 \wedge \ldots \wedge dx_{m-1}$ differs from $\int_{\mathbb{R}^{m-1}} f_m(x_1, \ldots, x_{m-1}, 0) dx_1 \ldots dx_{m-1}$ by a $(-1)^m$ sign. We obtain

$$\int_{\partial M} \omega = (-1)^{2m-1} \int_{\mathbb{R}^{m-1}} f_m(x_1, \ldots, x_{m-1}, 0) dx_1 \ldots dx_{m-1},$$

which concludes the proof. ∎

Stokes' theorem provides a simple way to show that a form is not exact.

COROLLARY 6.7.2
Assume that M is an m-dimensional oriented manifold (without boundary). If

$$\int_M \omega \neq 0,$$

then ω is not exact.

On any compact oriented manifold (without boundary), there is always an m-form ω whose support is contained in some coordinate neighborhood with $\int_M \omega > 0$. If $(x_1, \ldots, x_m)$ are local coordinates on V that agree with the orientation of M, one can define $\omega = f dx_1 \wedge \ldots \wedge dx_m$, where $f > 0$ on V and vanishes outside V. Such a form is not exact by the previous corollary. From the Poincaré lemma we obtain:

COROLLARY 6.7.3
A compact orientable manifold (without boundary) is not contractible to a point.

6.7. STOKES' THEOREM

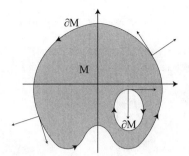

FIGURE 6.7.2
An oriented region in $\mathbb{R}^2$.

The classical Green's, Gauss' and Stokes' formulas from calculus follow from the general form of Stokes' theorem.

Example 6.7.4
(i) Suppose that M is a bounded region of $\mathbb{R}^2$, whose orientation agrees with that of $\mathbb{R}^2$. The orientation on ∂M is the positive orientation induced by M. Suppose that f and g are smooth functions on M. Green's formula asserts

$$\int_{\partial M} f dx + g dy = \int_M d(f dx + g dy) = \int_M \left(\frac{\partial g}{\partial x} - \frac{\partial f}{\partial y} \right) dx \wedge dy$$
$$= \int_M \left(\frac{\partial g}{\partial x} - \frac{\partial f}{\partial y} \right) dx dy.$$

Figure 6.7.2 shows how to orient the boundary of a region M.

(ii) Suppose that M is a bounded region in space whose orientation is consistent with that of $\mathbb{R}^3$. The outward normal direction of the boundary surface ∂M induces an orientation on ∂M. Suppose that f, g, and h are smooth functions on M. Gauss' formula asserts

$$\int_{\partial M} f dy dz + g dz dx + h dx dy = \int_M d(f dy \wedge dz + g dz \wedge dx + h dx \wedge dy)$$
$$= \int_M \left(\frac{\partial f}{\partial x} + \frac{\partial g}{\partial y} + \frac{\partial h}{\partial z} \right) dx dy dz.$$

(iii) Suppose that M is a bounded region of an oriented embedded surface S in $\mathbb{R}^3$. The positive orientation on ∂M is defined by the 'right-hand rule': when the fingers of the right hand point in the positive direction of the boundary, the thumb points in the direction of the positive normal vector to S. Let f, g, and h be smooth functions on S. Stokes' formula asserts

$$\int_{\partial M} f\,dx + g\,dy + h\,dz = \int_M d(f\,dx + g\,dy + h\,dz)$$
$$= \int_M \left[\left(\frac{\partial h}{\partial y} - \frac{\partial g}{\partial z}\right) dy\,dz \right.$$
$$\left. + \left(\frac{\partial f}{\partial z} - \frac{\partial h}{\partial x}\right) dz\,dx + \left(\frac{\partial g}{\partial x} - \frac{\partial f}{\partial y}\right) dx\,dy\right].$$

□

As an application of Stokes' theorem, we give an alternative proof of the no-retraction theorem — Theorem 2.2.6.

THEOREM 6.7.5 (No-retraction theorem)
There is no continuous retraction $r : \bar{B}^n \to S^{n-1}$.

PROOF Assume, by contradiction, that such a retraction r exists.
Step 1. Using r, we produce a smooth mapping $h : \mathbb{R}^n \to \mathbb{R}^n \setminus \{0\}$ such that $h(x) = x$ outside of some neighborhood of 0.
  We define $R : \bar{B}^n \to \mathbb{R}^n \setminus \{0\}$ by

$$R(x) = \begin{cases} r(2x) & \text{if } \|x\| < 1/2, \\ x & \text{otherwise}. \end{cases}$$

This function is continuous, since $R(x) \to x$ as $\|x\| \to 1/2$ from either direction. Let $f : \mathbb{R}^n \to \mathbb{R}$ be a smooth even function (i.e. $f(-x) = f(x)$ for all x), whose support is contained in $\|x\| < 1/4$, and satisfying $\int_{\mathbb{R}^n} f(x)\,dx = 1$. Define

$$h(x) = \int_{\mathbb{R}^n} R(y) f(x - y)\,dy.$$

It is not difficult to show that, although R is only continuous, h is as smooth as f is. For $\|x\| > 3/4$, we can have $(x - y) \in \mathrm{supp}(f)$ only if $\|y\| > 1/2$, so

$$h(x) = \int_{\mathbb{R}^n} R(y) f(x-y)\,dy = \int_{\mathbb{R}^n} y f(x-y)\,dy$$
$$= x \int_{\mathbb{R}^n} f(x-y)\,dy - \int_{\mathbb{R}^n} (x-y) f(x-y)\,dy.$$

Since $y \to yf(y)$ is an odd function (i.e., $(-y)f(-y) = -(yf(y))$ for all y), we have

$$\int_{\mathbb{R}^n} (x-y)f(x-y)dy = -\int_{\mathbb{R}^n} yf(y)dy = 0.$$

Hence $h(x) = x$ for all $\|x\| > 3/4$.

Step 2. *We show that the existence of the function h constructed at Step 1 leads to a contradiction.*

Consider the solid angle form σ in $\mathbb{R}^n \setminus \{0\}$, discussed in Example 6.6.3. On the domain $\|x\| > 3/4$ on which $h(x) = x$, we have $h^*\sigma = \sigma$. The solid angle form satisfies

$$\int_{S^{n-1}} \sigma = \text{Area}(S^{n-1}) > 0.$$

On the other hand, using Stokes' theorem and the fact that σ is closed, we have

$$\int_{S^{n-1}} \sigma = \int_{S^{n-1}} h^*\sigma = \int_{\partial \bar{B}^n} h^*\sigma$$
$$= \int_{\bar{B}^n} d(h^*\sigma) = \int_{\bar{B}^n} h^*(d\sigma) = 0.$$

Contradiction. ∎

More applications of Stokes' theorem can be found in Abraham, Marsden and Ratiu (1988).

6.8 De Rham cohomology

Loosely speaking, the Poincaré lemma says that the 'shape' of a manifold decides whether or not every closed form is also exact. Stokes' theorem describes a duality between the boundary operator acting on manifolds with boundary, and the exterior derivative operator, acting on differential forms on that manifold. By combining these ideas, one can obtain quite detailed information on the the shape of a manifold in terms of the relationship between its closed forms and its exact forms.

For each $k \geq 0$, the closed k-forms form a vector space, and the exact k-forms form a subspace of the closed k-forms, since $d \circ d = 0$. In order to measure the extent to which the closed k-forms fail to be exact, we consider the quotient space

$$H^k(M) = \text{closed } k\text{-forms on } M/\text{exact } k\text{-forms on } M.$$

Two closed k-forms ω_1 and ω_2 are in the same equivalence class if and only if if there exists a $(k-1)$-form η with $\omega_2 - \omega_1 = d\eta$. Denote by $[\omega]$ the equivalence class of ω. The quotient space $H^k(M)$ is a vector space over the real numbers with the induced operations

$$[\omega_1] + [\omega_2] = [\omega_1 + \omega_2], \text{ and } a[\omega_1] = [a\omega_1].$$

DEFINITION 6.8.1
The vector space $H^k(M)$ is called the k-th de Rham cohomology group. The dimension β_k of the vector space $H^k(M)$ is called the k-th Betti number (provided it is finite).

In many instances, the Betti numbers are symmetric.

THEOREM 6.8.2 (Poincaré duality)
If M is a compact connected orientable m-dimensional manifold, then $H^k(M)$ and $H^{m-k}(M)$ have the same dimension.

See Spivak (1999) for a proof.

Example 6.8.3
(i) A closed 0-form on a manifold is a smooth function of zero derivative, so it is constant on each connected component of M. Since no 0-form is exact, we conclude that $H^0(M) \simeq \mathbb{R}^q$, where q is the number of connected components.

(ii) If M is a manifold contractible to a point (for example an open ball $B(0, r) = \{x \in \mathbb{R}^n \mid \|x\| < r\}$ in $\mathbb{R}^n$, or $\mathbb{R}^n$ itself), then all cohomology groups of order greater than zero are trivial, that is $H^k(M) = 0$ for all $k \geq 1$. □

If two manifolds are diffeomorphic, their corresponding de Rham cohomology groups are isomorphic. For any smooth map $f : M \to N$ we have $d \circ f^* = f^* \circ d$, implying that f^* maps closed forms to closed forms and exact forms to exact forms. Thus, it induces a homomorphism $f^{*k} : H^k(N) \to H^k(M)$, for each $k \geq 0$. We denote by $H^*(M)$

6.8. DE RHAM COHOMOLOGY

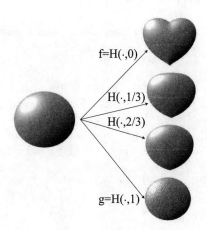

FIGURE 6.8.1
Homotopic maps.

and $H^*(N)$ the cohomology groups of an unspecified order, and by f^* the induced homomorphism. The following properties hold

$$\text{id}^* = \text{id},$$
$$(g \circ f)^* = f^* \circ g^*,$$

where $f : M \to N$ and $g : N \to P$ are smooth maps. In particular, if f is a diffeomorphism from M to N, then $f^* : H^*(N) \to H^*(M)$ is a linear isomorphism. This shows that the de Rham cohomology groups are differentiable invariants of a manifold. As we will see in Section 6.10, they are in fact topological invariants.

Now we investigate the effect of homotopies on the induced homomorphisms on the de Rham cohomologies. Two smooth maps $f, g : M \to N$ between two manifolds are said to be (smoothly) homotopic if there exists a smooth homotopy $h : M \times [0, 1] \to N$ such that

$$h(\cdot, 0) = f, \quad \text{and} \quad h(\cdot, 1) = g.$$

See Figure 6.8.1. Note that M is contractible to a point if and only if the identity mapping on M is homotopic to a constant mapping.

THEOREM 6.8.4 (Homotopy property of de Rham cohomology)
If $f, g : M \to N$ are smoothly homotopic, then the induced homomorphisms $f^*, g^* : H^*(N) \to H^*(M)$ are equal, $f^* = g^*$.

PROOF We need to prove that if ω is a closed form on M, then $f^*\omega \sim g^*\omega$, or $f^*\omega - g^*\omega$ is an exact form. Recall the family of embeddings $i_t : M \to M \times [0,1]$, $i_t(x) = (x,t)$, defined in the proof of the Poincaré lemma. By the first part of the proof of this lemma, we have $i_1^* - i_0^* = d \circ P + P \circ d$. Since $f = h \circ i_0$ and $g = h \circ i_1$, we have

$$g^*\omega - f^*\omega = (h \circ i_1)^*\omega - (h \circ i_0)^*\omega = i_1^*(h^*\omega) - i_0^*(h^*\omega)$$
$$= (d \circ P)(h^*\omega) + (P \circ d)(h^*\omega) = d(P(h^*\omega)) + P(h^*(d\omega))$$
$$= d(P(h^*\omega)),$$

since ω is closed. ∎

Example 6.8.5
We compute the cohomology groups of the n-dimensional sphere ($n \geq 1$):

$$H^k(S^n) = \begin{cases} \mathbb{R} & \text{if } k = 0, \\ 0 & \text{if } 0 < k < n, \\ \mathbb{R} & \text{if } k = n. \end{cases} \qquad (6.8.1)$$

That $H^0(S^n) = \mathbb{R}$ follows from the connectedness of the sphere. For the higher order groups, we proceed inductively.

Consider the mapping

$$[\omega] \in H^n(S^n) \to \int_{S^n} \omega \in \mathbb{R}. \qquad (6.8.2)$$

We will prove that this map is an isomorphism, and that every closed k-form on S^n, with $0 < k < n$, is exact.

The above mapping is obviously well defined and linear. It is also non-trivial, since for a volume form ω on S^n we have that $\int_{S^n} \omega \neq 0$. To prove that it is an isomorphism, we only need to verify its injectivity.

Case 1. Let $n = 1$.

Let ω be a 1-form on S^1, with $\int_{S^1} \omega = 0$. Then $\omega = f(\theta)d\theta$, for some periodic function f of period 2π. For all t we have

$$0 = \int_{S^1} \omega = \int_t^{t+2\pi} f(\theta)d\theta = F(t+2\pi) - F(t),$$

6.8. DE RHAM COHOMOLOGY

where F is an antiderivative of f. Therefore F is also 2π-periodic. We conclude that ω is exact with $\omega = dF$, for some 0-form F on S^1, so $[\omega] = 0$. The statement regarding $0 < k < n$ is vacuously true.

Case 2. Let $n > 1$. Assume that the mapping defined in (6.8.2) is an isomorphism in the case of S^{n-1}, and that that every closed k-form on S^{n-1}, with $0 < k < n-1$, is exact.

Let ω be a an n-form on

$$S^n = \{x \in \mathbb{R}^{n+1} \mid x_1^2 + \cdots + x_{n+1}^2 = 1\},$$

with $\int_{S^n} \omega = 0$. Let

$$H_N = \{x \in S^n \mid x_{n+1} \geq 0\}, \text{ and } H_S = \{x \in S^n \mid x_{n+1} \leq 0\},$$

be the closed upper and lower hemisphere, respectively, regarded as manifolds with boundary. The two hemispheres intersect along the equator, and they induce opposite orientations on it. The equator is an embedded copy of S^{n-1}, and so we will identify it with S^{n-1}. Consider the neighborhoods of the hemispheres

$$U_N = \{x \in S^n \mid x_{n+1} > -\epsilon\}, \text{ and } U_S = \{x \in S^n \mid x_{n+1} < \epsilon\},$$

for some small $\epsilon > 0$. See Figure 6.8.2. Since U_N and U_S are contractible to points, the Poincaré lemma implies that there exist $(n-1)$-forms η_N on U_N and η_S on U_S such that $\omega = d\eta_N$ on U_N and $\omega = d\eta_S$ on U_S. The difference $\eta_N - \eta_S$ is a closed $(n-1)$-form on $U_N \cap U_S$. Stokes' theorem yields

$$0 = \int_{S^n} \omega = \int_{H_N} d\eta_N + \int_{H_S} d\eta_S = \int_{S^{n-1}} \eta_N - \int_{S^{n-1}} \eta_S$$
$$= \int_{S^{n-1}} (\eta_N - \eta_S).$$

By the induction step, we obtain that $\eta_N - \eta_S$ is exact on S^{n-1}. Now, consider the mapping $r : U_N \cap U_S \to S^{n-1}$ that sends each point $p \in U_N \cap U_S$ along the meridian circle all the way to the equator. This map is homotopic to the identity map on $U_N \cap U_S$, and coincides with the identity on S^{n-1}. By the homotopy property of de Rham cohomology, we have $\text{id}^* = r^*$, so $(\eta_N - \eta_S) - r^*(\eta_N - \eta_S)$ is an exact form on $U_N \cap U_S$. The form $r^*(\eta_N - \eta_S)$ must be exact on $U_N \cap U_S$, because it is the image of the exact form $\eta_N - \eta_S$ on S^{n-1} through the homomorphism r^*. Hence $\eta_N - \eta_S$ is exact on $U_N \cap U_S$, so $\eta_N - \eta_S = d\gamma$ for some $(n-2)$-form γ on $U_N \cap U_S$.

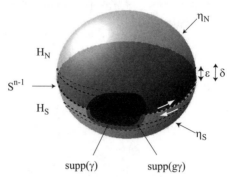

FIGURE 6.8.2
The cohomology groups of the sphere.

Since η_N and $\eta_S + d\gamma$ agree on $U_N \cap U_S$, we now only need to extend γ to the whole sphere. For $\delta > 0$ only slightly greater than ϵ, consider a bump function g on the sphere that is equal to 1 on $U_N \cap U_S$ and to 0 on $|x_{n+1}| > \delta$. Thus, $g\gamma$ coincides with γ on $U_N \cap U_S$, and can be extended by zero to the whole sphere. Define an $(n-1)$-form μ to be equal to η_N on U_N and to $\eta_S + d(g\gamma)$ on U_S. We have $d\mu = \omega$, so ω is exact. This shows that the mapping defined in (6.8.2) is an isomorphism for $k = n$.

Now let ω be a closed k-form on S^n, with $0 < k < n$. Let η_N and η_S be $(k-1)$-forms defined as before. As $d\eta_N = d\eta_S$ on $U_N \cap U_S$, the $(k-1)$-form $\eta_N - \eta_S$ is closed on $U_N \cap U_S$, so, in particular, is closed when restricted to the equator S^{n-1}. The induction step implies that it is exact there. By an argument similar to the above, it follows that ω is exact on S^n. □

We can give another proof to the no-retraction theorem.

THEOREM 6.8.6 (No-retraction theorem)
There exists no continuous retraction $r : \bar{B}_1^n \to S_1^{n-1}$.

PROOF Assume that such a retraction exists. If $i : S^{n-1} \to \bar{B}^n$ is the canonical inclusion, then $r \circ i = \mathrm{id}_{S^{n-1}}$. Passing to cohomology, we have $i^* \circ r^* = \mathrm{id}_{H^*(S^{n-1})}$. However, $H^{n-1}(S^{n-1}) = \mathbb{R}$, and $H^{n-1}(\bar{B}^n) = 0$, since the ball is contractible to a point. Thus $i^* = 0$, which is a contradiction. ∎

6.9 Singular homology

There is a more geometric way to describe the 'shape' of a manifold than through de Rham cohomology groups. We begin this section by briefly describing some other groups, called singular homology groups, that can be associated with a manifold, or, more general, with a topological space.

Let $\{p_0, \ldots, p_k\}$ be $k+1$ points in $\mathbb{R}^m$ with the property that the vectors $p_i - p_0$, $i = 1, \ldots, k$ are linearly independent. The smallest convex set $[p_0, \ldots, p_k]$ containing $\{p_0, \ldots, p_k\}$ is called a k-simplex in $\mathbb{R}^m$. See Figure 6.9.1. The points p_i are called the vertices of the simplex. The $(k-1)$-simplices $[p_0, \ldots, p_{i-1}, \widehat{p}_i, p_{i+1}, \ldots, p_k]$, where $\widehat{}$ designates an omitted vertex, are the $(k-1)$-faces of $[p_0, \ldots, p_k]$. Similarly, we can define its $(k-2)$-faces, $(k-3)$-faces, and so on.

Example 6.9.1
The simplex Δ_k of vertices $e_0 = (0, \ldots, 0)$, $e_1 = (1, 0, \ldots)$, $e_2 = (0, 1, \ldots)$, $\ldots$, $e_k = (0, \ldots, 1, \ldots, 0)$, is called the standard k-simplex. It consists of all $x \in \mathbb{R}^n$ with $0 \leq x_i \leq 1$ and $\sum_{i=1}^{k} x_i \leq 1$. □

Each point x of a simplex $[p_0, \ldots, p_k]$ can be uniquely written as

$$x = \sum_{i=0}^{k} x_i p_i,$$

with $0 \leq x_i \leq 1$ for all i, and $\sum_{i=0}^{k} x_i = 1$. The coefficients $(x_i)_{i=0,\ldots,k}$ are called the barycentric coordinates of p.

A given ordering of the vertices $(p_0, \ldots, p_k)$ of a k-simplex $[p_0, \ldots, p_k]$ determines an orientation on that simplex. Two orderings of the same set of vertices determine the same orientation provided that they differ by an even permutation. Any m-simplex $[p_0, \ldots, p_m]$ in $\mathbb{R}^m$ has a canonical orientation, defined by the sign of the determinant of the matrix formed by $(p_1 - p_0, \ldots, p_m - p_0)$. An oriented simplex $[p_0, \ldots, p_k]$ determines an orientation on each of its $(k-1)$-faces: the orientation of $[p_0, \ldots, p_{i-1}, \widehat{p}_i, p_{i+1}, \ldots, p_k]$ differs, by definition, from the orientation of $[p_0, \ldots, p_k]$ by a factor of $(-1)^i$. One can inductively define an orientation on faces of any order of a simplex.

One can always subdivide a given simplex $[p_0, \ldots, p_k]$ into arbitrarily small simplices of the same dimension. One way to do this is by using

264   6. TENSORS AND DIFFERENTIAL FORMS

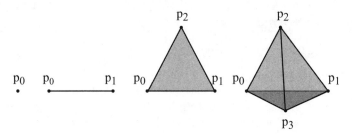

FIGURE 6.9.1
Simplices.

barycentric subdivision. The barycenter of the simplex $[p_0, \ldots, p_k]$ is the point b whose barycentric coordinates are all equal, that is, $x_i = 1/(k+1)$ for each i. The barycentric subdivision of the 0-simplex $[p_0]$ consists of $[p_0]$ alone. The barycentric subdivision of the 1-simplex $[p_0, p_1]$ consists of $[p_0, b]$ and $[b, p_1]$. Inductively, the k-subsimplices of the barycentric subdivision of $[p_0, \ldots, p_k]$ are given by all combinations $[b, q_0, \ldots, q_{k-1}]$, where b is the barycenter of $[p_0, \ldots, p_k]$, and $[q_0, \ldots, q_{k-1}]$ are the $(k-1)$-simplices obtained by the barycentric subdivision of all $(k-1)$-dimensional faces of $[p_0, \ldots, p_k]$. By repeated barycentric subdivision, one can systematically break any given simplex into subsimplices of arbitrarily small diameter. See Figure 6.9.5.

Let M be a smooth m-dimensional manifold. A singular k-simplex in M is a smooth map $s : \Delta_k \to M$. Let $S_k(M)$ be the collection of all formal finite sums $c = \sum_j a_j s_j$, where a_j are real numbers, and s_j are singular k-simplices. The elements of $S_k(M)$ are called the singular k-chains in M. One can naturally define addition and scalar multiplication of singular k-chains. These operations make $S_k(M)$ into a vector space.

For each $k \geq 1$, we define the boundary ∂s_k of a singular k-simplex s to be the singular $(k-1)$-chain

$$\sum_{i=0}^{k}(-1)^i s^i,$$

where, for each i, the $(k-1)$-simplex s^i (the i-th face of s) is given by

$$s^i = s\,|_{[p_0,\ldots,p_{i-1},\widehat{p}_i,p_{i+1},\ldots,p_k]}\,.$$

Example 6.9.2
A singular 0-simplex is only a point in M; its boundary is not defined.

6.9. SINGULAR HOMOLOGY

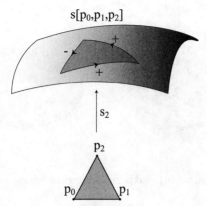

FIGURE 6.9.2
The boundary of a singular 2-simplex.

A singular 1-simplex is a path $s : [p_0, p_1] \to M$, and its boundary ∂s is the formal sum $s(p_1) - s(p_0)$. A singular 2-simplex $s : [p_0, p_1, p_2] \to M$ is a triangular region of M, and ∂s is a sum of three paths with signs. The sign of an edge is chosen positive if the counterclockwise direction agrees with the natural orientation of that edge and is chosen negative otherwise. See Figure 6.9.2. □

We extend the boundary operator linearly to singular chains

$$\partial : S_k(M) \to S_{k-1}(M).$$

It is easy to verify that the boundary of the boundary of a chain is always zero, that is, $\partial \circ \partial = 0$.

One can integrate differential forms on singular simplices on a manifold. If s_0 is a singular 0-simplex (which is a point in M), and ω is a 0-form, we set

$$\int_s \omega = \omega(s(p_0)).$$

If s is a singular k-simplex in M, $k \geq 1$, and ω is a k-form, we define

$$\int_s \omega = \int_{\Delta_k} s^* \omega.$$

Since the right-hand side integral is calculated over a polyhedral body in $\mathbb{R}^m$, we do not need to assume that M is orientable. Integration of differential forms extends by linearity to singular chains.

FIGURE 6.9.3
Manifolds without and with holes covered by singular chains.

In this context, Stokes' theorem becomes:

THEOREM 6.9.3 (Stokes' theorem for singular chains)
Suppose that c is singular k-chain on a manifold M, and ω is a differential $(k-1)$-form on M. Then

$$\int_c d\omega = \int_{\partial c} \omega.$$

See Warner (1971) for a proof. Roughly speaking, one applies Stokes' theorem — Theorem 6.7.1— to each simplex in the chain and then sums the results.

Singular chains can be used to describe the 'shape' of a manifold. We start with an intuitive description. In a region where the manifold has no 'holes', every closed singular chain is the boundary of some chain of higher dimension contained in the manifold. Near a 'hole', one can always find a closed singular chain that is not the boundary of any chain of higher dimension contained in that manifold. See Figure 6.9.3.

We call a singular n-chain c_n closed if $\partial c_n = 0$, and exact if $c = \partial \widetilde{c}$, for some singular $(k+1)$-chain $\widetilde{c}$. The closed singular k-chains form a vector space. Since $\partial \circ \partial = 0$, the exact singular k-chains form a subspace of the closed singular k-chains. We define the (smooth) k-th singular homology group of M as

$H_k(M)$ = closed singular k-chains/exact singular k-chains.

We denote the equivalence class of a singular k-chain c by $[c]$. Each singular homology group has a vector space structure, as a quotient of

vector spaces. Any smooth mapping $f : M \to N$ between two manifolds induces a homomorphism $f_* : H_*(M) \to H_*(N)$ between their homology groups, defined by
$$f_*([c]) = [f \circ c],$$
for $[c] \in H_*(M)$.

REMARK 6.9.4 The simplest way to compute the homology of a manifold is by using a triangulation. A triangulation of a manifold is a collection of singular simplices of dimension equal to the dimension of the manifold such that the intersection of any two simplices is either disjoint or a face of each, and any point in the manifold has a neighborhood that intersects only finitely many simplices. Any smooth manifold is triangulable. Given a triangulation, one can use its simplices and their faces of various orders to compute the homology groups. The homology groups are independent of the triangulation used in computation (see Hatcher (2002) for a proof). ∎

Example 6.9.5
We represent the torus $\mathbb{T}^2$ as a singular 2-chain as in Figure 6.9.4. That is, we take a rectangle with the opposite sides identified, and we break it into two triangles. We put the standard 2-simplex in correspondence with each of these two triangles, obtaining two singular 2-simplices s and s'. Due to identifications, there is only one distinct vertex p, and only three distinct edges α, β, and γ. We have

$$S_0(\mathbb{T}^2) = \{ap \,|\, a \in \mathbb{R}\},$$
$$S_1(\mathbb{T}^2) = \{a\alpha + b\beta + c\gamma \,|\, a, b, c \in \mathbb{R}\},$$
$$S_2(\mathbb{T}^2) = \{as + bs' \,|\, a, b \in \mathbb{R}\}.$$

There are no singular 3-chains. All chains in $S_0(\mathbb{T}^2)$ are closed and exact, so $[p]$ is a generator for $H_0(\mathbb{T}^2)$, and thus $H_0(\mathbb{T}^2) = \mathbb{R}$. All chains in $S_1(\mathbb{T}^2)$ are closed, and they are generated by α, β, γ, or, equivalently, $\alpha, \beta, \alpha + \beta - \gamma$. Since $\partial s = \partial s' = \alpha + \beta - \gamma$, all exact 1-chains are generated by $\alpha + \beta - \gamma$. It follows that $H_2(\mathbb{T}^2)$ is generated by $[\alpha]$ and $[\beta]$, and thus $H_1(\mathbb{T}^2) = \mathbb{R} \times \mathbb{R}$. The closed chains in $S_2(\mathbb{T}^2)$ are generated by $s - s'$, and there are no exact ones. Hence $H_2(\mathbb{T}^2)$ is generated by $[s] - [s']$, and thus $H_2(\mathbb{T}^2) = \mathbb{R}$. □

For an extended treatment of singular homology, see Hatcher (2002).

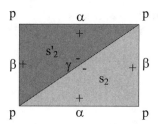

FIGURE 6.9.4
The torus represented as a singular 2-chain.

REMARK 6.9.6 We can define singular homology groups with coefficients in some other abelian groups besides $\mathbb{R}$. For example, one can define $S_k(M)$ as the collection of all formal finite sums $c = \sum_j a_j s_j$, where a_j are integer numbers, and s_j are singular k-simplices. Starting from here, the homology groups

$$H_k(M; \mathbb{Z}) = \text{closed singular } k\text{-chains/exact singular } k\text{-chains},$$

are vector spaces over $\mathbb{Z}$. Most of the theory of singular homology with real coefficients carries over to singular homology with integer coefficients (see Hatcher (2002)). ∎

We end this section with another proof of the Brouwer fixed point theorem which relies on the theory of simplices. We start with a combinatorial lemma due to Sperner, which concerns labeling of triangulations. Suppose that $[p_0, \ldots, p_k]$ is a k-simplex. Label the vertex p_i by i for $0 \leq i \leq k$. Consider a finite triangulation of $[p_0, \ldots, p_k]$, that is, a partition into finitely many k-simplices with the property that the intersection of any two simplices of the triangulation is either empty or a face of each. The vertices of the triangulation are labeled by numbers from $\{0, \ldots, k\}$, so that the following conditions are satisfied:

(i) if a vertex p lies on a face of $[p_0, \ldots, p_k]$ labeled by $i_0, \ldots, i_s$, then p is is assigned one of the labels $i_0, \ldots, i_s$;

(ii) if a vertex p does not lie on a face of $[p_0, \ldots, p_k]$, then p is labeled by an arbitrary number from $\{0, \ldots, k\}$.

Such a labeling is referred as a Sperner labeling. A k-simplex is said to be completely labeled if its $k + 1$ vertices are labelled with all numbers from $\{0, \ldots, k\}$.

6.9. SINGULAR HOMOLOGY

LEMMA 6.9.7 *(Sperner's lemma)*
If the vertices of a triangulation of a k-simplex $[p_0, \ldots, p_k]$ have a Sperner labeling, then the number of simplices from the triangulation that are completely labeled is odd. See Figure 6.9.5.

PROOF The proof goes by induction on the dimension $k \geq 1$ of the simplex.

For $k = 1$, let $[p_0, p_1]$ be a 1-simplex with p_0 labeled as 0, and p_1 labeled as 1. Consider a subdivision of $[p_0, p_1]$ into 1-simplices, and a labeling of all interior vertices by $\{0, 1\}$, according to the first labeling condition. We count the number of vertices labelled 0 in the following way: we imagine the 1-simplices from the partition separated from one another, and then count the number of 0 labels. This means that the 0 label at p_0 is counted once, but any vertices with the 0 label that are in the interior of the interval $[p_0, p_1]$ contribute twice to the count. The count must be an odd number. Now let a be the number of 1-simplices from the partition whose vertices are both labeled 0, and b be the number of 1-simplices from the partition with one vertex labeled 0 and the other vertex labeled 1. Then our count must be equal to $2a + b$. Since the count is odd, b is odd.

Now assume that the statement of the lemma is true for all $(k-1)$-simplices, $k \geq 2$. We consider a k-simplex $[p_0, \ldots, p_k]$ with a triangulation whose vertices have a Sperner labeling. We now count the number of $(k-1)$-faces of simplices from the partition that are labelled with all numbers from $\{0, \ldots, k-1\}$ by imagining that the k-simplices are separated from one another and then counting simplex by simplex. This means that a face which lies in the boundary of $[p_0, \ldots, p_k]$, and therefore lies in only one simplex of the partition, contributes once to the count if it is completely labeled. The conditions on the labeling ensure that such faces can lie only in $[p_0, \ldots, p_{k-1}]$. The number of these faces is odd by the induction hypothesis. On the other hand, a face which is in the interior of $[p_0, \ldots, p_k]$, and therefore lies in two simplices from the partition, either contributes twice to the count, if it is completely labeled, or it does not contribute at all if it is not completely labelled. The number of these faces is even. Thus the total count is odd.

Observe that if a k-simplex has a $(k-1)$-face labelled $\{0, \ldots, k-1\}$, then the remaining vertex is either labelled with k, or with some i in $\{0, \ldots, k-1\}$. In the former case, the k-simplex has one $(k-1)$-face labelled $\{0, \ldots, k-1\}$. In the later case there are two such $(k-1)$-faces $\{0, \ldots, k-1\}$. Now let a be the number of k-simplices from the partition

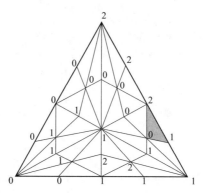

FIGURE 6.9.5
Subdivision of a triangle into triangles.

with two $(k-1)$-faces labelled $\{0, \ldots, k-1\}$, and b be the number of k-simplices from the partition whose vertices are labelled with all numbers from $\{0, \ldots, k\}$. Then the count made above must be equal to $2a + b$. Thus b must be odd because the total count is odd. ∎

Note that the above proof shows that the number of completely labelled simplices with the same orientation as the original simplex is exactly one plus the number of completely labelled simplices with the opposite orientation.

We use Sperner's lemma to prove the Brouwer fixed point theorem. It is also easy to use the Brouwer theorem to prove Sperner's lemma (see Yoseloff (1974)).

THEOREM 6.9.8 (Brouwer fixed point theorem)

Any continuous mapping of the closed unit ball into itself has a fixed point.

PROOF Since the closed unit n-dimensional ball is homeomorphic to an n-simplex, we can assume that f is a continuous mapping of an n-simplex $[p_0, \ldots, p_n]$ into itself. Let $(x_0, \ldots, x_n)$ denote the barycentric coordinates of $x \in [p_0, \ldots, p_n]$, and $(f_0(x), \ldots, f_n(x))$ the barycentric coordinates of $f(x) \in [p_0, \ldots, p_n]$. For each $i \in \{0, \ldots, n\}$, we define

$$A_i = \{x \in [p_0, \ldots, p_n] \mid f_i(x) \leq x_i\}.$$

A point x is in A_i if f maps x 'away' from p_i. It is impossible for a point to be mapped 'away' from all vertices $p_0, \ldots, p_n$, for the sum of the barycentric coordinates of a point is always equal to 1. Hence the closed sets A_i cover all of $[p_0, \ldots, p_n]$.

For every $m \geq 1$, we triangulate the simplex into finitely many small n-simplices with diameters less than $1/K$. For example, one can use repeated barycentric subdivisions to achieve this. If at the m-th of the subdivision there is a vertex z with $f_i(z) = z_i$ for all i, then z is a fixed point of f and there is nothing left to prove. If not, for each vertex z there exists a smallest i with $f_i(z) < z_i$. Label the vertex z by i. According to this labeling, the points $p_0, \ldots, p_n$ are labeled by $0, \ldots, n$, respectively. Moreover, if a vertex z lies on a face $[q_0, \ldots, q_s]$ of $[p_0, \ldots, p_k]$, and this face is already labeled by $i_0, \ldots, i_s$, respectively, then $z_i = 0$ and so $f_i(z) \geq z_i$ for all $i \neq i_0, \ldots, i_s$. It follows that $f_i(z) < z_i$ for some $i \in \{i_0, \ldots, i_s\}$, thus z is assigned one of the labels $\{i_0, \ldots, i_s\}$. This agrees with the first Sperner labeling condition.

Due to Sperner's lemma, at each step m of the subdivision we can find and fix a completely labeled simplex from the triangulation. Each of its vertices carries a label from $\{0, \ldots, n\}$. Call z_i^m the vertex of this completely labeled subsimplex carrying the label i. Note that $z_i^m \in A_i$ for all i. By compactness, all these sequences $(z_i^m)_{i \geq 0}$ contain convergent subsequences. Since the diameters of the corresponding completely labeled subsimplices tend to zero, these subsequences approach the same limit x. The sets A_i being closed, we have $x \in \bigcap_{i=0}^n A_i$, so $f_i(x) \leq x_i$ for all i. Since the barycentric coordinates of any point add up to 1, we must have $f_i(x) = x_i$ for all i, thus $f(x) = x$. ∎

6.10 The de Rham theorem

The de Rham theorem states that the information provided by the singular homology groups is essentially the same as the information provided by the de Rham cohomology groups.

THEOREM 6.10.1 (De Rham theorem)
Let M be a smooth m-manifold, $H^k(M)$ be the k-th de Rham cohomology group, $H_k(M)$ the k-th singular homology group, and $(H_k(M))^*$ the vector space of linear functionals on $H_k(M)$. Then, for each $0 \leq k \leq m$,

the mapping

$$[\omega] \in H^k(M) \to \alpha_k([\omega]) \in (H_k(M))^*,$$

given by

$$\alpha_k([\omega])[c] = \int_c \omega$$

is a linear isomorphism.

We will prove only the following special case of the de Rham theorem, and recommend Warner (1971) for a proof of the general theorem.

THEOREM 6.10.2
Let M be a compact and connected orientable m-dimensional manifold. The mapping

$$[\omega] \in H^m(M) \to \int_M \omega \in \mathbb{R}$$

is a linear isomorphism, and so $H^m(M) \simeq \mathbb{R}$.

PROOF It is clear that the above mapping is well defined, non-trivial and linear. We only need to verify the injectivity of the map. In other words, we need to show that an m-form ω on M must be exact if $\int_M \omega = 0$.

Step 1. We claim that if Ω is an m-form with compact support on $\mathbb{R}^m$ and $\int_{\mathbb{R}^m} \Omega = 0$, then Ω is the exterior derivative of a compactly supported $(m-1)$-form.

Consider the stereographic projection $p : S^m \to \mathbb{R}^m$. See Figure 6.10.1. The pull-back $p^*\Omega$ of Ω has its support in S^m disjoint from some small, contractible neighborhood U_N of the north pole. By the change of variable formula, we have

$$\int_{S^m} p^*\Omega = \int_{\mathbb{R}^m} \Omega = 0.$$

As in the proof of Example 6.8.5, we derive that $p^*\omega = d\eta$, for some $(m-1)$-form η on the sphere. Since $d\eta = 0$ on U_N, by the Poincaré lemma, $\eta = d\gamma$ for some $(m-2)$-form γ on U_N. Using a bump function, we can extend γ to an $(m-1)$-form on S^m, with support disjoint from that of $p^*\Omega$. Consider now the $(m-1)$-form $(\eta - d\gamma)$: its exterior derivative equals $p^*\Omega$ and its support is disjoint from U_N. Thus $(p^{-1})^*(\eta - d\gamma)$

6.10. THE DE RHAM THEOREM

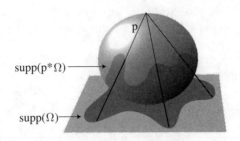

FIGURE 6.10.1
Pull-back of a form through stereographic projection.

is a compactly supported $(m-1)$-form on $\mathbb{R}^m$ whose exterior derivative is equal to Ω. This completes Step 1.

Due to compactness, we can choose a finite number of local coordinate systems $\phi_i : V_i \to \mathbb{R}^m$, $i = 1, \ldots, n$, such that $V_1, \ldots, V_n$ is an open cover of M.

Step 2. Consider an m-form ω on M whose support is contained in a single coordinate neighborhood V_i. We claim that ω must be exact if $\int_M \omega = 0$.

Since $0 = \int_{V_i} \omega = \int_{\mathbb{R}^m} \phi_i^* \omega$, Step 1 tells us that $\phi_i^* \omega$ is the exterior derivative of a compactly supported $(m-1)$-form on $\mathbb{R}^m$. It follows immediately that ω is exact.

Step 3. Suppose that $V_i \cap V_j \neq \emptyset$. We claim that, for any m-form α whose support lies in V_i, there exist an m-form β with support in V_j and an $(m-1)$-form η such that $\beta = \alpha + \eta$.

Let β_0 be the product of a volume form on M and a bump function whose support lies in the open set $V_i \cap V_j$. Then $\int_M \beta_0 > 0$, and we can choose $\beta = c\beta_0$, where

$$c = \int_M \alpha \bigg/ \int_M \beta_0.$$

It is clear that β is supported in V_j. Moreover, $\int_M (\beta - \alpha) = 0$, so by Step 2 there is an $(m-1)$-form η such that $\beta - \alpha = d\eta$.

Step 4. We claim that, for any m-form α whose support lies in V_i and any V_j, there exist an m-form β with support in V_j and an $(m-1)$-form η such that $\beta = \alpha + \eta$.

Since M is connected, there is a path from V_i to V_j. Cover this path with coordinate neighborhoods $V_{i_1} = V_i, V_{i_2}, \ldots, V_{i_k} = V_j$, such that consecutive neighborhoods in the sequence have nonempty intersection. Now do the obvious inductive argument using Step 3.

Step 5. Finally we claim that every m-form ω on M with $\int_M \omega = 0$ is exact.

Let $g_1, \ldots, g_n$, be a partition of unity that is subordinate to the covering $V_1, \ldots, V_n$. Since $g_i \omega$ is supported in V_i, we can use Step 4 to choose m-forms $\beta_1, \ldots, \beta_n$ and $(m-1)$-forms $\eta_1, \ldots, \eta_n$ such that β_i is supported in V_1 and $\beta_i = g_i \omega + \beta_i$ for $i = 1, \ldots, n$. Then

$$\omega = g_1 + \cdots + g_n \omega = \beta_1 + \cdots + \beta_n + d(\eta_1 + \cdots + \eta_n),$$

and hence

$$0 = \int_M \omega = \int_M (\beta_1 + \cdots + \beta_n).$$

Since $\beta_1 + \cdots + \beta_n$ is supported in V_1, Step 2 says that $\beta_1 + \cdots + \beta_n$ is exact. Hence ω is exact.

This completes the proof. ∎

A consequence of the de Rham theorem is that Poincaré duality also holds for homology groups. If M is a compact connected orientable m-dimensional manifold, then $H_k(M)$ and $H_{m-k}(M)$ have the same dimension.

REMARK 6.10.3 For non-compact manifolds, one can define a different type of cohomology, the compactly supported cohomology. The k-th compactly supported de Rham cohomology group $H_c^k(M)$ of M consists of the quotient of the vector space of the compactly supported closed k-forms, over the vector space of exterior derivatives of compactly supported $(k-1)$-forms. Note that the exterior derivatives of compactly supported $(k-1)$-forms are not the exact compactly supported k-forms. The compactly supported de Rham cohomology is, in general, different from the regular one: for example $H^n(\mathbb{R}^n) \simeq 0$, while $H_c^n(\mathbb{R}^n) \simeq \mathbb{R}$. One can show that for a connected non-compact m-manifold M, $H_c^m(M) \simeq \mathbb{R}$ if M is orientable and $H_c^m(M) \simeq 0$ if M is non-orientable, while $H^m(M) \simeq 0$ whether M is orientable or not (see Abraham, Marsden and Ratiu (1988)). ∎

6.10. THE DE RHAM THEOREM

We prove that if two volume forms give the same total volume, then one is the pull-back of the other through some diffeomorphism (see Moser (1965)).

THEOREM 6.10.4 *(Moser's theorem)*
If two compact and connected orientable manifolds are diffeomorphic and have the same total volume, then there is a diffeomorphism that preserves the volume element.

PROOF Suppose that $f : M \to N$ is an orientation preserving diffeomorphism, and ω_M and ω_N are volume forms on M and N, respectively. We need to prove that if $\int_M \omega_M = \int_N \omega_N$, then $\omega_M = g^*\omega_N$ for some (other) diffeomorphism $g : M \to N$. We can reduce this to the case of two volume forms on the same manifold by taking the pull back $f^*\omega_N$ of ω_N. By the change of variable formula, we have

$$\int_M \omega_M = \int_N \omega_N = \int_M f^*\omega_N.$$

The form $\delta = f^*\omega_N - \omega_M$ satisfies $\int_M \delta = 0$, so, by the previous theorem, we have $\delta = d\eta$ for some $(m-1)$-form η on M. The form $\omega_t = \omega_M + t\delta = (1-t)\omega_M + tf^*\omega_N$ is a volume form for all $t \in [0,1]$.

Note that the mapping $X_p \to (\omega_t)_p(X_p, \cdot)$, assigning to each vector in T_pM an $(m-1)$-form, is linear. It is also injective, since ω_t is a volume form. Since T_pM and $\Lambda^{m-1}(T_pM)$ have the same dimension, this mapping is actually an isomorphism. Hence, at each p, there exists a unique tangent vector $(X_t)_p$ such that $(\omega_t)_p((X_t)_p, \cdot) = -\eta$. For a given t, the vector $(X_t)_p$ depends smoothly on p; see Exercise 6.11.4 below. Thus, for each $t \in [0,1]$, there is a smooth vector field X_t on M, such that $\omega_t(X_t, \cdot) = -\eta$. Using the result of Exercise 6.11.3 from below, this translates into $i_{X_t}(\omega_t) = -\eta$. By Theorem 2.3.3, there exists a smooth flow $(s, p) \in \mathbb{R} \times M \to (\phi^t)(s, p) \in M$, $s \in \mathbb{R}$, with $d\phi^t/ds = X_t$ and $(\phi^t)_0 = \text{id}$, for all t. Using Exercise 6.11.3 again, we have

$$\frac{d}{dt}((\phi^t)^*\omega_t) = (\phi^t)^*(L_X\omega_t) + (\phi^t)^*\left(\frac{d}{dt}\omega_t\right)$$

$$= (\phi^t)^*(i_X d\omega_t + d i_X \omega_t) + (\phi^t)^*(\delta) = \frac{d}{dt}(-d\eta + \delta)$$

$$= 0.$$

Hence, $(\phi^t)^*(\omega_t)$ is constant, and so $(\phi^1)^*(\omega_1) = (\phi^0)^*(\omega_0)$, or $(\phi^1)^*(f^*\omega_N) = \omega_M$. The desired diffeomorphism is thus $f \circ \phi^1$. ∎

6.11 Exercises

6.11.1 Prove that the tangent bundle and the cotangent bundle on a manifold are isomorphic.

Hint: Consider a Riemannian metric $\langle \cdot, \cdot \rangle$, and define the function $(x, v) \in TM \to (x, \langle \cdot, v \rangle) \in TM^*$, and show that it is an isomorphism.

6.11.2 Let $\pi : E \to B$ be a fiber bundle with fiber space F. For every open neighborhood V of a point $x \in B$, the open set $\pi^{-1}(V)$ is called a tubular neighborhood of the fiber $\pi^{-1}(x)$. We say that π is tubular if for every open set $U \subset E$ containing a fiber, there is a tubular neighborhood of that fiber contained in U. That is, $\pi^{-1}(x) \subseteq \pi^{-1}(V) \subset U$.

(i) Prove that π is tubular if and only if π is a closed map.

(ii) Show that π tubular implies that π is a quotient map, i.e., a map with the property that $V \subseteq B$ is open if and only if $\pi^{-1}(V) \subseteq E$ is open.

(iii) Give an example of a quotient map that is not tubular.

6.11.3 The Lie derivative of differential forms is defined similarly to the Lie derivative of vector fields: if ω is a form and X is a vector field on a manifold M, the Lie derivative of ω along X at a point p is given by

$$(L_X(\omega))_p = \lim_{t \to 0} \frac{(\phi^t)^* \omega_{\phi^t(p)} - \omega_p}{t},$$

where ϕ^t is the local flow defined by X in a neighborhood of p. Show the following properties:

(i) $L_X(f) = X(f)$, provided f is a 0-form (smooth function) on M;

(ii) $L_X(\omega_1 + \omega_2) = L_X(\omega_1) + L_X(\omega_2)$ and $L_X(f\omega) = X(f)\omega + f L_X(\omega)$;

(iii) $L_X(d\omega) = d(L_X(\omega))$;

(iv) Cartan's magic formula: $L_X = i_X \circ d + d \circ i_X$ where $i_X : \Lambda^k(TM) \to \Lambda^{k-1}(TM)$ is the 'contraction' operator defined by

$$(i_X \omega)_p(V_1, \ldots, V_{k_1}) = \omega_p(X_p, V_1, \ldots, V_{k-1}).$$

6.11. EXERCISES

(v) With the notation from the proof of Theorem 6.5.9, show that

$$(P\Omega)(x) = \int_0^1 L_{\partial/\partial t}\Omega(x,t)dt.$$

Use this and (iv) to give another proof of equation (6.5.3).

6.11.4 Show that if ω is a smooth volume form and α is a smooth $(m-1)$-form on an m-dimensional manifold, then the vector field X defined by the equation $i_X\omega = \alpha$ is smooth.

6.11.5 Show that the form

$$\omega = \left(\frac{-y}{x^2+y^2}\right) dx + \left(\frac{x}{x^2+y^2}\right) dy,$$

defined on the punctured plane $\mathbb{R}^2 \setminus \{(0,0)\}$ is closed but is not exact.

Hint: Integrate this form along a circular path around the origin.

6.11.6 Show that the solid angle form σ on $\mathbb{R}^{n-1} \setminus \{0\}$ defined in Example 6.6.3 is not exact.

6.11.7 Show that, for any $n \geq 2$, we have

$$\int_{S^{n-1}} \sigma = \begin{cases} \dfrac{2\pi^{n/2}}{(n/2-1)!} & \text{if } n \text{ is even,} \\ \dfrac{2^{(n+1)/2}\pi^{(n-1)/2}}{(n-2)!!} & \text{if } n \text{ is odd,} \end{cases}$$

where σ is the solid angle form defined in Example 6.6.3.

Hint: Define the annulus

$$A_n = \{x \in \mathbb{R}^n \,|\, a < \|x\| < b\},$$

where $a < b$ are real numbers. Define the diffeomorphism $f : (a,b) \times S^{n-1} \to A_n$ by $f(t,x) = tx$. Show $f^*(e^{-\|x\|^2} dx_1 \wedge \ldots \wedge dx_n) = t^{n-1} e^{-t^2}(dt \wedge \sigma)$. Derive

$$\int_{\mathbb{R}^n} e^{-\|x\|^2} dx_1 \wedge \ldots \wedge dx_n = \int_a^b t^n e^{-t^2} dt \cdot \int_{S^{n-1}} \sigma,$$

and pass $a \to 0$, $b \to \infty$ to deduce the desired formula.

6.11.8 If $(q_0,\ldots,q_n)$ are non-negative integers obtained by permuting $(0,\ldots,n)$, we define $\text{sign}(q_0,\ldots,q_n)$ as the sign of the permutation; if $(q_0,\ldots,q_n)$ are not obtained by permutating $(0,\ldots,n)$, then we set $\text{sign}(q_0,\ldots,q_n)$ to be 0. For example $\text{sign}(0,1) = 1 = -\text{sign}(1,0)$, and $\text{sign}(0,0) = \text{sign}(0,2) = 0$. If $[p_0,\ldots,p_n]$ is an n-simplex whose vertices are labeled by $q_0,\ldots,q_n$, then we define $\text{sign}[p_0,\ldots,p_n] = \text{sign}(q_0,\ldots,q_n)$. This definition extends linearly to labeled chains.

Prove the following version of Sperner's lemma for chains. If c is an n-chain whose vertices are labeled from $\{0,\ldots,n\}$, then $\text{sign}(c) = (-1)^n \text{sign}(\partial(c))$. In particular, if the vertices of a triangulated surface with boundary are assigned labels from $\{0,1,2\}$, then the number of completely labeled triangles counted with orientation equals the number of edges labeled $\{0,1\}$ on the boundary counted with orientation.

Hint: Reduce the problem to the case of a simplex s. In this case, one needs only to show that

$$\text{sign}(q_0,\ldots,q_n) = (-1)^n \sum_{j=0}^{n} (q_0,\ldots,q_{j-1},\widehat{q_j},q_{j+1},\ldots,q_n).$$

Write the left side as the determinant of some 'permutation' matrix and evaluate its determinant.

6.11.9 Use the above version of Sperner's lemma to prove the fundamental theorem of algebra.

Hint: Divide the complex plane into three sectors T_i, given by

$$T_i = \{z \,|\, 2\pi i/3 \leq \arg(z) < 2\pi(i+1)/3\},$$

$i = 0,1,2$. Consider a large disk in the complex plane on which the polynomial $p(z) \sim z^n$. Assign to each point z a label i provided $p(z) \in T_i$. Triangulate the disk and proceed as in the proof of the Brouwer fixed point theorem.

Chapter 7

Fixed Points and Intersection Numbers

7.1 Introduction

Finding fixed points is closely related to solving equations. Indeed, x is a solution of $f(x) = y$ if and only if x is a fixed point of the mapping $F(x) = f(x) + x - y$. Let us consider an equation of the type $f(x) = y$, where y is a given point in the plane, and f is a continuous mapping of a disk D into the plane, which distorts the disk in some complicated way. We are interested in conditions on the behavior of f restricted to the boundary of D that guarantee that $f(D)$ contains y. For this purpose, we define the 'winding number' which counts how many times f wraps the boundary of D around y.

We start by describing the winding number of a closed curve relative to a point. Let $\gamma : [0,1] \to \mathbb{R}^2$ be a smooth closed curve (not necessarily simple), and p be a point not on γ. We define the mapping $u_\gamma : [0,1] \to S^1$ by

$$u_\gamma(t) = \frac{\gamma(t) - p}{\|\gamma(t) - p\|}, \tag{7.1.1}$$

and count the number of complete turns made by $u_\gamma(t)$ around the circle as t goes from 0 to 1. The counterclockwise turns are counted as positive, and the clockwise turns are counted as negative. See Figure 7.1.1. Let $\theta(t)$ be the angle coordinate. Then $u_\gamma(t) = (\cos\theta(t), \sin\theta(t))$. The function θ is not uniquely determined, since we can add or subtract multiples of 2π, but its derivative is well defined. The angle swept out by $u_\gamma(t)$ is $\int_0^1 \theta'(t)dt$. Thus, the signed number of complete turns is given by $W(\gamma, p) = (1/2\pi)\int_0^1 \theta'(t)dt$; it is referred to as the winding number

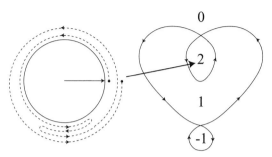

FIGURE 7.1.1
The winding numbers of various points relative to a given curve.

of γ relative to p. For example, if $\gamma(t) = (\cos(2\pi nt), \sin(2\pi nt))$, with n an integer, then $W(\gamma, p) = n$ for every point p inside the circle, and $W(\gamma, p) = 0$ for every point p outside the circle.

Going back to the definition of $W(\gamma, p)$, let us assume, for simplicity, that $p = 0$. Recall the angle form on $\mathbb{R}^2 \setminus \{0\}$ given by

$$\sigma_p = d\theta = \frac{-xdy + ydx}{x^2 + y^2}.$$

Then

$$W(\gamma, p) = \frac{1}{2\pi} \int_\gamma d\theta = \frac{1}{2\pi} \int_0^1 \gamma^* d\theta.$$

Since $d\theta$ is exact, the above integral depends only on the values of θ at the extremities, which can only differ by an integer multiple of 2π. This implies that the winding number is always an integer. It is easy to see that if γ and γ' are homotopic relative to $\mathbb{R}^2 \setminus \{0\}$, then, by the homotopy property of the de Rham cohomology (Theorem 6.8.4) we have $\gamma^* d\theta = (\gamma')^* d\theta$, so

$$W(\gamma, p) = W(\gamma', p).$$

We can also define the winding number for continuous curves. We chose a partition $0 = t_0 < t_1 < \cdots < t_{n-1} < t_n = 1$ of $[0, 1]$ such that $\gamma([t_i, t_{i+1}])$ is contained in a sector of angle less than 2π, as $i = 0, \ldots, n - 1$. Within each sector containing $\gamma([t_i, t_{i+1}])$, we can define a continuous function θ_i that measures the angle swept out by $\gamma(t)$ counterclockwise. Then we define

$$W(\gamma, p) = \frac{1}{2\pi} \sum_{i=0}^{n-1} \theta_i(\gamma(t_{i+1})) - \theta_i(\gamma(t_i)).$$

7.1. INTRODUCTION

This definition is independent of the choice of a partition with the specified properties, and it extends the definition of the winding number of a smooth curve.

We now describe a combinatorial method to compute the winding number $W(\gamma, p)$ of a closed curve γ in $\mathbb{R}^2 \setminus \{0\}$ about $p = 0$. We divide $\mathbb{R}^2$ into three sectors T_i, given by

$$T_q = \{(r, \theta) \mid 2\pi q/3 \leq \theta < 2\pi(q+1)/3\},$$

for $q = 0, 1, 2$. To each point $\gamma(t)$ we assign a label q by the rule $\gamma(t) \in T_q$. We can choose a partition $0 = t_0 < t_1 < \cdots < t_{n-1} < t_n = 1$ of $[0,1]$ such that each segment $\gamma[t_i, t_{i+1}]$ has its endpoints $\gamma(t_i)$ and $\gamma(t_{i+1})$ on two different prongs separating the sectors, and $\gamma[t_i, t_{i+1}]$ does not intersect the third prong. For example, for a piece of the curve that intersects a prong infinitely many times as in Figure 7.1.2, we can choose t_1 as the first instant of time when γ meets that prong, and then choose $t_2 > t_1$ as the first instant of time when γ meets a different prong. In this way, the polygonal line $\gamma(t_0)\gamma(t_1)\ldots\gamma(t_n)$ has its vertices labeled by $q_0, \ldots, q_n$, where the labels are from $\{0, 1, 2\}$, and any two successive labels in the sequence are different. Recall from Exercise 6.11.8 that $\text{sign}(q_i, q_{i+1})$ is defined to be the sign of the permutation (q_i, q_{i+1}) provided $\{q_i, q_{i+1}\} = \{0, 1\}$, and 0 otherwise. The winding number is then given by

$$W(\gamma, p) = \sum_{i=0}^{n-1} \text{sign}(q_i, q_{i+1}).$$

We now show how to use the winding number to investigate the existence of solutions of an equation $f(x) = y$, where f is a continuous mapping of a disk into the plane. Let γ be the restriction of f to the boundary of the disk D. We claim that if $W(\gamma, y) \neq 0$, then the equation $f(x) = y$ has at least one solution $x \in D$.

We outline the proof. Assume, for simplicity, that $y = 0$. We triangulate the polygon $\gamma(t_0)\gamma(t_1)\ldots\gamma(t_n)$, and give labels to all vertices of the triangles according to the rule described earlier. By the version of the Sperner's lemma from Exercise 6.11.8, there is always a completely labeled triangle, i.e., with labels $\{0, 1, 2\}$. Choosing finer and finer triangulations, a compactness argument shows that there is a point x in D that has a completely labeled triangle in any sufficiently small neighborhood. For such a point x, we must have $f(x) = 0$. The reader is invited to compare this argument with the proof of the fundamental theorem of algebra, outlined in Exercise 6.11.9.

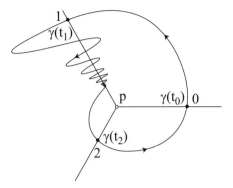

FIGURE 7.1.2
Combinatorial computation of the winding number.

In this chapter, we will extend the concept of winding number to higher dimensions, develop tools to identify and classify fixed points, and provide applications.

7.2 The Brouwer degree

In this section we present a homotopy invariant associated with a smooth mapping between manifolds of the same dimension. We will show how this invariant can be used to define the winding number in higher dimensional situations.

Let $f : M \to N$ be a smooth mapping between two oriented manifolds M and N, with M compact, N connected, and $\dim(M) = \dim(N)$. Let $\operatorname{sign}(df)_x$ be $+1$ if $(df)_x : T_x M \to T_{f(x)} N$ is orientation preserving, and be -1 if $(df)_x$ is orientation reversing. Let $y \in N$ be a regular value of f. Regular values always exist due to Sard's theorem 1.12.1. Then $f^{-1}(y)$ is a 0-dimensional submanifold of M (possibly empty). Since M is compact, $f^{-1}(0)$ is finite. We define the degree of f at y by

$$\deg(f, y) = \sum_{x \in f^{-1}(y)} \operatorname{sign}(df)_x. \qquad (7.2.1)$$

If $f^{-1}(y) = \emptyset$, we let $\deg(f, y) = 0$. If $y \in f(M)$ is a regular value, by the inverse function theorem 1.9.1, all points y' in a sufficiently small neighborhood of y are regular values as well. Moreover, $f^{-1}(y')$ has the same number of points as $f^{-1}(y)$. Indeed, if $f^{-1}(y) = \{x_1, \ldots, x_k\}$, then each x_i has a connected neighborhood U_i diffeomorphic to some neighborhood V_i of y. The closed set $f(M \setminus (U_1 \cup \ldots \cup U_k))$ is disjoint from y, and the points outside $f(M \setminus (U_1 \cup \ldots \cup U_k))$ have all of their preimages in $U_1 \cup \ldots \cup U_k$. We can choose the neighborhood U_i so that they are pairwise disjoint. Then $V = (V_1 \cap \ldots \cap V_k) \setminus f(M \setminus (U_1 \cup \ldots \cup U_k))$ is a neighborhood of y with the property that each point $y' \in V$ has exactly k preimages, one in each U_i. Since sign(df) is constant on each of the connected sets U_i, we see that $\deg(f, y) = \deg(f, y')$ for all $y' \in V$. Thus, $\deg(f, y)$ is a locally constant function of the regular values y. The argument described above is known as 'the stack of records argument'.

Now we would like to define the degree of a map as a global, homotopy invariant. This is preceded by some preparations.

LEMMA 7.2.1 (Extension lemma)
Assume that M is a compact, connected, orientable $(m+1)$-dimensional manifold with boundary ∂M, N is a connected, orientable m-dimensional manifold, $f : \partial M \to N$ is a smooth mapping, and y is a regular value for f. If f can be extended to a smooth mapping $F : M \to N$, then $\deg(f, y) = 0$.

PROOF Case 1. Assume that y is a regular value for both F and $f = F|_{\partial M}$.

In this case, $f^{-1}(y)$ is a finite number of points in ∂M and $F^{-1}(y)$ is a compact 1-dimensional manifold. There is a finite number of components of $F^{-1}(y)$, each of which is either diffeomorphic to a circle or is the image of a smooth map $\alpha : [0, 1] \to M$, according to Remark 1.11.5. The endpoints $\alpha(0)$ and $\alpha(1)$ must lie in ∂M, since otherwise y could not be a regular value of F. The circle components of $F^{-1}(y)$ cannot intersect ∂M, for otherwise y could not be a regular value of f. We see that the points of $f^{-1}(y)$ occur in pairs, each pair being the endpoints $\alpha(0)$ and $\alpha(1)$ of a smooth curve $\alpha : [0, 1] \to M$ that lies in $F^{-1}(y)$. We now show that

$$\text{sign}\left((df)_{\alpha(0)}\right) + \text{sign}\left((df)_{\alpha(1)}\right) = 0 \qquad (7.2.2)$$

for each such pair.

We can choose smooth vector fields $V_1(t), \ldots, V_m(t)$ along α so that $V_i(0)$ and $V_i(1)$ are tangent to ∂M for each i and $\alpha'(t), V_1(t), \ldots, V_m(t)$

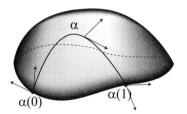

FIGURE 7.2.1
The extension lemma.

is a basis for $T_{\alpha(t)}M$ for each $t \in [0,1]$. Since $F \pitchfork \{y\}$, the vectors $dF(\alpha'(t)), dF(V_1(t)), \ldots, dF(V_m(t))$ must span $T_y N$ for each t. But $dF(\alpha'(t)) = 0$, and $T_y N$ is m-dimensional, so this is possible only if $dF(V_1(t)), \ldots, dF(V_m(t))$ is a basis for $T_y N$ for each t. It follows that the bases $dF(V_1(0)), \ldots, dF(V_m(0))$ and $dF(V_1(1)), \ldots, dF(V_m(1))$ have the same sign. On the other hand, the bases $V_1(0), \ldots, V_m(0)$ of $T_{\alpha(0)}(\partial M)$ and $V_1(1), \ldots, V_m(1)$ of $T_{\alpha(1)}(\partial M)$ must have opposite orientations in ∂M because one of the vectors $\alpha'(0)$ and $\alpha'(1)$ points into M and the other points out of M. This justifies (7.2.2). Summing these equations over all curve segments in $F^{-1}(y)$ yields $\deg(f, y) = 0$.

Case 2. Assume that y is a regular value only for f.

There exists a neighborhood U of y whose points are all regular values for f. By Sard's theorem 1.12.1, the regular values of F form an open dense set in N. Hence there exists a point $y' \in U$ that is a regular value for both F and f. From Case 1, it follows that $\deg(f, y') = 0$. Since the degree is a locally constant function of the regular values, we obtain $\deg(f, y) = 0$. ∎

The converse of the above lemma is also true, but we will not use it. See Hirsch (1976) for a proof.

PROPOSITION 7.2.2

If $g_0, g_1 : M \to N$ are smoothly homotopic maps that have a common regular value y, then $\deg(g_0, y) = \deg(g_1, y)$.

PROOF Let $F : M \times [0,1] \to N$ be a smooth homotopy with $F(\cdot, 0) = g_0$ and $F(\cdot, 1) = g_1$. The domain $M \times [0,1]$ of F is a region with boundary in $M \times \mathbb{R}$. See Figure 7.2.2. The orientation of the

7.2. THE BROUWER DEGREE

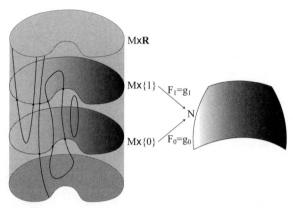

FIGURE 7.2.2
A homotopy between g_0 and g_1.

product determines the orientation of the boundary, so

$$\partial(M \times [0,1]) = M \times \{1\} - M \times \{0\}.$$

For any regular value y of $F|_{\partial(M \times [0,1])}$, we have

$$\begin{aligned}\deg(F|_{\partial(M\times[0,1])}, y) &= \deg(F(\cdot,1), y) - \deg(F(\cdot,0), y) \\ &= \deg(g_1, y) - \deg(g_0, y).\end{aligned}$$

Since $F|_{\partial(M \times [0,1])}$ is only a restriction of F, the previous lemma implies that $\deg(F|_{\partial(M\times[0,1])}, y) = 0$, hence $\deg(g_1, y) = \deg(g_0, y)$. ∎

Now we define $\deg(f, y)$, when y is not necessarily a regular value of f, by $\deg(g, y)$, where g is a map homotopic to f such that y is a regular value of g. The above proposition says that this definition is independent of the choice of a homotopic map.

PROPOSITION 7.2.3
If $f : M \to N$ is as above, then $\deg(f, y)$ is independent of $y \in N$.

PROOF We need to show that $\deg(f, y)$ is a locally constant function of the values $y \in N$. This is true if y is a regular value of f, by the stack of records argument. This is also true if y is a regular value for some g homotopic to y. To prove that $\deg(f, y)$ is locally constant, we only need to show that there exists a smooth g as C^0-close to f as we wish, such that y is a regular value of g. Let $\psi_\alpha : U_\alpha \to V_\alpha$ be a local coordinate system near $y \in N$, where y is not a regular value of f and V_α is relatively compact in N. Choose a compact neighborhood $V \subseteq V_\alpha$ of y. Let C be a constant vector field on V, which is the image under $d\psi_\alpha$ of a constant vector field on U_α. We can extend C to a smooth vector field on N by setting it equal to 0 outside of V_α, and smoothing out with some bump function on $V_\alpha \setminus V$. Let ϕ^t be the flow obtained by integrating this vector field. Since the vector field has compact support, ϕ^t is globally defined. The mapping $F : M \times [0,1] \to N$ given by $F(x,t) = \phi^t(f(x))$ defines a smooth homotopy with $F_0 = f$. In local coordinates, the mapping $\phi^t(f(x))$ is a translation in some neighborhood of y. By Sard's theorem 1.12.1, there exists an arbitrarily small C such that $\phi^1 \circ f$ has y as a regular value. We define $g = \phi^1 \circ f$. The map g is homotopic to f and has y as a regular value, which proves our claim. The conclusion of the proposition follows from the compactness of N. ∎

Now we can define the degree of a map.

DEFINITION 7.2.4
Let f be a smooth map from the compact, oriented m-dimensional manifold M to the connected, oriented m-dimensional manifold N. The degree $\deg(f)$ of f is defined as the degree $\deg(f, y)$ at any regular value y of f.

The definition of the degree has built in the homotopy invariance, so we immediately obtain:

PROPOSITION 7.2.5
If $f, g : M \to N$ are smoothly homotopic, then $\deg(f) = \deg(g)$.

PROOF By the above arguments, we have $\deg(f, y) = \deg(g, y)$, provided y is a regular value for both f and g. Such common regular values exist due to Sard's theorem 1.12.1. ∎

REMARK 7.2.6 Let us analyze the role of the conditions imposed on the manifolds in the definition of the Brouwer degree. The compact-

7.2. THE BROUWER DEGREE

ness of M ensures that the number of terms in (7.2.1) is finite. If N is not connected, then the degree is constant on each connected component of N. If M is not compact, one can define a 'local' degree $\deg(f, D, y)$ at each regular value y, and for each open subset D of M with $y \notin \partial D$ and $f^{-1}(y) \cap D$ a compact set, by

$$\deg(f, D, y) = \deg(f|_D, y).$$

The definition of the degree also relies on the orientability of M and N. If M and N are not orientable, then one can define the degree mod 2 of a map f by

$$\deg_2(f, y) = \text{number of points of } f^{-1}(y) \pmod{2},$$

where y is a regular value of f. It can be shown that the mod 2 degree is independent of the choice of a regular value y of f (see Milnor (1965)). ∎

Proposition 7.2.5 enables one to extend the definition of the degree to continuous maps between M and N. If $f : M \to N$ is a continuous map, we define $\deg(f)$ as the degree of any smooth map $g : M \to N$ which is continuously homotopic to f. The homotopy property remains valid if $f, g : M \to N$ are only continuously homotopic.

REMARK 7.2.7 An alternative definition of the degree of a map $f : M \to N$ can be given in terms of volume forms:

$$\deg(f) = \frac{\int_M f^*\omega}{\int_N \omega},$$

where ω is a volume form on N. To show that this is an equivalent definition, let y be a regular value of f, $f^{-1}(y) = \{x_1, \ldots, x_k\}$, V be a neighborhood of y, and U_i be a neighborhood of x_i diffeomorphic to V, for $i = 1, \ldots, k$. We start with a volume form η on N with support contained in V. Then $f^*\eta$ has its support contained in $\bigcup_{i=1}^n U_i$, and, due to the change of variable formula (6.6.2), we have $\int_{U_i} f^*\eta = \pm \int_V \eta$, depending on whether $f|_{U_i}$ is orientation preserving or not. We have

$$\int_M f^*\eta = \sum_{i=1}^k \int_{U_i} f^*\eta = \sum_{i=1}^k \text{sign}(df)|_{U_i} \int_V \eta = \deg(f) \int_N \eta.$$

If ω is an arbitrary volume form on N (N is assumed to be compact), then we can re-scale η so that $\int_N \omega = \int_N \eta$, or $\int_N (\omega - \eta) = 0$. By Step 1 of the proof of Theorem 6.10.2, we have $\omega - \eta = d\nu$, for some form $(m-1)$-form ν on N. Stokes' theorem 6.7.1 yields

$$\int_M f^*\omega = \int_M f^*\eta + \int_M f^*(d\nu) = \int_M f^*\eta + \int_M d(f^*\nu)$$
$$= \int_M f^*\eta + \int_{\partial M} f^*\nu = \deg(f)\int_N \eta = \deg(f)\int_N \omega.$$

Using this alternative definition of the degree, the homotopy property of the Brouwer degree follows directly from the homotopy property of the de Rham cohomology (Theorem 6.8.4): if f and g are homotopic, then $f^* = g^*$, thus $\int_M f^*\omega = \int_M g^*\omega$. ∎

REMARK 7.2.8 Another alternative definition of the degree can be obtained in terms of singular homology groups. The m-th cohomology group of a compact, connected, orientable m-manifold is 1-dimensional, as shown by Theorem 6.9.6. By the de Rham theory, the m-th singular homology group is also 1-dimensional. We can define singular homology groups with integer coefficients (see Remark 6.9.6). Fixing orientations, we have $H_m(M) \simeq \mathbb{Z}$ and $H_m(N) \simeq \mathbb{Z}$. A map $f : M \to N$ induces an automorphism $f_m : H_m(N) \to H_m(M)$, hence an automorphism of $\mathbb{Z}$. Any automorphism of $\mathbb{Z}$ is of the form $n \in \mathbb{Z} \to d \cdot n \in \mathbb{Z}$, for some unique $d \in \mathbb{Z}$. De Rham theory leads to $d = \deg(f)$. ∎

Example 7.2.9
(i) Consider the antipodal map $A : S^n \to S^n$ defined by $A(x) = -x$. Each $y \in S^n$ is a regular value, and $A^{-1}(y) = \{-y\}$. Since A is orientation preserving if n is odd, and orientation reversing if n is even, we have $\deg(A) = +1$ for n odd, and $\deg(A) = -1$ for n even.

(ii) Think of S^n ($n \geq 2$) as the set of all points $(z, x_3, \ldots, x_{n+1})$ with $z \in \mathbb{C}$ and $x_3, \ldots, x_{n+1} \in \mathbb{R}^n$ satisfying $|z|^2 + x_3^2 + \ldots + x_{n+1}^2 = 1$. We define a family of mappings of the sphere $r_k : S^n \to S^n$, $k \in \mathbb{Z}$, by

$$r_k(z, x_3, \ldots, x_{n+1}) = \begin{cases} \left(|z|\left(\dfrac{z}{|z|}\right)^k, x_3, \ldots, x_{n+1}\right), & \text{if } z \neq 0, \\ (0, x_3, \ldots, x_{n+1}), & \text{if } z = 0. \end{cases}$$

It is easy to check that $\deg(r_k) = k$, for each $k \in \mathbb{Z}$. Moreover, it can be shown that these mappings are, up to homotopy, all of the possible mappings of the sphere. ☐

7.2. THE BROUWER DEGREE

Example 7.2.10
The winding number of a closed curve $\gamma : [0,1] \to \mathbb{R}^2$ relative to a point p not on γ was introduced in the preceding section. Alternately, we can think of γ as a map from S^1 to $\mathbb{R}^2$. This induces a map $u_\gamma : S^1 \to S^1$, defined by (7.1.1). Since the winding number at p counts how many times $u_\gamma(t)$ moves around p counterclockwise minus how many times it moves around p clockwise, it follows that the winding number of γ at p equals the degree of the mapping u_γ. □

This example suggests how to define the winding number of a smooth mapping $f : M \to \mathbb{R}^m$ at a point $y \notin f(M)$, where M is a compact oriented m-dimensional manifold. Let $u_f : M \to S^{m-1}$ be given by

$$u_f(x) = \frac{f(x) - y}{\|f(x) - y\|}.$$

We define the winding number $W(f, y)$ of f relative to y as $\deg(u_f, y)$.

A simple yet powerful consequence of the definition of the degree of a map $f : M \to N$, is that if $\deg(f) \neq 0$, then f is surjective, i.e., $f(M) = N$. Otherwise, since non-values are regular values, we would have that $\deg(f) = \deg(f, y) = 0$ for any $y \in N \setminus f(M)$. We use this to give a proof to the Fundamental Theorem of Algebra (compare this with Exercise 6.11.9).

THEOREM 7.2.11 (The Fundamental Theorem of Algebra)
Any non-constant polynomial with complex coefficients has at least one complex root.

PROOF Using stereographic projection, we can extend the complex polynomial to a smooth map $f : S^2 \to S^2$. A point in S^2 is critical if and only if the complex derivative f' is zero at that point. Since the polynomial is non-constant, there exists a regular value $w \in S^2$, with $f^{-1}(w)$ non-empty. Let z be an arbitrary point in $f^{-1}(w)$. If $f = f_1 + if_2$, the fact that f is differentiable in the complex variable z is equivalent to the Cauchy-Riemann equations (see Rudin (1966)):

$$\frac{\partial f_1}{\partial x} = \frac{\partial f_2}{\partial y} \text{ and } \frac{\partial f_1}{\partial y} = -\frac{\partial f_2}{\partial x}.$$

This implies that the determinant of

$$(df)_z = \begin{pmatrix} \dfrac{\partial f_1}{\partial x} & \dfrac{\partial f_1}{\partial y} \\ \dfrac{\partial f_2}{\partial x} & \dfrac{\partial f_2}{\partial y} \end{pmatrix}$$

equals $(\partial f_1/\partial x)^2 + (\partial f_2/\partial x)^2$, and so is positive. Since this holds for any $z \in f^{-1}(w)$, it implies that $\deg(f, w) \neq 0$. It follows that f is surjective, and the equation $f(z) = 0$ has at least one solution.

Division with remainder applied repeatedly yields that any complex polynomial of algebraic degree n has n complex roots (not necessarily distinct). Therefore $\deg(f, w)$ is the number of roots of the equation $f(z) = w$, counted with multiplicity. ∎

As an application of the Brouwer degree, we give another proof of the hairy ball theorem.

THEOREM 7.2.12 (Hairy ball theorem)
On any even dimensional sphere S^n there is no continuous vector field of non-zero tangent vectors.

PROOF Let V be a nowhere zero continuous vector field on S^n. By replacing $V(x)$ to $V(x)/\|V(x)\|$, if necessary, we can assume that V is made of unit tangent vectors, i.e., $x \cdot V(x) = 0$ and $V(x) \cdot V(x) = 1$ for all $x \in M$. We construct the homotopy $F : S^n \times [0, 1] \to S^n$ by

$$F(x, t) = x \cos(\pi t) - V(x) \sin(\pi t).$$

The fact that $F(x, t)$ takes values on the sphere S^n can be easily verified:

$$(x \cos(\pi t) - V(x) \sin(\pi t)) \cdot (x \cos(\pi t) - V(x) \sin(\pi t))$$
$$= (x \cdot x) \cos^2(\pi t) + (V(x) \cdot V(x)) \sin^2(\pi t) = 1.$$

Note that $F(x, 0) = x$ and $F(x, 1) = -x$, so the identity and the antipodal map are homotopic. However, the degree of the identity is $+1$, but the degree of the antipodal map is $(-1)^{n-1}$. By Proposition 7.2.5 the degrees of $F(\cdot, 0)$ and of $F(\cdot, 1)$ must be equal, so S^n can carry a non-vanishing vector field only if n is odd. ∎

7.3 The oriented intersection number

The oriented intersection number of two oriented submanifolds of complementary dimensions in an ambient manifold is the signed number of intersection points. Each intersection point is counted with a positive sign if the combined orientations of the submanifolds agree with the orientation of the manifold, and is counted with a negative sign otherwise. The oriented intersection number can also be defined for transverse maps.

Suppose that $f : M \to N$ is a smooth mapping transverse to P, where M is a compact oriented manifold, N is an oriented manifold, and P is a closed submanifold of N with $\dim M + \dim P = \dim N$. Then $f^{-1}(P)$ is a finite collection of points in M. For any point $x \in f^{-1}(P)$, we have $(df)_x(T_xM) + T_{f(x)}P = T_{f(x)}N$. Thus for each $x \in f^{-1}(P)$ the image under df of a basis for T_xM and a basis for $T_{f(x)}P$ together form a basis for $T_{f(x)}N$. We define the oriented intersection number $I(f,P)_x$ to be $+1$ or -1 depending on whether we obtain a positively or negatively oriented basis for $T_{f(x)}N$ when we use positively oriented bases for T_xM and $T_{f(x)}P$ and put the basis of $T_{f(x)}P$ after the image of T_xM.

DEFINITION 7.3.1
The oriented intersection number of f with P is

$$I(f,P) = \sum_{x \in f^{-1}(P)} I(f,P)_x.$$

If M and N have equal dimensions and $P = \{y\}$, the condition of f being transverse to y is equivalent to y being a regular value for f. Thus $I(f, \{y\}) = \deg(f, y)$. The oriented intersection number is a homotopy invariant. Following the ideas in the proof of Proposition 7.2.5, one can show:

PROPOSITION 7.3.2
If $f, g : M \to N$ are both transverse to P and smoothly homotopic, then $I(f, P) = I(g, P)$.

For any smooth map $f : M \to N$ we can define the intersection number $I(f, P)$ of f with P as the intersection number $I(g, P)$ of a

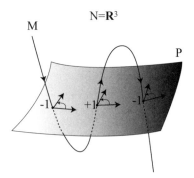

FIGURE 7.3.1
The intersection number of two submanifolds of $\mathbb{R}^3$. Here $I(M, P) = 1$.

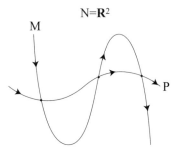

FIGURE 7.3.2
The intersection number of two submanifolds of $\mathbb{R}^2$. Here $I(M, P) = 1$ and $I(P, M) = -1$.

homotopy deformation g of f which is transverse to P. The above proposition implies that $I(f, P)$ is well defined.

For M and P compact submanifolds of complementary dimensions of N, we define the oriented intersection number $I(M, P)$ as the oriented intersection number of the canonical embedding $i : M \hookrightarrow P$ with P. See Figure 7.3.1. Since P is also compact, we can also compute $I(P, M)$. These numbers are not equal, in general. In fact,

$$I(M, P) = (-1)^{\dim(M)\dim(P)} I(P, M).$$

See Figure 7.3.2.

7.4 The fixed point index and the index of a vector field

In this section, we apply the techniques developed so far to analyze the behavior of a map or a flow near a fixed point. We are interested in fixed points that are stable under small perturbations.

A fixed point x_0 of a smooth mapping $f : M \to M$ of a manifold M is said to be isolated if there is an open neighborhood V of x_0 that contains no other fixed point of f besides x_0. Suppose first that $M = \mathbb{R}^m$, and $D \subseteq M$ is a diffeomorphic copy of an m-dimensional closed disk, with x_0 in the interior of D, and $D \subseteq V$. Define the map $v_{f,D} : \partial D \to S^{m-1}$ by

$$v_{f,D}(x) = \frac{f(x) - x}{\|f(x) - x\|}.$$

We show that the degree of the map $v_{f,D}$ does not depend on the choice of D. Suppose that D and D' are two disks as above. Let $D_0 \subseteq D \cap D'$ be another disk containing x_0. We claim that $\deg(v_{f,D}) = \deg(v_{f,D_0})$. The 'shell' $\Delta = D \setminus \text{int}(D_0)$ is a region with boundary, on which the mapping

$$x \to \frac{f(x) - x}{\|f(x) - x\|}$$

is well defined. The restriction to $\partial \Delta$ of this mapping has degree zero, due to Lemma 7.2.1. The boundary components carry opposite orientations, that is $\partial \Delta = \partial D - \partial D_0$, so we obtain $\deg(v_{f,D}) - \deg(v_{f,D_0}) = 0$, which proves our claim. This allows us to define the fixed point index $\text{Ind}_{x_0}(f)$ of f at x_0 as the degree of $v_{f,D}$ for any choice of a disk D as above.

We now want to define the index of a mapping f of a manifold M. Consider two local coordinate systems near a fixed point x_0 of f, such that x_0 corresponds to 0 in both systems. With an abuse of notation, assume that f is the expression of the mapping with respect to the first coordinate system, and $\theta \circ f \circ \theta^{-1}$ is the expression of the mapping with respect to second coordinate system, where the coordinate change mapping $\theta : D \to \mathbb{R}^m$ satisfies $\theta(0) = 0$. For simplicity, assume that 0 is the only fixed point for both f and $\theta \circ f \circ \theta^{-1}$ within some closed disk D.

Case 1. The diffeomorphism θ is orientation preserving.

We claim that θ is homotopic to the identity on D. Define $F : U \times [0,1] \to \mathbb{R}^m$ by

$$F(x,t) = \begin{cases} \theta(tx)/t & \text{if } t \neq 0, \\ (d\theta)_0(x) & \text{if } t = 0. \end{cases}$$

This is a smooth homotopy. Indeed, by the fundamental theorem of calculus applied to $t \to \theta(tx)$, and by the substitution $s = ut$, we obtain

$$\theta(tx) = \theta(tx) - \theta(0) = \int_0^t \frac{d}{ds}(\theta(sx)) ds = \int_0^t (d\theta)_{(sx)}(x) ds$$
$$= t \int_0^1 (d\theta)_{(utx)}(x) du = tg(tx),$$

where the smooth function g satisfies $g(0) = (d\theta)_0(x)$. Hence $F(x,t) = g(tx)$ for all $t \in [0,1]$, $F_1 = \theta$, and $F_0 = (d\theta)_0$. Since g is smooth, θ and $(d\theta)_0$ are smoothly homotopic. Since θ is orientation preserving, the linear mapping $(d\theta)_0$ is orientation preserving, so is smoothly homotopic to the identity mapping. Thus θ and the identity mapping are smoothly homotopic. It results that f and $\theta \circ f \circ \theta^{-1}$ are smoothly homotopic. The homotopy can be chosen so that 0 remains an isolated fixed point at all times. By the homotopy property, the degrees of the mappings induced by f and $\theta \circ f \circ \theta^{-1}$ must be equal.

Case 2. The diffeomorphism θ is orientation reversing.

By a similar argument, $\theta \circ f \circ \theta^{-1}$ and $\theta_0 \circ f \circ \theta_0^{-1}$ are smoothly homotopic, where θ_0 is given by $(x_1, \ldots, x_{m-1}, x_m) \to (x_1, \ldots, x_{m-1}, -x_m)$. It is easy to check that the degrees of the mappings induced by f and $\theta_0 \circ f \circ \theta_0^{-1}$, restricted to some small disk, are equal.

This discussion legitimizes the following:

DEFINITION 7.4.1
Assume that $\phi : U \to V \subseteq M$ is a local parametrization, with $x_0 \in V$, and that $f(x) \neq x$ for all $x \neq x_0$ in V. Then the fixed point index $\mathrm{Ind}_{x_0}(f)$ of f at x_0 is, by definition, the index of $\phi^{-1} \circ f \circ \phi$ at $\phi^{-1}(x_0)$.

Note that the index of f at x_0 can be expressed as the winding number of the mapping $v_{f,D}$ relative to 0,

$$\mathrm{Ind}_{x_0}(f) = W(v_{f,D}, 0). \tag{7.4.1}$$

If x_0 is not a fixed point of f, then we are still be able to compute $\mathrm{Ind}_{x_0}(f)$ through the same procedure. Assuming that we are in $\mathbb{R}^m$, we

7.4. THE FIXED POINT INDEX

have $f(x) \neq x$ on some neighborhood of x_0, and $(f(x)-x)/\|f(x)-x\|$ is well defined and continuous on some neighborhood of x_0. By continuity, there exists a sufficiently small disk D about x_0 such that

$$\left\| \frac{f(x)-x}{\|f(x)-x\|} - \frac{f(x_0)-x_0}{\|f(x_0)-x_0\|} \right\| < 1,$$

for all $x \in \partial(D)$. This means that the values of $v_{f,D}$ are all contained in some hemisphere of the unit sphere S^{m-1}, hence $\deg(v_{f,D}) = 0$. Thus $\operatorname{Ind}_{x_0}(f) = 0$. We conclude that if the index of a point is well defined and non-zero, then that point is a fixed point.

If the index of a point is zero, it does not mean that the point is not fixed. For example, the map $f : \mathbb{R} \to \mathbb{R}$ defined by $f(x) = x^2$ has an isolated fixed point at $x = 0$ with $\operatorname{Ind}_{x_0}(f) = 0$.

Fixed points with non-zero index persist under small perturbations:

PROPOSITION 7.4.2
Let M be a Riemannian manifold, $f : M \to M$ be a smooth mapping, and x_0 be an isolated fixed point of f of index $\operatorname{Ind}_{x_0}(f) \neq 0$. Let V be an isolating neighborhood of x_0. Then there exists $\epsilon > 0$ such that any smooth mapping $g : M \to M$ with $\|f(x) - g(x)\| < \epsilon$ for all $x \in V$, has a fixed point in V.

PROOF Using a local parametrization, we can assume that $M = \mathbb{R}^m$. Let $D \subseteq V$ be a closed disk as in the definition of the fixed point index, and $\delta = \min\{\|f(x) - x\| \mid x \in \partial D\} > 0$. Set $\epsilon = \delta/2$ and let g be a mapping with $\|g(x) - f(x)\| < \delta/2$ on U. Since we have

$$\|g(x) - x\| \geq \|f(x) - x\| - \|f(x) - g(x)\| > \delta/2 > 0,$$

it results that the mapping $v_{g,D} : \partial D \to S^{n-1}$ is well defined. We claim that $F : \partial D \times [0,1] \to S^{n-1}$ given by

$$F(x,t) = \frac{tv_{f,d} + (1-t)v_{g,D}}{\|tv_{f,D} + (1-t)v_{g,D}\|}$$

defines a homotopy between $v_{f,D}$ and $v_{g,D}$. We only need to verify that the denominator never vanishes. The denominator would be equal to 0 only if $v_{f,D}$ and $v_{g,D}$ are antipodal points in S^{m-1}, which would require that $f(x), x, g(x)$ are collinear with x between $f(x)$ and $g(x)$. However this is impossible because $\|f(x) - g(x)\| < \delta/2$ while $\|f(x) - x\| > \delta$. The homotopy invariance of the degree yields $\deg(v_{g,D}) = \deg(v_{f,D}) \neq 0$, which implies that g has a fixed point in V. Note that the fixed point of g may be neither unique nor isolated. ∎

In order to guarantee that isolated fixed points of non-zero index survive as isolated fixed points of the same index throughout a perturbation, we need to assume some non-degeneracy condition. A fixed point x_0 of $f : \mathbb{R}^m \to \mathbb{R}^m$ is called non-degenerate if the derivative of $x \to f(x) - x$ at x_0 is an invertible mapping.

DEFINITION 7.4.3
A fixed point x_0 of a mapping $f : M \to M$ is called non-degenerate if the linear mapping $(df)_{x_0} - \mathrm{id} : T_{x_0}M \to T_0 M$ is invertible.

It is easy to see that the non-degeneracy condition of a fixed point x_0 is equivalent to each of the following conditions:

- $(df)_{x_0}(v) \neq v$ provided $v \neq 0$;

- 1 is not an eigenvalue of $(df)_{x_0}$;

- the diagonal $\Delta = \{(x,x) \mid x \in M\}$ is transverse to the graph of f at x_0 in $M \times M$.

To check that the last conditions are equivalent, note that $T_{(x_0,x_0)}(\mathrm{graph}f) = \mathrm{graph}(df)_{x_0}$, and assume that the graph of f is not transverse to Δ at x_0, that is

$$T_{(x_0,x_0)}\Delta + \mathrm{graph}(df)_{x_0} \neq T_{(x_0,x_0)}(M \times M).$$

Since the two subspaces on the left side are of complementary dimensions in the tangent space on the right side, there must exist a non-zero vector $(v,w) \in T_{(x_0,x_0)}\Delta \cap \mathrm{graph}(df)_{x_0}$. On the one hand, we need to have $w = (df)_{x_0}(v)$, and on the other hand, $v = w$. This implies that 1 is not an eigenvalue of $(df)_{x_0}$. Conversely, if 1 is not an eigenvalue of $(df)_{x_0}$, then the graph of f is not transverse to Δ at x_0.

PROPOSITION 7.4.4
The fixed point index $\mathrm{Ind}_{x_0}(f)$ at a non-degenerate fixed point x_0 equals the fixed point index of $(df)_{x_0}$ at 0, which is given by

$$\mathrm{Ind}_0((df)_{x_0}) = \mathrm{sign}\det((df)_{x_0} - I) = (-1)^{\mathrm{card}\{\lambda_i \in \mathbb{R} \mid \lambda_i < 1\}},$$

7.4. THE FIXED POINT INDEX

where $\{\lambda_i\}_i$ are all of the eigenvalues of $(df)_{x_0}$, counted with multiplicity. The eigenvalues $\lambda_i \in \mathbb{C} \setminus \mathbb{R}$ do not count in the above formula.

PROOF For simplicity, we assume that we are already in $\mathbb{R}^m$ and that $x_0 = 0$.

Step 1. We claim that $\mathrm{Ind}_0(f) = \mathrm{Ind}_0((df)_0)$.
The differentiability of f at x_0 implies

$$f(x) = (df)_0(x) + R(x),$$

where $R(x)/x \to 0$ as $x \to 0$.

We want to show that $v_{f,D}$ and $v_{(df)_0,D}$ are homotopic mappings for some small closed disk D centered at 0, with a homotopy F given by

$$F(x,t) = \frac{tv_{f,D} + (1-t)v_{(df)_0,D}}{\|tv_{f,D} + (1-t)v_{(df)_0,D}\|}.$$

This reduces to show that it is impossible to have $v_{f,D}$ and $v_{(df)_0,D}$ antipodal, or, equivalently, to have $f(x)$, x, and $(df)_0(x)$ collinear with x between $f(x)$ and $(df)_0(x)$. Since 1 is not an eigenvalue of $(df)_0$, $(df)_0 - I$ is invertible, therefore $\|((df)_0 - I)^{-1}(x)\| \leq M\|x\|$ for some $M > 0$, or $\|((df)_0(x) - x\| \geq (1/M)\|x\|$ for all x. There exists a disk D centered at 0 such that $\|(df)_0(x) - f(x)\| < (1/M)\|x\|$ for all $x \in D$. So the distance between $f(x)$ and $(df)_0(x)$ is shorter than the distance between x and $(df)_0(x)$, hence x cannot lie between $f(x)$ and $(df)_0(x)$.

Step 2. We claim that $\mathrm{Ind}_0((df)_0) = \mathrm{sign}\det((df)_0 - I)$.

Since $((df)_0 - I)$ is invertible, $v_{(df)_0,D}(x) = v_{(df)_0,D}(y)$ yields $y = x$. Thus $v_{(df)_0,D}$ is a diffeomorphism of ∂D onto S^{n-1}. Its degree is 1 if $(df)_0(x) - I$ is orientation preserving, and -1 otherwise.

Step 3. We claim that $\mathrm{sign}\det((df)_0 - I) = (-1)^{\mathrm{card}\{\lambda_i \,|\, \lambda < 1\}}$.

The eigenvalues λ_i of the matrix $(df)_0$ are the roots of the polynomial $\det((df)_0 - tI)$. From $\det((df)_0 - tI) = \Pi_i(\lambda_i - t)$, we obtain $\det((df)_0 - I) = \Pi_i(\lambda_i - 1)$. Each $\lambda_i > 1$ contributes with a positive factor, and also each pair of complex conjugated eigenvalues contributes with a positive factor $(\lambda_i - 1)(\bar{\lambda}_i - 1)$ to $\Pi_i(\lambda_i - 1)$. This shows that the sign of $\det((df)_0 - I)$ is positive if there is an even number of $\lambda_i < 1$, and is negative otherwise. ∎

Example 7.4.5

Hyperbolic fixed points are non-degenerate. For a surface map, the fixed point index of a repelling fixed point is 1, the fixed point index of

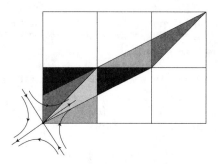

FIGURE 7.4.1
The image of the torus under A. The diagram in the corner shows the fixed point and a few nearby trajectories.

an attracting fixed point is also 1, and the fixed point index of a saddle point is -1.

For example, consider the mapping f_A of the torus $\mathbb{T}^2 \simeq \mathbb{R}^2/\mathbb{Z}^2$ induced by the linear transformation

$$A = \begin{pmatrix} 2 & 1 \\ 1 & 1 \end{pmatrix}.$$

Since the entries of the matrix are integers, $A(x_1', x_2') \sim A(x_1, x_2)$ provided $(x_1', x_2') \sim (x_1, x_2)$ (recall that $(x_1', x_2') \sim (x_1, x_2)$ means $x_1' - x_1$ and $x_2' - x_2$ are both integers). The determinant of A is 1, so the inverse A^{-1} also has integer entries. Thus A induces a diffeomorphism of the torus. The origin 0 is an isolated fixed point of f_A, and the eigenvalues of A are $\lambda_{1,2} = (3 \pm \sqrt{5})/2$. Since $\lambda_1 < 1$ and $\lambda_2 > 1$, the fixed point index of f_A at 0 is -1. See Figure 7.4.1. □

PROPOSITION 7.4.6
Hyperbolic fixed points persist under smooth homoptopy.

PROOF Assume that x_0 is hyperbolic fixed point of a map $f : M \to M$. Let $F : M \times [0, 1] \to M$ be a homotopy with $F_0 = F(\cdot, 0) = f$. The index of $(dF_t)_x$ and the eigenvalues of $(dF_t)_x$ vary continuously with $t \in [0, 1]$ and with x near x_0. Thus, for all sufficiently small t, F_t has a hyperbolic fixed point near x_0. ∎

The fixed point index can also be expressed as an intersection number.

7.4. THE FIXED POINT INDEX

PROPOSITION 7.4.7
The fixed point index of f at a non-degenerate fixed point x_0 is equal to the intersection number $I(\Delta, \operatorname{graph}(f))_{x_0}$ of the diagonal Δ of $M \times M$ with the graph of f at x_0.

PROOF Let $(v_1, \ldots, v_m)$ be a positively oriented basis for $T_{x_0} M$. A positively oriented basis for $T_{(x_0, x_0)}(M \times M)$ is

$$((v_1, 0), \ldots, (v_m, 0), (0, v_1), \ldots, (0, v_m)).$$

Let $D = (df)_{x_0}$. We have $I(\Delta, \operatorname{graph}(f)) = \pm 1$ depending on whether

$$((v_1, v_1), \ldots, (v_m, v_m), (v_1, D(v_1)), \ldots, (v_m, D(v_m)))$$

forms a positively or a negatively oriented basis of $T_{(x_0, x_0)}(M \times M)$. By subtracting the first m vectors from the corresponding last m vectors, we obtain a new basis

$$((v_1, v_1), \ldots, (v_m, v_m), (0, (D-I)(v_1)), \ldots, (0, (D-I)(v_m))),$$

with the same orientation as the original one. Since $(D-I)$ is invertible, $(D-I)(v_1), \ldots, (D-I)(v_m)$ is a basis of $T_{x_0} M$. Hence, using linear combinations of the last m-vectors, we obtain a new basis

$$((v_1, 0), \ldots, (v_m, 0), (0, (D-I)(v_1)), \ldots, (0, (D-I)(v_m))),$$

with the same orientation as the original one. This basis is positively oriented if and only if $D - I$ is orientation preserving, that is, $\operatorname{sign} \det(D - I) > 0$. Thus

$$I(\Delta, \operatorname{graph} f)_{x_0} = \operatorname{sign} \det(D - I) = \operatorname{Ind}_{x_0}(f).$$

∎

Now we discuss the analogue of the fixed point index in the case of a vector field with an isolated zero. Suppose first that

$$X : V \to \mathbb{R}^m$$

is a smooth vector field on an open subset V of $\mathbb{R}^m$ and $x_0 \in V$ is an isolated zero of the vector field. There exists a disk $D \subseteq V$ containing x_0 in its interior such that $X(x) \neq 0$ for all $x \neq x_0$ in D. The way in

300   7. FIXED POINTS AND INTERSECTION NUMBERS

which the direction of $X(x)$ changes as x varies near x_0 is measured by the map $v_{X,D} : \partial D \to S^{m-1}$ given by

$$v_{X,D}(x) = \frac{X(x)}{\|X(x)\|}.$$

The index of X at x_0 is defined to be the degree of this map:

$$\text{Ind}_{x_0}(X) = \deg(v_{X,D}).$$

This definition does not depend on the choice of the disk D.

In the case of a vector field X on a manifold M, we define the index of X at an isolated zero x_0 through a local parametrization.

DEFINITION 7.4.8
Assume that $\phi : U \to V \subseteq M$ is a local parametrization, and that $X(x) \neq 0$ for all $x \neq x_0$. Then the index $\text{Ind}_{x_0}(X)$ of X at x_0 is, by definition, the index of $d\phi^{-1} \circ X \circ \phi$ at $\phi^{-1}(x_0)$.

The fact that this definition is independent of a local parametrization can be argued as in the case of the fixed point index.

The index of a planar vector field at an isolated zero x_0 is the number of times $X(x)$ rotates about 0 as as x is moved counterclockwise around a small circle about x_0. Since a smooth vector field can always be integrated (at least locally) to a flow, we can represent the vector field by its corresponding phase portrait. See Figure 7.4.2.

The index of a vector field can also be expressed in the language of intersection theory. A smooth vector field X is a smooth section of the tangent bundle TM. The zero section $X_0 : M \to TM$, defined by $X_0(x) = 0 \in T_x M$ for all $x \in M$, represents an embedding of M in TM. The zeroes of X correspond to the intersections between the zero section X_0, regarded as an m-dimensional submanifold in TM, and the section X, viewed as a smooth function from M to TM. The index of the vector field X at an isolated zero x_0 is the same as the intersection number

$$I(X_0, X)_{x_0}.$$

A zero x_0 of a vector field $X : M \to TM$ is said to be non-degenerate if the derivative mapping of $d\phi^{-1} \circ X \circ \phi : U \to \mathbb{R}^m$ is invertible at $\phi^{-1}(x_0)$. This is equivalent to the fact that the zero section X_0 is transverse to the graph of X at $(x_0, 0)$. In terms of the local coordinates $(X_1, \ldots, X_m)$

7.4. THE FIXED POINT INDEX

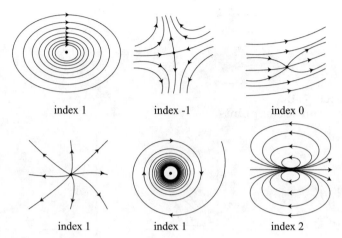

FIGURE 7.4.2
Indices of zeroes of plane vector fields.

of X, this condition says that the determinant of the matrix associated with dX at x_0 is non-zero. Then the index of X at x_0 is given by

$$\text{Ind}_{x_0}(X) = \text{sign}\det((dX)_{x_0}) = (-1)^{\text{card}\{\lambda_i \in \mathbb{R} \mid \lambda_i < 0\}}, \quad (7.4.2)$$

where $\{\lambda_i\}$ are all of the eigenvalues of $(dX)_{x_0}$, counted with multiplicity. All of these facts follow immediately from the properties of the fixed point index, and their proofs are left to the reader.

The zeroes of a vector field X correspond to the fixed points of the induced flow ϕ. Assume that for a fixed point x_0 of the flow, there exist an isolating neighborhood V and a time increment t_0 such that the flow does not have any periodic orbits in V of period less than t_0. The time discretization of the flow ϕ^{t_0} has no other fixed points in V besides x_0. Then

$$\text{Ind}_{x_0}(X) = \text{Ind}_{x_0}(\phi^{t_0}).$$

Example 7.4.9

We provide an easy method to compute the index at a zero x_0 of a vector field X on a surface. Since the computation of the index is local, we can assume that the surface is just an open set in the plane. Let ϕ be the (local) flow obtained by integrating X. The point x_0 is called a center if it has a neighborhood where all flow lines are closed, inside one another, and contain the zero in their interior. The point x_0 is called a

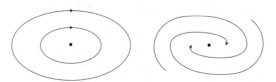

FIGURE 7.4.3
Center and focus points.

FIGURE 7.4.4
Elliptic, parabolic, and hyperbolic sectors.

focus if it has a neighborhood where all flow lines tend to x_0 as $t \to \infty$, or all flow lines tend to x_0 as $t \to -\infty$. See Figure 7.4.3.

One can show that a zero x_0 of a generic vector planar field X is either a focus, or a center, or it has a small circular neighborhood D that can be divided into finitely many sectors of one of the types described below. A sector is called elliptic if all interior flow lines return to x_0. A sector is called parabolic if either all interior flow lines move away from x_0 or all interior flow lines head towards x_0. A sector is called hyperbolic if all interior flow lines leave the sector in both positive and negative time. See Figure 7.4.4. It is obvious that if x_0 is a center or a focus, the index is ± 1, depending on orientation. We now calculate the index in the case when x_0 has a circular neighborhood that can be divided into e elliptic sectors, p parabolic sectors, and h hyperbolic sectors. When the point x crosses a sector of angle θ, the corresponding vector $X(x)$ turns by an angle of $\theta + \pi$ if the sector is elliptic, by an angle of θ if the sector is parabolic, and by an angle of $\theta - \pi$ is the sector is hyperbolic. Summing up the turns along a complete rotation around a circle, $X(x)$ turns by $2\pi + e\pi - h\pi$. Thus

$$\operatorname{Ind}_{x_0}(X) = 1 + \frac{e-h}{2}.$$

As an example, one can use this method to compute the indices of the zeroes of the vector fields shown in Figure 7.4.2. □

7.5 The Lefschetz number

When a mapping is slightly perturbed, some of its fixed points may disappear and some new fixed points may be created. The Lefschetz number is a homotopy invariant associated with the collection of all isolated fixed points of a mapping. Since a fixed point of f is an intersection point of the diagonal $\Delta = \{(x,x) \,|\, x \in M\}$ with the graph of f, it is only natural to use intersection theory to account for all fixed points. The intersection number of two submanifolds that are not transverse is defined as the intersection number of some homotopy deformation of the submanifolds which makes them transverse.

DEFINITION 7.5.1
Let $f : M \to M$ be a smooth mapping of a compact orientable manifold M. The Lefschetz number $\mathcal{L}(f)$ of f is the intersection number $I(\Delta, \text{graph}(f))$ of the diagonal Δ with the graph of f in $M \times M$.

The Lefschetz number provides an effective method for detecting fixed points.

THEOREM 7.5.2 *(Lefschetz fixed point theorem)*
Let $f : M \to M$ be a smooth map of a compact orientable manifold. If $\mathcal{L}(f) \neq 0$, then f has a fixed point.

PROOF We prove the contrapositive. If f has no fixed point, then $\Delta \cap \text{graph}(f) = \emptyset$, yielding $\mathcal{L}(f) = 0$. ∎

Since the intersection number is a homotopy invariant, we immediately obtain:

PROPOSITION 7.5.3
The Lefschetz number is a homotopy invariant.

See Figure 7.5.1. We emphasize that for homotopy invariance we need the manifold to be compact. For manifolds with boundary, we need to require that the homotopy does not move the boundary points.

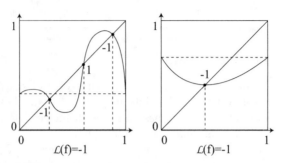

FIGURE 7.5.1
The Lefschetz numbers of a mapping of $S^1 = [0,1]/0 \sim 1$ and of a homotopic transformation of it. The indices of the fixed points are marked along the diagonal.

Consider the particular case when the graph of f is transverse to the diagonal Δ. Each fixed point x of f is isolated and non-degenerate, so $I(\Delta, \operatorname{graph}(f))_x = \operatorname{Ind}_x(f)$. Thus

$$\mathcal{L}(f) = \sum_{x \text{ fixed point of } f} \operatorname{Ind}_x(f).$$

We now show that the above formula is valid even under the weaker assumption that f has only isolated fixed points, but its graph is not necessarily transverse to the diagonal.

PROPOSITION 7.5.4
If $f : M \to M$ is a smooth mapping of a compact orientable manifold M, with all of its fixed points isolated, then

$$\mathcal{L}(f) = \sum_{x \text{ fixed point of } f} \operatorname{Ind}_x(f).$$

PROOF We only need to prove that for some homotopy deformation g of f, whose graph is transverse to Δ, we have

$$\sum_{y \text{ fixed point of } g} \operatorname{Ind}_y(g) = \sum_{x \text{ fixed point of } f} \operatorname{Ind}_x(f).$$

It suffices to show that if x is an isolated fixed point of f and V is an isolating neighborhood of x, if g is homotopic to f with $\Delta \pitchfork \operatorname{graph}(g)$

7.5. THE LEFSCHETZ NUMBER

and $g = f$ outside of V, then

$$\sum_{y \text{ fixed point of } g \text{ in } V} \mathrm{Ind}_y(g) = \mathrm{Ind}_x(f).$$

Such a g exists due to the transversality theorem —Theorem 1.14.4.

For simplicity, we assume that $M = \mathbb{R}^m$. Let D be a diffeomorphic copy of a disk in V, containing x and all of the fixed points $\{y_1, \ldots, y_k\}$ of g in V. For each $i = 1, \ldots, k$, let $D_i \subseteq D$ be a diffeomorphic copy of the disk, such that $y_i \in D_i$ and $D_i \cap D_j = \emptyset$ if $i \neq j$. On the one hand, we have $\mathrm{Ind}_x(f) = \deg(v_{f,D})$. On the other hand, the mapping

$$z \to \frac{g(z) - z}{\|g(z) - z\|}$$

is well defined on the entire manifold with boundary $D \setminus \bigcup_{i=1}^{k} \mathrm{int}(D_i)$, so the restriction of this mapping to the boundary has degree 0, due to Lemma 7.2.1. The boundary component ∂D has orientation opposite to all other boundary components ∂D_i, $i = 1, \ldots, k$, hence

$$\deg(v_{g,D}) = \sum_{i=1}^{k} \deg(v_{g,D_i}) = \sum_{i=1}^{k} \mathrm{Ind}_{y_i}(g).$$

However, f and g coincide on the boundary of D, so $\deg(v_{f,D}) = \deg(v_{g,D})$, yielding the conclusion of the proposition. ∎

As an application to Lefschetz fixed point theorem, we give another proof to the Brouwer fixed point theorem. Let $f : B^n \to B^n$ be a smooth map of the unit ball into itself. The map f is homotopic to any constant map $h(x) = x_0$, with $x_0 \in B^n$. An example of such a homotopy is $F(t, x) = (1 - t)f(x) + th(x)$, for $t \in [0, 1]$. The mappings h_{*k} induced by h on the homology groups $H_k(B^n)$ are all zero for $k \geq 1$. Since the ball is connected, $H_0(B^n) = \mathbb{R}$, and h_0 is the identity mapping. It follows that $\mathcal{L}(h) = 1$, so, by the homotopy property, $\mathcal{L}(f) = 1 \neq 0$. The Lefschetz fixed point theorem implies that f has at least one fixed point.

REMARK 7.5.5 The Lefschetz number can also be defined for continuous maps on topological spaces, through homotopy deformations. In order for the Lefschetz number of a mapping to be well defined, one usually assumes that the topological space is a compact Euclidean neighborhood retract (compact ENR). A topological space M is said to be an ENR if there exists a homeomorphism $h : M \to h(M)$ from M onto a subset $h(M)$ of an open neighborhood U in some Euclidean space, for which there is a retraction $r : U \to h(M)$ of U onto $h(M)$. If M is a compact ENR, then one can define the index of a map $f : M \to M$ at a fixed point x of f as the index of $h \circ f \circ h^{-1} \circ r$ at $h(x)$. Only finitely many homology groups of a compact ENR are non-trivial, and they all are free and finitely generated. Hence, the map $f_{*k} : H_k(M) \to H_k(M)$ induced by f on the k-dimensional homology is represented by a (finite) matrix for each k. The Lefschetz number of f is defined by

$$\mathcal{L}(f) = \sum_{k=0}^{n} (-1)^k \text{trace}(f_{*k}).$$

This definition is equivalent to the earlier one in the case of a compact manifold. For more details, see Dold (1980) and R.F. Brown (1971).

7.6 The Euler characteristic

One of the oldest results in topology is the Euler formula: for any polyhedron, the number of vertices minus the number of edges plus the number of faces always equals 2. Using this formula one can show, for example, that there are five Platonic solids. There is a natural generalization of Euler's formula to higher dimensional objects.

DEFINITION 7.6.1
Let M be a compact m-dimensional smooth manifold, and $\mathcal{T}$ be a finite triangulation of M. The Euler characteristic of $\mathcal{T}$ is defined as

$$\chi(\mathcal{T}) = \sum_{i=0}^{m} (-1)^i s_i,$$

where s_i represents the number of i-dimensional singular simplices in the triangulation. This number is independent of the choice of a triangulation. We define the Euler characteristic $\chi(M)$ of M as the Euler characteristic $\chi(\mathcal{T})$ of some triangulation $\mathcal{T}$ of M.

7.6. THE EULER CHARACTERISTIC

Every smooth manifold is triangulable (see Remark 6.9.4). For any compact manifold, we can always find a finite triangulation. We will prove in Theorem proposition:eulercharhomo (and also in Theorem 7.6.6) that $\chi(\mathcal{T})$ is independent of the choice of a triangulation. For the time being, we take this fact for granted.

Example 7.6.2

(i) Using the triangulations of the sphere and of the torus shown in Figure 6.9.3, we obtain $\chi(S^2) = 2$ and $\chi(\mathbb{T}^2) = 0$. In general, one can show that the Euler characteristic for a surface of genus g is $2 - 2g$.

(ii) An n-dimensional sphere can be represented as the n-dimensional simplicial chain formed by the faces of an n-simplex. There are $\binom{n+2}{1}$ vertices, $\binom{n+2}{2}$ edges, $\binom{n+2}{3}$ 3-dimensional faces, ..., and $\binom{n+2}{n+1}$ n-dimensional faces. Using the binomial formula, we obtain

$$\chi(S^n) = \binom{n+2}{1} - \binom{n+2}{2} + \cdots + (-1)^n \binom{n+2}{n+1}$$
$$= \binom{n+2}{0} + (-1)^{n+2}\binom{n+2}{n+2} - (1-1)^n$$
$$= 1 + (-1)^{n+2}.$$

In conclusion, $\chi(S^n) = 2$ if n is even and $\chi(S^n) = 0$ if n is odd. □

The Euler characteristic can be expressed in terms of the homology groups of M.

PROPOSITION 7.6.3
Let M be a compact m-dimensional smooth manifold. Then

$$\chi(M) = \sum_{i=0}^{m}(-1)^i \operatorname{rank}(H_i(M)), \qquad (7.6.1)$$

where $\operatorname{rank}(H_i(M)) = \beta_i$ (the i-th Betti number) is the dimension of the homology group $H_i(M)$, regarded as a vector space.

PROOF Consider a finite triangulation of the compact manifold M. Denote by S_i the vector space of singular i-chains on M, by C_i the vector space of closed singular i-chains on M, and by E_i the vector space of exact singular i-chains on M. The vector spaces S_i, C_i and E_i

depend on the triangulation, and they are finite dimensional, since the triangulation is finite. Let H_i be the i-th singular homology group on M. The homology groups H_i — and implicitly their dimensions — computed via triangulations are independent of the choice of a triangulation (see Remark 6.9.4).

We have $E_i \subseteq C_i \subseteq S_i$, $\partial S_i = E_{i-1}$, $C_i = \ker(\partial)$, and $H_i = C_i/E_i$. From linear algebra, if $f : V \to W$ is a linear homomorphism between the vector spaces V and W, we have $\dim(V) = \dim(\ker(f)) + \dim(\mathrm{im}(f))$. Applying this property to the boundary mapping $\partial : S_i \to C_{i-1}$, and to the canonical projection $\pi : C_i \to H_i$ (whose kernel is E_i), we obtain

$$\dim(S_i) = \dim(C_i) + \dim(E_{i-1}),$$
$$\dim(C_i) = \dim(H_i) + \dim(E_i).$$

Since $\dim(E_m) = 0$, $\dim(H_0) = \dim(C_0) - \dim(E_0)$, and $\dim(S_0) = \dim(C_0)$,

$$\chi(M) = \sum_{i=0}^{m}(-1)^i s_i = \sum_{i=0}^{m}(-1)^i \dim(S_i),$$

$$= \dim(S_0) + \sum_{i=1}^{m}(-1)^i \left(\dim(C_i) + \dim(E_{i-1})\right)$$

$$= \dim(S_0) + \sum_{i=1}^{m}(-1)^i \left(\dim(H_i) + \dim(E_i) + \dim(E_{i-1})\right)$$

$$= \dim(S_0) - \dim(E_0) + \sum_{i=1}^{m}(-1)^i \dim(H_i) = \sum_{i=0}^{m}(-1)^i \dim(H_i).$$

∎

REMARK 7.6.4 This proof shows that

$$\chi(\mathcal{T}) = \sum_{i=0}^{m}(-1)^i \dim(H_i),$$

where the quotient groups $H_i = C_i/E_i$ are defined by a triangulation $\mathcal{T}$. We emphasize the fact that the dimensions $\dim(H_i)$, $i = 0,\ldots,m$, are independent of the triangulation shows that $\chi(\mathcal{T})$ is also independent of the triangulation. This implies that the Euler characteristic computed via triangulations equals the Euler characteristic computed via homology. ∎

7.6. THE EULER CHARACTERISTIC

REMARK 7.6.5 The Euler characteristic can also be expressed in terms of the cohomology groups.

$$\chi(M) = \sum_{i=0}^{m}(-1)^i \mathrm{rank}(H^i(M)).$$

From the de Rham theorem, $H^i(M)$ is isomorphic to the space $(H_i(M))^*$ of linear functionals on $H_i(M)$ for each i. Since $\dim((H_i(M))^*) = \dim(H_i(M))$, $\dim(H^i(M)) = \dim(H_i(M))$. ∎

Now we define the Euler characteristic from the vantage point of intersection theory. A first definition is

$$\chi(M) = I(\Delta, \Delta), \qquad (7.6.2)$$

where $\Delta = \{(x,x) \,|\, x \in M\} \subseteq M \times M$. Here the manifold M is assumed to be oriented. A practical way to use this definition is to find some smooth function $f : M \to M$, homotopic to the identity, and compute its Lefschetz number. Since the graph of the identity is Δ, by the homotopy invariance of the Lefschetz number we obtain

$$\chi(M) = \mathcal{L}(f), \text{ provided } f \text{ is homotopic to the identity.} \qquad (7.6.3)$$

A second definition is in terms of the intersection number of vector fields on the manifold. The orientation on M naturally induces an orientation on TM. Each smooth vector field $X : M \to TM$ is a section of TM. Let X_0 be the zero section, assigning the 0 vector to all points in M. We define

$$\chi(M) = I(X_0, X_0). \qquad (7.6.4)$$

We think of the first factor X_0 as an embedded copy of M in TM, and of the second factor X_0 as a smooth function from M to TM. Since any smooth vector field X on M is homotopic to the zero section, the homotopy property of the oriented intersection number implies

$$\chi(M) = I(X_0, X). \qquad (7.6.5)$$

It is advantageous to choose X transverse to X_0, meaning that the mapping $(dX)_{x_0}$ is invertible at every zero x_0 of X. Then the Euler characteristic is computed as

$$\chi(M) = I(X_0, X) = \sum_{x_0 \text{ zero of } X} I(X_0, X)_{x_0}$$

$$= \sum_{x_0 \text{ zero of } X} \mathrm{Ind}_{x_0}(X).$$

These two intersection theory-based definitions of the Euler characteristic are equivalent to each other. By the tubular neighborhood theorem 6.3.2, the mapping $(x,0) \to (x,x) \in \Delta \times \Delta$ can be extended to a diffeomorphism from a neighborhood of X_0 in TM onto a neighborhood of $\Delta \times \Delta \in M \times M$, and so $I(X_0, X_0) = I(\Delta, \Delta)$.

It is not obvious that the above intersection theory-based definitions of the Euler characteristic are equivalent to the earlier ones. In Theorem 7.6.6 we will show that the Euler characteristic via intersection theory equals the Euler characteristic via triangulations, and so, in particular, that the Euler characteristic via triangulations is independent of the choice of a triangulation. The following theorem, which will be proved in Section 8.6, says that the Euler characteristic via intersection theory equals the Euler characteristic via homology.

THEOREM 7.6.6 (Poincaré-Hopf theorem)
If X is a smooth vector field with isolated zeroes on a compact oriented manifold M, then the sum of the indices of the zeroes of X is $\chi(M)$ — defined through homology.

This yields to an alternative proof of the fact that the Euler characteristic via triangulations equals the Euler characteristic via homology (compare to Remark 7.6.5).

We first illustrate the usage of the intersection theory-based Euler characteristic definitions in some examples.

Example 7.6.7
(i) We compute again the Euler characteristic of the sphere S^n. We consider a field X of tangent vectors to the sphere, radiating from the south pole towards the north pole, as in Figure 7.6.1. We can define X by $X(x) = N - (N \cdot x)x$, where N is the position vector in $\mathbb{R}^{n+1}$ of the north pole. The vectors at the north pole and at the south pole are zero. Note that, in local coordinates, $dX = I$ at the south pole and $dX = -I$ at the north pole, so $I(X_0, X)_S = \text{Ind}_S(X) = 1$ and $I(X_0, X)_N = \text{Ind}_N(X) = (-1)^n$. Then

$$\chi(M) = I(X_0, X) = I(X_0, X)_S + I(X_0, X)_N$$
$$= 1 + (-1)^n.$$

We can also compute the Euler characteristic by using the flow ϕ obtained by integrating the vector field X. Let ϕ^t be the time t map for

7.6. THE EULER CHARACTERISTIC

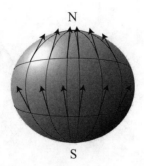

FIGURE 7.6.1
The Euler characteristic of a sphere.

some small t. The map ϕ^t pushes each point of the sphere further north, except for the south and north poles, which stay fixed. It is clear that ϕ^t is homotopic to $\phi^0 = \text{id}$. We have $\text{Ind}_S(\phi^t) = 1$ and $\text{Ind}_N(\phi^t) = (-1)^n$. Then

$$\chi(M) = \mathcal{L}(\phi^t) = \text{Ind}_S(\phi^t) + \text{Ind}_N(\phi^t) = 1 + (-1)^n.$$

(ii) The Euler characteristic of any odd-dimensional, compact, oriented manifold is zero. Indeed, if M is odd-dimensional so is Δ, so, switching the order in (7.6.2), we obtain

$$I(\Delta, \Delta) = (-1)^{\dim M} I(\Delta, \Delta),$$

which implies $I(\Delta, \Delta) = 0$.

Alternatively, we can show that $\chi(M) = 0$ for an odd-dimensional manifold M by using (7.6.1) and the Poincaré duality theorem 6.8.2.
□

We now prove the equivalence between the triangulation-based definition of the Euler characteristic and intersection theory-based definition.

THEOREM 7.6.8 (Poincaré-Hopf theorem)
If X is a smooth vector field with isolated zeroes on a compact, connected, orientable, smooth manifold M, then the sum of the indices of the zeroes of X equals $\chi(M)$ — defined through triangulations.

312   7. FIXED POINTS AND INTERSECTION NUMBERS

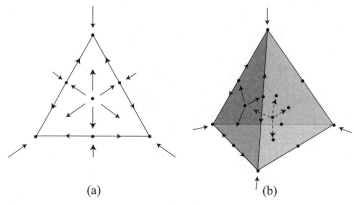

FIGURE 7.6.2
(a) Vector field on 2-dimensional simplex. (b) Vector field on 3-dimensional simplex (the vector field is shown only in the center and on one face).

PROOF Consider a triangulation $\mathcal{T}$ of M into finitely many m-dimensional singular simplices. On each singular m-simplex, we construct a vector field that has a zero of index $(-1)^m$ in the center of the singular simplex (i.e., the point corresponding to the barycenter in the standard m-simplex), and a zero of index $(-1)^{m-j}$ in the center of each $(m-j)$-dimensional face of the simplex. In particular, it has a zero of index 0 at each vertex of the simplex. This is illustrated in Figure 7.6.2 in the case of a 2-dimensional simplex and in the case of a 3-dimensional simplex. The vector fields on neighboring m-simplices are chosen so that the resulting vector field Y on M is smooth. Thus, each i-dimensional singular simplex in the triangulation corresponds to a zero of Y of index $(-1)^i$.

The intersection number of the zero section of the tangent bundle with a vector field on M is independent of the choice of a vector field. We have

$$I(X_0, Y) = \sum_{y_0 \text{ zero of } Y} \text{Ind}_{y_0}(Y) = \sum_{i=0}^{m} (-1)^i s_i,$$

where s_i denotes the number of i-dimensional simplices in the triangulation. Since $I(X_0, X) = I(X_0, Y)$,

$$I(X_0, X) = \sum_{x_0 \text{ zero of } X} \text{Ind}_{x_0}(X) = \sum_{i=0}^{m} (-1)^i s_i = \chi(\mathcal{T}).$$

7.7 The Gauss-Bonnet theorem

The Gauss-Bonnet theorem is one of the most important theorems in differential geometry. This theorem has a local version and a global version. The local version says that the curvature of a surface determines the angle sum of any geodesic triangle. The global version says that the curvature of the surface determines its topology.

A simple version of this result can be formulated in the case of the sphere with its standard metric. The local version of the theorem says that the angle sum of a geodesic triangle $\triangle ABC$ of the sphere S_r^2 of radius r is given by

$$\angle A + \angle B + \angle C = \pi + \frac{1}{r^2}\text{Area}(\triangle ABC).$$

A geodesic triangle of the sphere is formed by arcs of great circles. The the interior angles $\angle A$, $\angle B$, $\angle C$ of the triangle are measured between the tangents to the great circles at the contact points. See Figure 7.7.1.

To prove this statement, we should first note that $\triangle ABC$ is the intersection of three 'lunes' L_A, L_B, L_C. A lune is a region of the sphere enclosed between two great semi-circles. Each angle of the spherical triangle embraces some lune. The symmetric copies L'_A, L'_B, L'_C of the lunes L_A, L_B, L_C determine a symmetric copy $\triangle A'B'C'$ of $\triangle ABC$. It is not hard to see that the area of a lune equals $2r^2$ times the angle embraced by that lune, so $\text{Area}(L_A) = 2r^2 \angle A$, $\text{Area}(L_B) = 2r^2 \angle B$, and $\text{Area}(L_C) = 2r^2 \angle C$. The sum of the areas of the three pairs of lunes equals the area of the sphere plus twice $\text{Area}(\triangle ABC)$ plus twice the area of $\text{Area}(\triangle A'B'C')$, since each of these two triangles is covered simultaneously by three lunes. Due to the symmetries, we infer

$$2(2r^2 \angle A + 2r^2 \angle B + 2r^2 \angle C) = 4\pi r^2 + 4\text{Area}(\triangle ABC),$$

therefore

$$\angle A + \angle B + \angle C = \pi + \frac{1}{r^2}\text{Area}(\triangle ABC).$$

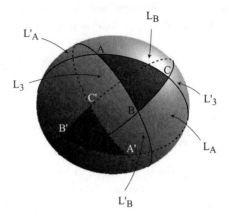

FIGURE 7.7.1
The Gauss-Bonnet theorem for spherical triangles.

Note that the factor $1/r^2$ of the area of the triangle represents the Gaussian curvature of the sphere of radius r. Thus, the angle sum of a spherical triangle exceeds the angle sum of a Euclidean triangle of equal area by an amount proportional to the Gaussian curvature of the sphere. We can immediately deduce that the sum of the exterior angles of a spherical polygon falls short from the sum of the exterior angles of a Euclidean polygon of equal area (which angle sum is always 2π) also by an amount proportional to the Gaussian curvature of the sphere. That is, if $A_1 \ldots A_n$ is a geodesic polygon, then

$$\sum_{i=1}^{n}(\pi - \angle A_i) = 2\pi - \frac{1}{r^2}\text{Area}(A_1 \ldots A_n). \tag{7.7.1}$$

The global version of the Gauss-Bonnet theorem for the sphere says that the Euler characteristic of the sphere multiplied by 2π equals the Gaussian curvature multiplied by the total area. Indeed, $2\pi\chi(S^2) = 4\pi$ and $(1/r^2)\text{Area}(S^2) = 4\pi$.

These types of relations hold in general, for compact orientable surfaces. The Gaussian curvature of a surface is not necessarily constant. In the general statement, the product of the Gaussian curvature by the area will be replaced by the integral of the Gaussian curvature over the corresponding surface.

7.7. THE GAUSS-BONNET THEOREM

We follow the outline in Morgan (1997). We begin with some simple results on geodesic curvature. Recall from Example 5.3.6 that for an arc-length parametrized curve c on a surface M, the geodesic curvature κ_g is given by $\|Dc'/dt\|$.

LEMMA 7.7.1
Assume that $X(t)$ is a parallel vector field along $c(t)$, and $\phi(t)$ is the angle from $X(t)$ to $c'(t)$, measured counter-clockwise relative to the positive orientation of the tangent plane. Then

$$\frac{d\phi}{dt} = \pm \kappa_g,$$

where the sign is plus or minus depending on whether c' is turning counter-clockwise or clockwise.

PROOF Assume that $\|X(t)\| = 1$. The vector $X \times N$, where N is the unit normal vector to M, is a unit tangent vector to the surface and is orthogonal to X. We have

$$c' = \cos\phi\, X + \sin\phi\, (N \times X),$$
$$c'' = -\sin\phi \frac{d\phi}{dt} X + \cos\phi \frac{dX}{dt} + \cos\phi \frac{d\phi}{dt}(X \times N) + \sin\phi \frac{d(N \times X)}{dt}.$$

The covariant derivative D/dt of a vector field on a surface is the tangential component to the surface of the usual derivative d/dt of the vector field. By taking the tangential component to the surface of each side of the above equation, we obtain

$$\frac{Dc'}{dt} = -\sin\phi \frac{d\phi}{dt} X + \cos\phi \frac{DX}{dt} + \cos\phi \frac{d\phi}{dt}(X \times N) + \sin\phi \frac{D(N \times X)}{dt}$$
$$= -\sin\phi \frac{d\phi}{dt} X + \cos\phi \frac{d\phi}{dt}(N \times X),$$

since X and $N \times X$ are both parallel vector fields along c. Applying the Pythagorean theorem, we conclude

$$\kappa_g = \left\|\frac{Dc'}{dt}\right\| = \left|\frac{d\phi}{dt}\right|.$$

∎

The next result is a version of Hopf's theorem of turning tangents in the particular case of the sphere. See do Carmo (1976) for the general case.

LEMMA 7.7.2
Let S_r^2 be the sphere of radius r with the standard metric. If R is a positively oriented, embedded disk in S_r^2 with ∂R a smooth curve, then

$$\int_{\partial R} \kappa_g = 2\pi - \frac{1}{r^2}\mathrm{Area}(R). \tag{7.7.2}$$

If R is a positively oriented, embedded polygon in S_r^2, with ∂R a piecewise smooth curve, then

$$\int_{\partial R} \kappa_g + \sum_{i=1}^{n}(\pi - \angle A_i) = 2\pi - \frac{1}{r^2}\mathrm{Area}(R), \tag{7.7.3}$$

where $\pi - \angle A_i$ are the exterior angles between the smooth arcs at the corners of the boundary, measured counter-clockwise relative to the positive orientation of the tangent plane. See Figure 7.7.2.

PROOF Assume first that ∂R is a smooth closed curve c parametrized by arc length. We approximate R by a polygon whose edges $\gamma_0, \ldots, \gamma_{k-1}$ are geodesic segments. The (internal) angle between γ_j and γ_{j+1} is denoted by α_j, and so the external angle is $\pi - \alpha_j$.

Fix a tangent vector $X(0)$ at $c(0) = \gamma_0(0)$ and parallel transport it along $\gamma_0 \ldots \gamma_{k-1}$. We obtain a parallel vector field along $\gamma_0 \ldots \gamma_{k-1}$. The angle ϕ_i between $X|_{\gamma_j}$ and γ_j' stays constant on each edge γ_j, and makes a jump of $\phi_{j+1} - \phi_j = \pi - \alpha_j$ at each corner of the polygon. When we let the polygonal line $\gamma_0 \ldots \gamma_{k-1}$ approach c, in the limit we obtain a parallel vector field X along c (see Remark 4.2.9). The total angle jump $\sum_{j=1}^{k}(\pi - \alpha_j)$ approaches $\int_{\partial R}(d\phi/dt)dt$, where ϕ is the angle between X and c'. Applying (7.7.1) to each polygon and passing to limit we obtain

$$\int_{\partial R} \frac{d\phi}{dt} dt = 2\pi - \frac{1}{r^2}\mathrm{Area}(R).$$

Lemma 7.7.1 yields the conclusion.

If ∂R is a piecewise smooth closed curve, then the total angle jump $\sum_{j=1}^{k}(\pi - \alpha_j)$ approaches $\int_{\partial R}(d\phi/dt)dt + \sum_{i=1}^{n}(\pi - A_i)$, hence the formula (7.7.2) needs to be adjusted by a quantity of $\sum_{i=1}^{n}(\pi - \angle A_i)$. ∎

7.7. THE GAUSS-BONNET THEOREM

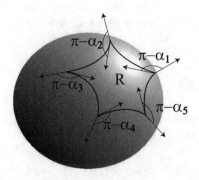

FIGURE 7.7.2
Hopf's theorem of turning tangents in the particular case of the sphere.

REMARK 7.7.3 As we know from Section 4.1, when we parallel transport a tangent vector along a simple closed curve back to the starting point, the final vector does not coincide, in general, with the original vector. The angle made by these two vectors is called the holonomy of the simple closed curve c, and is given by

$$\mathcal{H}(c) = 2\pi - \int_c \kappa_g,$$

if the curve is smooth, and by

$$\mathcal{H}(c) = 2\pi - \int_c k_g - \sum_{k=1}^{n}(\pi - \angle A_k),$$

if it is a curvilinear polygon. The holonomy does not depend on the particular choice of a tangent vector parallel transported along the curve, but only on the curve itself. Lemma 7.7.2 says that the holonomy of a simple closed piecewise smooth curve on the sphere of radius r is

$$\mathcal{H}(c) = \frac{1}{r^2}\text{Area}(R),$$

where R represents the region bounded by c.

An important property of holonomy is that it depends continuously on the curves in the C^1 topology. This means that if c_n is a sequence of smooth parametrized curves that approaches a smooth parametrized curve c in the C^1 topology, then $\mathcal{H}(c_n) \to \mathcal{H}(c)$. This property follows from the fact that if X_n is a parallel vector field along c_n and X_n approaches a vector field X along c, then X is a parallel vector field along c. See Remark 4.2.9. ∎

Let D be either the unit 2-dimensional disk or a polygon in $\mathbb{R}^2$, and $\psi : D \to S_r^2$ be a smooth map. We define the 'oriented area' of $\psi(D)$ by

$$\text{Area}(\psi(D)) = \int_D \psi^* \sigma',$$

where σ' is the standard area form on S_r^2 (see Example 6.6.3). In the particular case when ψ is an orientation preserving embedding, this is $+\psi(D)$, and in the case when ψ is orientation reversing, this is $-\psi(D)$.

THEOREM 7.7.4 (Gauss-Bonnet Theorem — local version)
Assume that R is a homeomorphic copy of a disk in a compact orientable surface M embedded in $\mathbb{R}^3$. Let K denote the Gaussian curvature of M.

If the boundary ∂R of R is a smooth curve parametrized by the arc length, then

$$\int_{\partial R} \kappa_g + \int_R K = 2\pi, \qquad (7.7.4)$$

where κ_g is the geodesic curvature of ∂R.

If the boundary ∂R of R is a piecewise smooth curve parametrized by the arc length, then

$$\int_{\partial R} \kappa_g + \sum_{i=1}^n (\pi - \angle A_i) + \int_R K = 2\pi, \qquad (7.7.5)$$

where $\pi - \angle A_i$ are the exterior angles between the smooth arcs at the corners of the boundary, measured counter-clockwise relative to the positive orientation of the tangent plane.

PROOF Consider the Gauss map $G : M \to S^2$ defined by $G(p) = N_p$, taking p to the positively oriented unit normal vector $N_p \in S^2$. We choose the orientation on M such that N_p points outwards S^2.

7.7. THE GAUSS-BONNET THEOREM

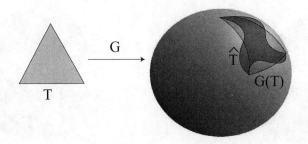

FIGURE 7.7.3
A geodesic triangle in S^2 approximating the image through G of a geodesic triangle in M.

Assume first that ∂R is represented by a smooth curve parametrized by arc length. We approximate R by a geodesic polygon in M, and triangulate the polygon into small geodesic triangles. The image $G(T)$ of a geodesic triangle T in M is not, in general, a geodesic triangle in S^2. However, by connecting the images of the vertices of T by unit speed geodesics in S^2, we obtain a geodesic triangle $\widehat{T}$ in S^2 which approximates $G(T)$. See Figure 7.7.3. In this way, we construct a network of geodesic triangles $\widehat{T}$ in S^2 such that the vertices of each $\widehat{T}$ are the images under G of the corresponding geodesic triangle T in M. See Figure 7.7.4.

For each geodesic triangle $\widehat{T}$ in S^2, by Lemma 7.7.2 we have

$$2\pi - \sum_{i=1}^{3} \angle A_i = \text{Area}(\widehat{T}), \tag{7.7.6}$$

where $\angle A_i$, $i = 1, \ldots, 3$, represent the angles of the geodesic triangle. The left side of the equality represents the holonomy of the boundary $\partial \widehat{T}$ of the geodesic triangle. The holonomy of $\partial \widehat{T}$ and the area of $\widehat{T}$ are counted positively if G preserves orientation (in the sense that the vertices of T traversed counter-clockwise correspond to the vertices of $\widehat{T}$ traversed counter-clockwise), and negatively otherwise. The above

320　7. FIXED POINTS AND INTERSECTION NUMBERS

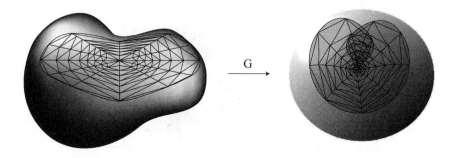

FIGURE 7.7.4
Triangulation of M and construction of a network of geodesic polygons in S^2.

relationship is valid even in the case when the geodesic triangle is degenerate.

We add the holonomies of the boundaries of the geodesic triangles. The parallel transports along the edges shared by adjacent geodesic triangles cancel, since each of these edges is traversed twice, with opposite orientations. Hence, the sum of the holonomies of the boundaries of the geodesic triangles equals the holonomy of the geodesic polygonal line approximating $G(\partial R)$.

Observe that the holonomy of $G(\partial R)$ equals the holonomy of ∂R. To see this, consider a parallel vector field X along c. Since the tangential component DX/dt of dX/dt is zero, it results that dX/dt is a multiple of N at each point. Define the vector field $\widehat{X}$ along $G(\partial R)$, where $\widehat{X}(t)$ is the parallel copy of $X(t)$ based at the corresponding image point in S^2. Since $d\widehat{X}/dt = dX/dt$, $\widehat{X}$ is a parallel vector field along the image of the boundary of R. Thus, the holonomy of $G(\partial R)$ is the same as the holonomy of ∂R, i.e.

$$2\pi - \int_{\partial R} \kappa_g = 2\pi - \int_{G(\partial R)} \kappa_g.$$

By (7.7.6), the holonomy of the geodesic polygonal line equals the

7.7. THE GAUSS-BONNET THEOREM

sum of the oriented areas of the geodesic triangles $\widehat{T}$. Since the geodesic polygonal line in S^2 formed by the geodesic triangles $\widehat{T}$ is C^1-close to the curve representing $G(\partial R)$, the sum of the oriented areas of the geodesic triangles is approximately equal to the oriented area of $G(R)$. To compute the oriented area of $G(R)$, note that the determinant of the Jacobi matrix of the Gauss map is the Gaussian curvature, so the change of variable formula implies

$$\text{Area}(G(R)) = \int_R \det(JG) = \int_R K.$$

We now pass to limit by letting the geodesic polygon approach R, and letting the size of the geodesic triangles T get smaller and smaller. As discussed in Remark 7.7.3, the holonomy of the geodesic polygon approaches the holonomy of $G(\partial R)$. The total area of the geodesic triangles approaches $\int_R K$. In the limit, we obtain

$$2\pi - \int_{\partial R} \kappa_g = \int_R K.$$

The case when the boundary of R is piecewise smooth follows similarly from (7.7.3). ∎

THEOREM 7.7.5 (Gauss-Bonnet theorem — global version)
Assume that M is a compact orientable surface M embedded in $\mathbb{R}^3$ and K is the Gaussian curvature of M. Then

$$\int_M K = 2\pi \chi(M). \tag{7.7.7}$$

PROOF Consider a triangulation of M with V vertices, E edges and F faces. Applying Theorem 7.7.4 to each face, we have

$$\int_\triangle K = 2\pi - \int_{\partial \triangle} \kappa_g - \sum_{i=1}^3 (\pi - \angle A_i). \tag{7.7.8}$$

We add these equations over all of the triangles. Each triangle is assumed to be positively oriented, and the boundary of each triangle is travelled counter-clockwise. This means that each edge of the triangulation, being shared by exactly two triangles, is travelled once clockwise and once counter-clockwise. The sum of $\int_{\partial \triangle} \kappa_g$ over all triangles equals 0. At each vertex, the sum of the adjacent angles $\angle A$ is 2π, so the sum of $-\sum_{i=1}^{3}(\pi - \angle A_i)$ over all triangles is $-\pi F + 2\pi V = 2\pi(V - F/2)$. In the proof of Theorem 7.6.8, we showed that for a general triangulation we have $V - F/2 = V - E + F = \chi(M)$. This shows that the sum of the right-hand side of (7.7.8) over all triangles is $2\pi\chi(M)$. It is clear that the sum of the left-hand side of (7.7.8) over all triangles is $\int_M K$. This completes the proof. ∎

The Gauss-Bonnet theorem remains valid for compact orientable 2-dimensional Riemannian manifolds (not necessarily embedded in $\mathbb{R}^3$). The proof relies on Hopf's theorem of turning tangents, and on Stokes' theorem, and can be found in Lee (1997).

Example 7.7.6
For a geodesic triangle $\triangle ABC$ in the hyperbolic plane, since $K = -1$, the Gauss Bonnet theorem implies

$$\angle A + \angle B + \angle C = \pi - \text{Area}(\triangle ABC).$$

☐

The global version of the Gauss-Bonnet theorem says that the curvature of a compact orientable 2-dimensional manifold M determines its topology, and, conversely, the topology of a manifold provides information about its curvature. For example, if $K > 0$, then M is a homeomorphic copy of a sphere, and if $K \leq 0$, then M is a surface of genus $g \geq 1$. Conversely, if M is homeomorphic to a sphere then $K > 0$ on some region of M; if M is homeomorphic to a flat torus then $K = 0$ or K reaches both positive and negative values; if M is a surface of genus $g \geq 1$ then $K < 0$ on some region of M.

REMARK 7.7.7 The Gauss-Bonnet theorem supplements our understanding of the intrinsic geometries of the standard surfaces. These geometries can be described through axiomatic systems. The parallel postulate of Euclidean geometry is equivalent to the statement that the angle sum of any triangle is 180°.

The parallel postulate can be negated in two ways. One way is by postulating the existence of infinitely many parallel lines to a given line. It turns out that the angle sum of any triangle is less than 180°. This corresponds to the hyperbolic plane model. The rest of the axioms from Euclidean geometry remain unchanged.

Another way to negate the parallel postulate is by postulating the existence of no parallel lines to a given line. This corresponds to spherical and elliptic geometry. In fact, in spherical and elliptic geometry there are no parallel lines at all. The angle sum of any triangle is greater than 180°. Some other axioms from Euclidean geometry become invalid. ∎

REMARK 7.7.8 The definition of the Gauss map $G : M \to S^{n-1}$ can be extended to orientable n-dimensional submanifolds of $\mathbb{R}^{n+1}$. The degree $\deg(G)$ of the Gauss map G is called the normal degree. The degree equals $\int_M G^*\omega / \int_{S^n} \omega$, where ω is the canonical volume form on S^n (see Remark 7.2.7). By the change of variable formula we obtain

$$\deg G = \frac{\int_M K}{\mathrm{Vol}_n(S^n)},$$

where Vol_n denotes the n-dimensional volume. The Gauss-Bonnet theorem for surfaces can be restated as

$$\deg(G) = \frac{\chi(M)}{2}.$$

This statement remains true for orientable n-dimensional submanifolds of $\mathbb{R}^{n+1}$ with n an even number. See Guillemin and Pollack (1974) for a proof. For odd-dimensional submanifolds, the above formula is false since the Euler characteristic is automatically 0. There is a generalization of the Gauss-Bonnet theorem to compact orientable even-dimensional Riemannian manifolds, due to Chern. For higher dimensions, curvature means the Riemannian curvature tensor. The Riemannian curvature tensor R can be used to define a scalar substitute $\mathcal{P}$ for the Gaussian curvature, called the Pfaffian, that can be integrated over M. The generalized Gauss-Bonnet theorem states that

$$\int_M \mathcal{P} = \frac{1}{2}\mathrm{Vol}_n(S^n)\chi(M).$$

A history of the Gauss-Bonnet theorem and some other extensions of it can be found in Gottlieb (1996). ∎

7.8 Exercises

7.8.1 Use the winding number to prove that the system of equations

$$-x\sin^2\left(\frac{1}{x^2+y^2+1}\right) = 0.1,$$

$$-y\sin^2\left(\frac{1}{x^2+y^2+2}\right) = 0.2,$$

has a solution inside the unit disk in $\mathbb{R}^2$.

7.8.2 Suppose that D is a disk in the complex plane, $f : \mathbb{C} \to \mathbb{C}$ is a holomorphic function, and y is a point not on $\partial(f(D))$. Prove that

$$\deg(f, D, y) = W(f|_{\partial D}, y) = \frac{1}{2\pi i}\int_{\partial D} \frac{f'(z)}{f(z) - y} dz.$$

7.8.3 Show by an example that the condition on $\mathrm{Ind}_{x_0}(f) \neq 0$ in Proposition 7.4.2 cannot be dropped.

7.8.4 Prove that the index of a vector field at an isolated zero equals the index of a time discretization of the flow induced by that vector field at the corresponding fixed point.

7.8.5 Prove the following theorem due to Perron and Frobenius. Any matrix A with all positive entries has a positive eigenvalue.

Hint: We regard A as a linear transformation of R^n. Then A maps the (closed) positive cone $\{(x_1, \ldots, x_n) \,|\, x_1 \geq 0, \ldots, x_n \geq 0\}$ to itself. The map $f(v) = Av/\|Av\|$ of the unit sphere into itself takes the region of the sphere within the closed positive cone into itself. This region is homeomorphic to a $(n-1)$-dimensional disk. Apply the Brouwer fixed point theorem.

7.8.6 (Borsuk-Ulam theorem). Let $f : S^n \to \mathbb{R}^n$ be a continuous map. There exists a pair of antipodal points on S^n that are mapped by f to the same point in $\mathbb{R}^n$.

7.8.7 Prove that whenever S^n is covered by the union of $n+1$ closed sets $A_1, \ldots, A_{n+1}$, then at least one of these sets must contain a pair of antipodal points x and $-x$. Find a cover of S^2 by four closed sets without antipodal pair of points.

7.8. EXERCISES

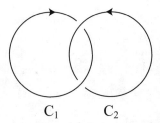

C_1 C_2

FIGURE 7.8.1
Linking number.

7.8.8 Show that if a map $f : S^n \to S^n$ has no fixed points, then $\deg(f) = (-1)^{n+1}$.

7.8.9 Let M_1 and M_2 be compact oriented submanifolds of $\mathbb{R}^n$, of dimensions m_1 and m_2, respectively. Assume that M_1 and M_2 are disjoint and $m_1 + m_2 = n - 1$. Define $f : M_1 \times M_2 \to S^2$ by

$$f(x_1, x_2) = \frac{x_1 - x_2}{\|x_1 - x_2\|},$$

where $M_1 \times M_2$ is given the product orientation. The linking number $\text{Lk}(M_1, M_2)$ is defined to be the degree of f.

(i) Let C_1 and C_2 be two two circles in $\mathbb{R}^3$ as shown in Figure 7.8.1. Find the linking number $\text{Lk}(C_1, C_2)$.

(ii) Show that $\text{Lk}(M_1, M_2) = (-1)^{m_1-1}(-1)^{m_2-1}\text{Lk}(M_2, M_1)$.

7.8.10 Show that in spherical geometry and in hyperbolic geometry, if two geodesic triangles are similar, then they are congruent. More precisely, if $\triangle ABC$ and $\triangle A'B'C'$ have $\angle A = \angle A'$, $\angle B = \angle B'$ and $\angle C = \angle C'$, then there exists an isometry that takes $\triangle ABC$ to $\triangle A'B'C'$.

7.8.11 Assume that M is an oriented Riemannian 2-dimensional manifold, and R is a homeomorphic copy of a disk in M. Assume that the boundary ∂R of R is represented by a smooth parametrized curve $c(t)$. Give an alternative proof of the local version of the Gauss-Bonnet theorem following the outline below.

(i) Show that if X and Y are smooth fields on M, then $\theta_{X,Y}(v) = \langle \nabla_v X, Y \rangle$ defines a 1-form on M.

(ii) Assume that X and Y are a positively oriented orthonormal basis on some neighborhood U of R. If X' and Y' are another positively oriented orthonormal basis on U, show that $\theta_{X,Y} = \theta_{X',Y'} + d\alpha$, where α is the angle between X' and X, positively measured.

(iii) Show that $d\theta_{X,Y} = KX \wedge Y$.

(iv) If X' and Y' are parallel along c, then $\int_c \theta_{X',Y'} = 0$.

(iv) Show that $\int_c d\alpha$ is the holonomy of c.

(v) Use Stokes' theorem to derive the local version of the Gauss-Bonnet theorem.

7.8.12 Assume that M is an oriented Riemannian 2-dimensional manifold, and R is a geodesic ball in M. Give an alternative proof of the local version of the Gauss-Bonnet theorem following the outline below.

(i) Assume that R is parametrized by $\alpha(s,t) = \exp_p(tv(s))$, where $v(s)$ is the unit vector at angle s from some fixed tangent vector $v_0 \in T_pM$ (the angle s is measured in agreement with the positive orientation on M), with $s \in [0, 2\pi]$ and $t \in [0,T]$, for some $T > 0$. Note that α represents a variation through geodesics. Consider the Jacobi field $Y_s(t) = \dfrac{\partial \alpha}{\partial s}(s,t)$ along the geodesic $t \to \alpha(s,t)$. Let $E(s,t)$ be a unit vector field on $R \setminus \{p\}$ that is tangent to the geodesic circles, such that the vectors $E(s,t)$ with t fixed agree with the positive orientation on M. Let $y_s(t) = \left\| \dfrac{\partial \alpha}{\partial s}(s,t) \right\| = \left\langle \dfrac{\partial \alpha}{\partial s}(s,t), E(s,t) \right\rangle$. Show that $y_s''(t) + K(\alpha(s,t))y_s(t) = 0$.

(ii) Show that $\left\langle \dfrac{D}{\partial t}\dfrac{D}{\partial t}\dfrac{\partial \alpha}{\partial s}(s,t), E(s,t) \right\rangle = \dfrac{d}{dt}\left\langle \dfrac{D}{\partial s}\dfrac{\partial \alpha}{\partial t}(s,t), E(s,t) \right\rangle$

and that $\displaystyle\int_0^T \int_0^{2\pi} y_s''(t)\,ds\,dt = \int_0^{2\pi} \left\langle \dfrac{D}{\partial s}\dfrac{\partial \alpha}{\partial t}(s,T), E(s,T) \right\rangle ds$.

(iii) Show that $\left\langle \dfrac{D}{\partial s}\dfrac{\partial \alpha}{\partial t}(s,T), E(s,T) \right\rangle = \theta_{X,Y}\left(\dfrac{\partial \alpha}{\partial s}(s,T) \right)$, where θ is the 1-form described in Exercise 7.8.5, X is the outward unit normal to ∂R, and Y is the unit tangent vector to ∂R.

(iv) Conclude the local version of the Gauss-Bonnet theorem.

Chapter 8

Morse Theory

8.1 Introduction

Morse theory can be viewed as a generalized version of the local maxima and minima tests from calculus: for a 1-variable function, the points where the graph changes most dramatically — the minimum and maximum points — are among the critical points of the function. This principle, translated to the setting of Morse theory, says that, for a generic real-valued function on a manifold, the changes in the topology of the level sets occur at the critical points of the function.

The following classical example illustrates this idea. Consider a torus sitting on top of the xy plane as in Figure 8.1.1. Let $f: M \to \mathbb{R}$ be the functional representing the height above the xy plane, and M^a be the sub-level set consisting of all points of the torus of height less than or equal to a.

The critical points p, q, r, s of f are the points where the tangent plane to the torus is horizontal. They are all nondegenerate: near each critical point, the manifold is not 'flat' in any direction. The points p and s are global extrema and q and r are saddle points. We will follow the changes of the topology of M^a as a varies.

- If $a < f(p)$, then M^a is the empty set.
- If $a = f(p)$, then M^a is a point.
- If $f(p) < a < f(q)$, then M^a is a disk, which is, from a homotopy point of view, the same as a point.
- If $a = f(q)$, then M^a is, from a homotopy point of view, the same as a cylinder, which can be obtained by attaching the endpoints of

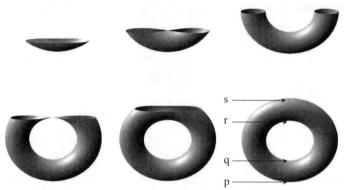

FIGURE 8.1.1
Sub-level sets of the height functional.

a 1-dimensional disk to the boundary of the disk from the previous step.

- If $f(q) < a < f(r)$, then M^a is still a cylinder.
- If $a = f(r)$, then M^a is, from a homotopy point of view, a torus with a cap removed, which can be obtained by attaching one endpoint of a 1-dimensional disk to one of the boundary circles of the cylinder from the previous step and attaching the other endpoint to the other boundary circle.
- If $f(r) \leq a < f(s)$, then M^a is still a torus with a cap removed.
- If $a \geq f(s)$, then M^a is torus, which can be obtained by attaching the boundary of a 2-dimensional disk to the boundary of the cut-off torus from the previous step.

We notice that the topology of M^a changes only when a passes through a critical value of f. Furthermore, the topology of the whole of M can be re-constructed by successively attaching disks of different dimensions. The attachment of those disks is done along the trajectories of the gradient flow. This flow is obtained by integrating the gradient vector field. In our example, the gradient flow lines follow the trajectories of an ink drop trickling on the surface of the torus. The critical points are the fixed points of the flow. The dimension of the attaching disk equals the number of directions in which the gradient flow leaves the critical point.

8.2 Functions with nondegenerate critical points

In this section, M will always denote a smooth m-dimensional Riemannian manifold without boundary. A critical point p of a smooth function $f : M \to \mathbb{R}$ is a point where $(df)_p = 0$. In local coordinates

$$\frac{\partial f}{\partial x_1}(p) = \cdots = \frac{\partial f}{\partial x_m}(p) = 0.$$

We can describe critical points as zeroes of the gradient vector field. The gradient grad f of a function on a surface was defined in the proof of Theorem 7.6.8 by the condition $\langle (\mathrm{grad} f)_p, v \rangle = (df)_p(v)$ for all tangent vectors v at p, where $\langle \cdot, \cdot \rangle$ is the Riemannian metric on the surface induced by the dot product. This will be the basis for our next definition.

DEFINITION 8.2.1
Let $f : M \to \mathbb{R}$ be a smooth function. The gradient vector field $\mathrm{grad} f$ is uniquely defined by the condition

$$\langle (\mathrm{grad} f)_p, V_p \rangle = (df)_p(V_p),$$

for every smooth vector field V on M.

In local coordinates, $\mathrm{grad} f$ is defined by

$$\mathrm{grad}\, f = \sum_{i=1}^{m} \left(\sum_{j=1}^{m} g^{ij} \frac{\partial f}{\partial x_j} \right) \frac{\partial}{\partial x_i},$$

where (g^{ij}) denotes the inverse of the matrix (g_{ij}) associated to the Riemannian metric on M.

The matrix of the second order partial derivatives of a function at a critical point can be defined intrinsically as follows:

DEFINITION 8.2.2
The Hessian of f at the critical point p is the symmetric bilinear form on the tangent space $\mathrm{Hess}_p(f) \colon T_pM \times T_pM \to \mathbb{R}$ defined by $\mathrm{Hess}_p(f)(v,w) = V(W(f))(p)$, where V and W are extensions of the tangent vectors v and w, respectively, to smooth vector fields in a neighborhood of p.

We prove that this definition is independent of the choice of vector field extensions. To this end, consider a local coordinate system $(x_1, \ldots, x_m)$, and assume that $V = \sum_{i=1}^m V_i(\partial/\partial x_i)$ and $W = \sum_{i=1}^m W_i(\partial/\partial x_i)$, with $V_i(p) = v_i$ and $W_i(p) = w_i$ for all i. We have

$$\mathrm{Hess}_p(f)(v,w) = \sum_{i,j=1}^m \left[V_i \frac{\partial W_j}{\partial x_i} \frac{\partial f}{\partial x_j} + V_i W_j \frac{\partial^2 f}{\partial x_i \partial x_j} \right](p)$$

$$= \sum_{i,j=1}^m v_i w_j \frac{\partial^2 f}{\partial x_i \partial x_j}(p),$$

since p is a critical point of f. This shows that the Hessian depends only on the initial tangent vectors v and w. We also see that the Hessian is symmetric in v and w. In local coordinates, the Hessian is represented by the matrix

$$\left(\frac{\partial^2 f}{\partial x_i \partial x_j}(p) \right)_{i,j}.$$

DEFINITION 8.2.3

A critical point p of f is called nondegenerate if the bilinear form $\mathrm{Hess}_p(f)$ is nondegenerate, that is, $\mathrm{Hess}_p(f)(v,w) = 0$ for all $w \in T_pM$ if and only if $v = 0$.

The Morse index of f at p is the maximal dimension of a subspace V of T_pM on which $\mathrm{Hess}_p(f)$ is negative definite, that is $\mathrm{Hess}_p(f)(v,v) < 0$ for all $v \in V$.

In terms of local coordinates, p is nondegenerate if and only if the Hessian matrix is non-singular, or, equivalently, all of its eigenvalues are different from 0. The Morse index of p equals the number of negative eigenvalues of $\mathrm{Hess}_p(f)$, counting multiplicities.

We can characterize nondegenerate critical points in terms of the gradient vector field. A critical point p of f is nondegenerate if and only if $\mathrm{grad} f$ is transverse to the zero section X_0 of TM at p. To see this, note that if p is a critical point, then $(\mathrm{grad} f)_p$ and

$$d(\mathrm{grad} f)_p : T_pM \to T_{(p,0)}(TM).$$

Let us use local coordinates $x_1, \ldots, x_m$ around p for which $\partial/\partial x_1, \ldots, \partial/\partial x_m$ is an orthonormal basis for T_pM. With respect to these coordinates and the corresponding local coordinates $x_1, \ldots, x_m, v_1, \ldots, v_m$

8.2. NONDEGENERATE CRITICAL POINTS

around $(p, 0)$ in TM, the matrix representing $d(\text{grad } f)_p$ is

$$\begin{pmatrix} \text{id} & \dfrac{\partial^2 f}{\partial x_i \partial x_j} \end{pmatrix}.$$

The matrix of the differential of the map $X_0 \colon M \to TM$ at p,

$$d(X_0)_p \colon T_p M \to T_{(p,0)}(TM)$$

is

$$d(X_0)_p = \begin{pmatrix} \text{id} & 0 \end{pmatrix}.$$

The transversality condition

$$d(\text{grad } f)_p(T_p M) + d(X_0)_p(T_p M) = T_{(p,0)}(TM)$$

is equivalent to the fact that $\text{Hess}_p(f)$ is non-singular.

The following lemma states that, relative to some special local coordinates, all level sets near a nondegenerate critical point are quadrics of the type $-y_1^2 - \cdots - y_l^2 + y_{l+1}^2 + \cdots + y_m^2 = \text{constant}$.

LEMMA 8.2.4 *(Morse's lemma)*
Let f be a smooth function on M, and p be a nondegenerate critical point of Morse index l. Then, in some open neighborhood of p, there are local coordinates $(y_1, \ldots, y_m)$ with $y_i(p) = 0$ for all i, such that, in these coordinates, the function f takes the form

$$f(y_1, \ldots y_m) = f(p) - y_1^2 - \cdots - y_l^2 + y_{l+1}^2 + \cdots + y_m^2. \tag{8.2.1}$$

PROOF The main is idea to mimic the standard diagonalization procedure of a quadratic form. The proof consists of three steps. For simplicity, we will assume that $p = 0$ and $f(0) = 0$.

Step 1. We show that

$$f(x_1, \ldots, x_m) = \sum_{i,j=1}^{m} x_i x_j h_{ij}(x_1, \ldots, x_m), \tag{8.2.2}$$

for some smooth functions h_{ij} with $h_{ij} = h_{ji}$ and $h_{ij}(0) = \dfrac{1}{2}\dfrac{\partial^2 f}{\partial x_i \partial x_j}(0)$.

We choose a neighborhood U of p so that it represents a disk in the local coordinates $x = (x_1, \ldots, x_m) \subseteq \mathbb{R}^m$. By the fundamental theorem of calculus,

$$f(x) = f(x) - f(0) = \int_0^1 \frac{d}{dt} f(tx_1, \ldots, tx_n) dt$$

$$= \sum_{i=1}^m x_i \int_0^1 \frac{\partial f}{\partial x_i}(tx_1, \ldots, tx_m) dt = \sum_{i=1}^m x_i g_i(x_1, \ldots, x_m),$$

where $g_i(x_1, \ldots, x_m) = \int_0^1 \frac{\partial f}{\partial x_i}(tx_1, \ldots, tx_m) dt$. Note that $g_i(0) = 0$ since $p = 0$ is a critical point. We repeat this procedure for each g_i, obtaining $g_i(x_1, \ldots, x_m) = \sum_{j=1}^m x_j h_{ij}(x_1, \ldots, x_m)$, where $h_{ij}(x_1, \ldots, x_m) =$
$\int_0^1 \frac{\partial g_i}{\partial x_j}(sx_1, \ldots, sx_m) ds = \int_0^1 \int_0^1 \frac{\partial^2 f}{\partial x_i \partial x_j}(tsx_1, \ldots, tsx_m) dt ds$. Combining these expressions, we obtain the equation (8.2.2). Note that $h_{ij}(0) = \int_0^1 \int_0^1 \frac{\partial^2 f}{\partial x_i \partial x_j}(0) dt ds = \frac{1}{2} \frac{\partial^2 f}{\partial x_i \partial x_j}(0)$, hence $(h_{ij}(0))_{i,j}$ is nonsingular.

Step 2. We show that there exist local coordinates $(y_1, \ldots, y_m)$ near p, with $y_i(p) = 0$ for all i, such that:

$$f(y_1, \ldots, y_m) = f(p) \pm y_1^2 \pm y_2^2 \pm \cdots \pm y_m^2. \tag{8.2.3}$$

We proceed by finite induction. Choosing a smaller neighborhood U of p if necessary, we suppose that there exist local coordinates $(u_1, \ldots, u_m)$ on U with $u_i(p) = 0$ for all i, such that

$$f(u_1, \ldots, u_m) = \pm u_1^2 \pm \cdots \pm u_{k-1}^2 + \sum_{i,j=k}^m H_{ij}(u_1, \ldots u_m) u_i u_j,$$

where the matrix $(H_{ij}(p))_{i,j}$ is symmetric and invertible.

The case $k = 1$ is clear by Step 1.

We want to verify the induction step. In order to do this, we first ensure that the matrix $(H_{ij})_{i,j \geq k}$ has a nonzero entry on the diagonal. If not, we can choose $i \neq j$ such that $H_{ij}(0) = H_{ji}(0)$ are nonzero. Now replace u_i by $u_i' + u_j'$ and u_j by $u_i' - u_j'$. Then the coefficients of $u_i'^2$ and $u_j'^2$ will be $2H_{ij}(0)$ and $-2H_{ij}(0)$ respectively. Once we have a nonzero diagonal entry in $(H_{ij})_{i,j \geq k}$, it can be moved to the top left corner by interchanging two of the coordinates.

8.2. NONDEGENERATE CRITICAL POINTS

Making U smaller, if necessary, we can assume that $\sqrt{|H_{kk}|}$ is smooth on U. Now we complete the square:

$$f(u_1,\ldots,u_m) = \pm u_1^2 \pm \cdots \pm u_{k-1}^2$$

$$\pm \left[|H_{kk}|u_k^2 + 2u_k \sum_{j\geq k+1} H_{jk}u_j + \left(\sum_{j\geq k+1} \frac{H_{jk}}{\sqrt{|H_{kk}|}} u_j\right)^2 \right.$$

$$\left. - \left(\sum_{j\geq k+1} \frac{H_{jk}}{\sqrt{|H_{kk}|}} u_j\right)^2 + \sum_{i,j=k+1}^{n} H_{ij}u_i u_j \right]$$

$$= \pm u_1^2 \pm \cdots \pm u_{k-1}^2 \pm \left(\sqrt{|H_{kk}|}u_k \pm \sum_{j\geq k+1} \frac{H_{jk}}{\sqrt{|H_{kk}|}} u_j\right)^2$$

$$\mp \left(\sum_{j\geq k+1} \frac{H_{jk}}{\sqrt{|H_{kk}|}} u_j\right)^2 \pm \sum_{i,j=k+1}^{m} H_{ij}u_i u_j.$$

Now we make the substitutions

$$v_i = \begin{cases} u_i, & \text{if } i \neq k, \\ \sqrt{|H_{kk}|}u_k \pm \sum_{j\geq k+1} \frac{H_{jk}}{\sqrt{|H_{kk}|}} u_j, & \text{if } i = k, \end{cases}$$

and obtain

$$f(v_1,\ldots,v_m) = \pm v_1^2 \pm \cdots \pm v_{k-1}^2 \pm v_k^2 + \sum_{i,j=k+1}^{m} H'_{ij}(v_1,\ldots,v_m)v_i v_j,$$

for some new coefficients H'_{ij}, with $i,j \geq k+1$.

The Jacobi matrix

$$\left(\frac{\partial v_i}{\partial u_j}\right)_{i,j} = \begin{pmatrix} 1 & 0 & & \cdots & & 0 & 0 \\ 0 & 1 & & \cdots & & 0 & 0 \\ & & \ddots & & & & \\ 0 & 0 & & \sqrt{|H_{kk}|} & & * & * \\ & & & & \ddots & & \\ 0 & 0 & & \cdots & & 1 & 0 \\ 0 & 0 & & \cdots & & 0 & 1 \end{pmatrix}$$

is invertible at p, so, by the inverse function theorem, the above substitution constitutes a legitimate coordinate change, possibly in some smaller neighborhood U of p. It also follows that $v_i(p) = 0$ for all i. The new coefficients H'_{ij} are symmetric in $i, j \geq k+1$. The non-degeneracy of p implies that the matrix $\left(H'_{ij}(p)\right)_{i,j}$ is invertible. This completes the induction step.

Step 3. We show that the number of negative coefficients in (8.2.3) is exactly l, so f has the same form as in (8.2.1).

We permute the coordinates $(y_1, \ldots, y_m)$ to get all the negative terms of (8.2.3) at the beginning. The Hessian form in represented in the $(y_1, \ldots, y_m)$ coordinates by the matrix

$$\begin{pmatrix} -2 & \ldots & & 0 & 0 & \ldots & 0 \\ & \ddots & & & & \ddots & \\ 0 & \ldots & -2 & 0 & \ldots & & 0 \\ 0 & \ldots & & 0 & 2 & \ldots & 0 \\ & \ddots & & & & \ddots & \\ 0 & \ldots & & 0 & 0 & \ldots & 2 \end{pmatrix}.$$

The number of negative entries on the diagonal equals the number of negative coefficients in (8.2.3). Since it also equals the number of negative eigenvalues of the Hessian, we conclude that (8.2.1) holds true.
∎

COROLLARY 8.2.5

The nondegenerate critical points of any smooth function $f : M \to \mathbb{R}$ are isolated. If M is compact, there are only finitely many nondegenerate critical points.

Example 8.2.6

(i) The function $f(x, y) = x^2 - y^2$ has the origin as a nondegenerate critical point. See Figure 8.2.1 (a).

(ii) The function $f(x, y) = x^2 + y^3$ has the origin as an isolated, but degenerate, critical point. See Figure 8.2.1 (b).

(iii) The function $f(x, y) = x^2 y^2$ has the origin as a degenerate, non-isolated, critical point. See Figure 8.2.1 (c). ☐

8.2. NONDEGENERATE CRITICAL POINTS

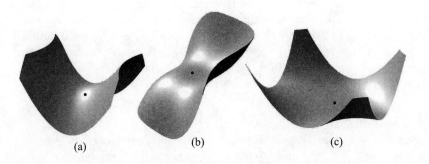

FIGURE 8.2.1
Critical points of functions.

DEFINITION 8.2.7
A smooth function $f: M \to \mathbb{R}$ is a Morse function if and only if all its critical points are nondegenerate.

As Morse functions are the principal tools in this chapter, we would like to know that there are many of them. We saw earlier in this section that the nondegeneracy of the critical points is equivalent to the transversality of the gradient vector field to the zero section of the tangent bundle. Since transversality is a generic condition, it follows that almost every smooth function is a Morse function.

THEOREM 8.2.8
The set of C^k-smooth Morse functions on a compact manifold M is open and dense in the set of C^k-smooth functions on M. The set of smooth Morse functions on a (non-compact) manifold M is dense in the set of smooth functions on M.

This theorem follows from the transversality theorem 1.14.4. The details of the proof can be found in Katok and Hasselblatt (1995).

As an example, we show directly that the 'height function' considered in Section 8.1 is a Morse function in most cases.

Example 8.2.9
Let M be a compact m-dimensional manifold embedded in $\mathbb{R}^{m+1}$, and u

be a unit vector in $\mathbb{R}^{m+1}$. The orthogonal projection $f_u \colon M \to \mathbb{R}$ of the the points of M into the line through u is described by the dot product $f_u(x) = u \cdot x$. The set of unit vectors u for which f_u is a Morse function is an open and dense set in S^m. The proof is left as an exercise. In the case when M is a surface in $\mathbb{R}^3$, the projection onto the z-axis is exactly the 'height function' from Section 8.1. □

Example 8.2.10
Suppose that four bodies of masses $m_1, \ldots, m_4$ are placed in general position at $x_1, \ldots, x_4$ in $\mathbb{R}^3$ (that is, they are at the vertices of a nondegenerate tetrahedron in $\mathbb{R}^3$). Assume that the positions of these four bodies stay fixed. The gravitational potential function of the configuration is
$$U(x) = -\frac{m_1}{\|x - x_1\|} - \cdots - \frac{m_4}{\|x - x_4\|}.$$
The critical points of U are the equilibrium points of the gravitational field, and an equilibrium point is nondegenerate if the critical point is so. Then for almost every x, the equilibrium points of the gravitational potential are nondegenerate and finite in number. The proof is left as an exercise. □

REMARK 8.2.11 Using Morse theory, one can prove the existence of a minimizing geodesic joining two points on a compact manifold. To explain this approach intuitively, we use the following physical model. We imagine a rubber band lying on a slippery surface between two fixed points. The position of the band is described by a parametric equation $x = c(t)$, $0 \leq t \leq 1$. The potential energy due to the elastic force is proportional to the quantity
$$E = \int_0^1 \left\| \frac{dc}{dt} \right\|^2 dt.$$
When we pull tight the band between the two fixed points, it will move towards an equilibrium position, corresponding to the minimum of the potential energy. This final position describes a geodesic connecting the two points. Thus, we can find minimizing geodesics by minimizing the energy functional. We can think of the energy as a Morse function defined on the space of all paths c connecting the two fixed points of the surface. This space can be modelled as some type of smooth manifold of infinite dimension. It is called a Hilbert manifold, since it is locally

diffeomorphic to an (infinite dimensional) Hilbert space. Geodesics turn out to be critical 'points' of the energy function on this Hilbert manifold. A treatment of Hilbert manifolds can be found in Klingenberg (1995). Morse theory on Hilbert manifolds is discussed in Palais (1963). Alternatively, one can use finite dimensional approximations to the infinite dimensional space of paths in order to obtain geodesics as energy minimizers (see Milnor (1963)).

Similar ideas can be used to prove the following theorem of Hadamard and Cartan: in every component of the space of all closed paths on a compact manifold, other than the component containing the constant map, there exists a non-trivial closed geodesic. The proof can be found in Bott (1982). ∎

8.3 The gradient flow

We fix a Morse function f. For every regular value a of f, the level set $f^{-1}(a)$ is a smooth $(m-1)$-dimensional submanifold of M.

From Theorem 2.3.3, the differential equation with initial condition

$$\frac{d\phi_x}{dt}(t) = -(\mathrm{grad} f)(\phi_x(t)),$$
$$\phi_x(0) = x, \qquad (8.3.1)$$

gives rise to a locally defined flow $\phi\colon (-\epsilon, \epsilon) \times U \to M$ on some open neighborhood U of x. This flow will be referred as the gradient flow. If M is a compact manifold, the gradient flow is globally defined. The critical points of f are the zeroes of the gradient vector field, and the index of the gradient vector field at a critical point is $(-1)^l$, where l is the Morse index of the critical point. Note that every fixed point p of the gradient flow ϕ is hyperbolic, and the Morse index of p is equal to the dimension of the unstable space $E^u(p)$ at p.

The next theorem describes the main properties of the gradient flow.

THEOREM 8.3.1
Let f be a Morse function on M.

(i) *The function f is strictly decreasing along the flow lines, except for the degenerate flow lines corresponding to the critical points where f is constant.*

(ii) *The gradient vector field is orthogonal to the level sets* $f^{-1}(a)$, *corresponding to regular values a of* f.

(iii) *If M is compact and $x \in M$, then $\lim_{t \to -\infty} \phi_x(t)$ and $\lim_{t \to +\infty} \phi_x(t)$ are critical points of f.*

PROOF (i) Let $t \in \mathbb{R} \longrightarrow \phi_x(t) \in M$ be a gradient flow line through the point x. We take the derivative of $(f \circ \phi_x)(t)$ with respect to t, and we use the definition of the gradient

$$\frac{d}{dt}(f(\phi_x(t)) = (df)_{\phi_x(t)}\left(\frac{d\phi_x}{dt}(t)\right) = \left\langle \operatorname{grad} f(\phi_x(t)), \frac{d\phi_x}{dt}(t) \right\rangle$$
$$= - \langle \operatorname{grad} f(\phi_x(t)), \operatorname{grad} f(\phi_x(t)) \rangle \leq 0. \qquad (8.3.2)$$

This shows that f is non-increasing along flow lines. If x is not a critical point, the inequality in (8.3.2) is strict.

(ii) Every tangent vector to $f^{-1}(a)$ at a point $x \in f^{-1}(a)$ is the derivative $(dc/dt)(0)$ of some curve $c : (-\epsilon, \epsilon) \to f^{-1}(a)$ with $c(0) = x$. Since $f(c(t)) = a$ for all t, by taking the derivative with respect to t, as above, we obtain $\langle \operatorname{grad} f, dc/dt \rangle = 0$, showing that the gradient vector field is orthogonal to the level sets.

(iii) Due to the compactness of the manifold, the gradient flow is globally defined, i.e., for every $x \in M$, $\phi_x(t)$ is defined for all $t \in \mathbb{R}$. Also, $t \in \mathbb{R} \to (f \circ \phi_x)(t) \in \mathbb{R}$ must be bounded, hence, by (8.3.2),

$$\lim_{t \to \pm\infty} \frac{d}{dt}(f(\phi_x(t)) = 0. \qquad (8.3.3)$$

By compactness and by Exercise 2.7.9, the ω-limit and the α-limit sets of x are non-empty, compact and connected, and by (8.3.3) they consist of critical points. Since the critical points are isolated (Corollary 8.2.5), the connectedness of the ω-limit implies that it consists of only one critical point, which equals $\lim_{t \to \infty} \phi_x(t)$. Similarly, the α-limit of x consists of only one critical point, which equals $\lim_{t \to -\infty} \phi_x(t)$. ∎

By the previous theorem, each point of the manifold flows towards some critical point, in forward or backward time. Therefore each critical point can be viewed as the crossing of two roads: one, consisting of the points that flow towards the point, and the other, consisting of the points that flow away from the point. These roads are the stable and unstable manifolds of p,

$$W^s(p) = \{x \in M \mid \lim_{t \to \infty} \phi_x(t) = p\},$$

8.3. THE GRADIENT FLOW

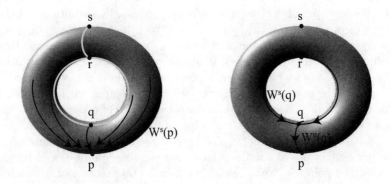

FIGURE 8.3.1
Stable and unstable manifolds of critical points.

$$W^u(p) = \{x \in M \mid \lim_{t \to -\infty} \phi_x(t) = p\},$$

defined in Section 2.6. Their existence follows from the stable manifold theorem, but in the case of the gradient vector field, one can prove their existence directly, using mainly the Morse lemma 8.2.4.

Example 8.3.2
In the example presented in Section 8.1, $W^s(p)$ is the torus with both the 'parallel' circle through q and r and the 'meridian' circle through r and s removed, while $W^u(p)$ is empty. For the critical point q, $W^s(q)$ is the 'parallel' circle through q and r with the point r removed, while $W^u(q)$ is the 'meridian' circle through q and p with the point p removed. See Figure 8.3.1. ◻

REMARK 8.3.3 Theorem 3.1 provides us with a topological decomposition of the manifold in terms of the gradient flow. The two basic structures of M are: the critical points, which are mapped by f into a nowhere dense subset of the real numbers (the critical values of f), and the regular points, whose trajectories always get trapped by the critical points. The function f is strictly decreasing along each of these trajectories.

This type of decomposition holds for general dynamical systems. To be precise, let us consider a flow ϕ acting on a compact metric space M. An (ϵ, τ)-pseudo-orbit (chain) from x to y is, by definition, a finite sequence of points $x_0 = x, x_1, \ldots, x_n = y$ such that $d(\phi^{t_j}(x_{i-1}), x_i) < \epsilon$ for $1 \leq i \leq n$, for some time increments $t_0, t_1, \ldots, t_{n-1}$, all greater than or equal to τ, and for some $n \geq 0$. An (ϵ, τ)-chain is essentially an orbit from x to y from an experimental point of view: after a time interval of at least τ, the position of $\phi^t(x)$ can only be measured with ϵ precision. The chain recurrent set $\mathcal{R}(\phi)$ is defined to be the collection of all points x for which there exists an (ϵ, τ)-chain from x to itself, for every pair of positive numbers ϵ and τ. The chain recurrent set contains all of the periodic orbits of the flow (including the fixed points), and maybe some other points as well. Conley's fundamental theorem of dynamical systems states that for any flow ϕ on a compact manifold M there exists a continuous function $\Gamma : M \to \mathbb{R}$ which is strictly decreasing along the flow lines which lie outside $\mathcal{R}(\phi)$, and such that the image of $\mathcal{R}(\phi)$ under Γ is a nowhere dense subset of the real numbers. See Robinson (1995) for details. ∎

8.4 The topology of level sets

For a real number a, the set M^a of all points x with $f(x) \leq a$ will be referred as the sub-level set of a. If a is a regular value of f, then M^a is an n-dimensional submanifold of M with boundary $\partial M^a = \{x \mid f(x) = a\}$. We will see that the topology of M^a changes only when a passes through a critical value of f. The comparison of the topologies will be done from a homotopy point of view. Two topological spaces M and N are said to be homotopy equivalent provided that there exist continuous mappings $f : M \to N$ and $g : N \to M$ such that $g \circ f$ is homotopic to the identity mapping on M, and $f \circ g$ is homotopic to the identity mapping on N. Intuitively, two spaces are homotopy equivalent provided that one space can be deformed into the other space without ripping or poking any holes. Homeomorphic spaces are obviously homotopy equivalent. The converse is not true: a disk and a point are homotopy equivalent without being homeomorphic. A natural class of homotopy equivalent spaces is obtained through deformation retractions. A subspace N is a deformation retract of M if there is a homotopy $r_t : M \to M$ with

8.4. THE TOPOLOGY OF LEVEL SETS

FIGURE 8.4.1
Attachments of a 1-dimensional disk to a cylinder. The 1-dimensional disk can be fattened up to a 'ribbon' whose ends are still glued to the boundary of the cylinder. On the left, the attachment is topologically equivalent to a torus with a 2-dimensional cap removed. On the right, the attachment is topologically equivalent to a cylinder with a 2-holed 2-dimensional cap glued to one end of the cylinder.

$r_0 = \mathrm{id}|_M$, $r_t|_N = \mathrm{id}|_N$ for all $t \in [0,1]$, and $r_1(M) \subseteq N$.

Before we state the theorem, we describe what it means to attach an l-dimensional disk D^l to a manifold with boundary M^a. This is a topological space, denoted by $M^a \cup D^l$, obtained by gluing the boundary of the l-dimensional disk D^l to the boundary ∂M^a of M^a, while the interior of the disk remains disjoint from the manifold. In Figure 8.4.1 we show a couple of different possible attachments of a 1-dimensional disk to the same manifold with boundary. It is important to keep in mind that the topology of the resulting space depends on the gluing.

THEOREM 8.4.1
Let f be a Morse function on M, and $a < b$ be real numbers with $f^{-1}[a,b]$ compact.

(i) If $f^{-1}[a,b]$ contains no critical point of f, then M^a is homotopy equivalent to M^b.

(ii) If $f^{-1}[a,b]$ contains exactly one critical point p of index l, then M^b is homotopy equivalent to M^a with an l-disk D^l attached.

(iii) If $f^{-1}[a,b]$ contains k critical points $p_1, p_2, \ldots, p_k$ of indices $l_1, l_2, \ldots, l_k$ respectively, corresponding to the same critical value, then M^b is homotopy equivalent to M^a with k disks $D^{l_1}, \ldots, D^{l_k}$ attached.

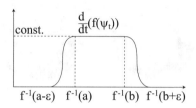

FIGURE 8.4.2
Rate of descent.

PROOF The idea is to push M^a along the gradient flow trajectories to M^b, in order to prove (i), or close enough to the critical level so that Morse's lemma can be used to prove (ii) and (iii).

(i) Since $f^{-1}[a,b]$ is compact, there exists a sufficiently small number $\epsilon > 0$ such that $f^{-1}[a-\epsilon, b+\epsilon]$ is still compact and free of critical points. Since the gradient flow trajectories are orthogonal to all of the level sets and contain no fixed points within $f^{-1}[a-\epsilon, b+\epsilon]$, we can define a smooth homotopy ψ_t of the level set $f^{-1}(b)$ into the level set $f^{-1}(b)$ that moves each point of $f^{-1}(b)$ along the gradient flow trajectory until it reaches $f^{-1}(a)$ in unit time, and at a constant descent rate, i.e.,

$$\frac{d}{dt}\bigl(f(\psi_t(x))\bigr) = \text{constant}, \quad \text{for all } t \in [0,1].$$

To extend this motion to the whole manifold, we are going to use the regions $f^{-1}[b, b+\epsilon]$ and $f^{-1}[a-\epsilon, a]$ as 'damping' zones. The motion speeds up from zero rate to constant rate while descending along the gradient flow trajectories within $f^{-1}[b, b+\epsilon]$, and slows down from constant rate to zero rate while descending along the gradient flow trajectories within $f^{-1}[a-\epsilon, a]$. This is shown in Figure 8.4.2.

Summarizing, the smooth deformation ψ_t satisfies the following properties:

- $\psi_t \colon M \to M$, for $t \in [0,1]$;

- ψ_t takes level sets into level sets, for each $t \in [0,1]$;

- $\psi_t =$ identity on the regions $f(x) \geq b + \epsilon$ and $f(x) \leq a - \epsilon$;

- $\psi_0 =$ identity;

- $\psi_1\bigl(f^{-1}(b)\bigr) \subset f^{-1}(a)$, $\psi_1\bigl(M^b\bigr) \subset M^a$.

8.4. THE TOPOLOGY OF LEVEL SETS

Since the inverse deformation is well defined by reversing the motion, it follows that ψ_1 is a diffeomorphism with $\psi_1(M^b) = M^a$. We conclude that M^a and M^b are homotopy equivalent.

(ii) Let $f(p) = c$, $a < c < b$. For simplicity, we assume that $c = 0$. By Morse's lemma, there exists an open neighborhood U of p and local coordinates $(y_1, \ldots, y_n)$ so that

$$f(x,y) = -|x|^2 + |y|^2, \qquad (8.4.1)$$

for all points $q = (x,y)$ in U, where we write $x = (y_1, \ldots, y_l)$, $y = (y_{l+1}, \ldots, y_m)$, $|x|^2 = y_1^2 + \cdots + y_l^2$, and $|y|^2 = y_{l+1}^2 + \cdots + y_m^2$. The gradient depends on the choice of the Riemannian metric. On $\mathbb{R}^m$ we consider the Euclidean metric. In order to be consistent, we choose a Riemannian metric on M that agrees on U with the metric induced by the Euclidean metric on $\mathbb{R}^m$ through the local parametrization, and we consider the gradient vector field with respect to this metric.

Step 1. We define a deformation of $f^{-1}[-\epsilon, \epsilon]$ into $f^{-1}(-\epsilon) \cup D^l$, for some l-disk D^l.

Let $\epsilon > 0$ be very small, such that U contains the box

$$B_0 = B^l(\sqrt{3\epsilon}) \times B^{m-l}(\sqrt{4\epsilon}),$$

where $B^l(\epsilon) = \{(x,0) \,|\, |x|^2 < \epsilon^2\}$ and $B^{m-l}(\epsilon) = \{(0,y) \,|\, |y|^2 < \epsilon^2\}$. The size of the box was chosen so that its 'corners' lie on the level set $f^{-1}(\epsilon)$. Define the l-dimensional disk to be attached by

$$D^l = \{(x,0) \,|\, |x|^2 \leq \epsilon\} = B^l(\sqrt{\epsilon}) \times 0.$$

Note that the boundary of D^l is represented by the points $(x,0)$ for which $|x|^2 = \epsilon$, so $f(x,y) = -\epsilon$. The boundary of this disk is glued to the level set $f^{-1}(-\epsilon)$. We now define a deformation of $f^{-1}[-\epsilon, \epsilon]$ into $f^{-1}(-\epsilon) \cup D^l$, relative to U.

Consider the smaller box

$$B_1 = B^l(\sqrt{2\epsilon}) \times B^{m-l}(\sqrt{3\epsilon})$$

located inside B_0. The 'corners' of this box also lie on the level set $f^{-1}(\epsilon)$. See Figure 8.4.3. The deformation of $f^{-1}[-\epsilon, \epsilon]$ into $f^{-1}(-\epsilon) \cup D^l$ will be obtained by patching together three deformations.

- In $B_1 \cap f^{-1}[-\epsilon, \epsilon]$, we construct a deformation that moves each point (x,y) at a constant descent rate along the interval joining (x,y) to $(x,0) \in D^l$, in the case when $|x|^2 < \epsilon$, or along the interval

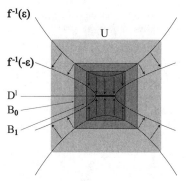

FIGURE 8.4.3
Attachments of an l-disk to $f^{-1}(-\epsilon)$.

joining (x, y) to the corresponding point $(x, y') \in f^{-1}(-\epsilon)$, in the case when $|x|^2 \geq \epsilon$. We choose the descent rate so that each point reaches $B_1 \cap f^{-1}(-\epsilon)$ in unit time. That means, the closer a point is to $\left(f^{-1}(-\epsilon) \cap B_1\right) \cup D^l$, the slower is its speed; hence, the points in $\left(f^{-1}(-\epsilon) \cap B_1\right) \cup D^l$ will be stationary.

- In $(U \backslash B_0) \cap f^{-1}[-\epsilon, \epsilon]$, we construct a deformation that moves each point (x, y) at constant speed along the trajectories of the gradient flow of f. We choose the descent rate along these trajectories so that each point reaches $f^{-1}(-\epsilon) \cap (U \backslash B_0)$ in unit time. The points of $f^{-1}(-\epsilon) \cap (U \setminus B_0)$ are stationary.

- In the 'damping zone' $f^{-1}[-\epsilon, \epsilon] \cap (B_0 \setminus B_1)$, we gradually push the direction of the gradient vector field of f towards the direction of the gradient vector field of g.

In the above, we also constructed a smooth vector field on $U \cap f^{-1}[-\epsilon, \epsilon]$, whose flow lines were used to define the appropriate deformations. Formally, this new vector field can be defined by

$$Y(x, y) = -\left((1 - \gamma(x, y))((\text{grad } f)(x, y)) + \gamma(x, y)((\text{grad } g)(x, y))\right),$$

where γ is a smooth 'damping' function whose graph is as in Figure 8.4.4. The function γ equals 1 on B_1 and equals 0 off B_0, hence the vector field Y agrees with $-\text{grad} f$ outside of B_0 and with $-\text{grad } g$ inside B_1. Note that the motion described above leaves the points of the set $f^{-1}(-\epsilon) \cup D^l$ fixed.

Step 2. We show that M^b is homotopy equivalent to $M^a \cup D^l$.

8.4. THE TOPOLOGY OF LEVEL SETS

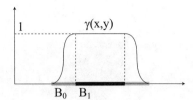

FIGURE 8.4.4
Damping function.

- We extend the deformation of $f^{-1}[-\epsilon, \epsilon]$ into $f^{-1}(-\epsilon) \cup D^l$ off U. We move each point at constant descent rate along the trajectories of $\text{grad} f$ so that it reaches $f^{-1}(-\epsilon) \cup D^l$ in unit time, in agreement with the motion described at Step 1. This motion defines a deformation retraction of $f^{-1}[-\epsilon, \epsilon]$ to $f^{-1}(-\epsilon)$, and implies the homotopy equivalence of M^ϵ to $M^{-\epsilon} \cup D^l$.

- Since $f^{-1}[\epsilon, b]$ and $f^{-1}[a, -\epsilon]$ do not contain any critical points, from part (i) we obtain M^b homotopy equivalent to M^ϵ and $M^{-\epsilon}$ homotopy equivalent to M^a.

We combine the homotopy equivalences from above and we conclude that M^b is homotopy equivalent to $M^a \cup D^l$.

(iii) We repeat the construction from (ii) near each critical point $p_1, \ldots, p_k$. The conclusion follows similarly. ∎

REMARK 8.4.2 (i) The condition that $f^{-1}[a, b]$ is compact is automatically satisfied provided that the manifold M is compact. However, if $f^{-1}[a, b]$ is not compact, the theorem is no longer true. For example, in Figure 8.4.5, the manifold M consists of a cylinder with one point p removed. The Morse function f is the 'height' function and has no critical points. Clearly, M^a and M^b are not homotopy equivalent. Even if $f^{-1}[a, b]$ is not compact but M possesses a complete Riemannian metric for which $\text{grad} f$ is bounded, then Theorem 8.4.1 (i) is still true.

(ii) Note that in Theorem 8.4.1 (i) we actually proved a stronger statement: M^b and M^a are diffeomorphic. Modifying slightly the motion along the gradient flow lines, one can prove that M^a is a deformation retract of M^b. See Milnor (1963) for details.

(iii) Note that in Theorem 8.4.1 (ii) and (iii) we obtain homotopy equivalences but not diffeomorphisms. Often $M^a \cup D^l$ cannot be represented as a smooth manifold (the 'dimension' varies from one point

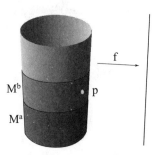

FIGURE 8.4.5
Sub-level sets that are not homotopy equivalent.

to another). Even in the cases when $M^a \cup D^l$ can be represented as a manifold, it is still possible that there is no diffeomorphism. See the application below and the subsequent remark for a counterexample. ∎

THEOREM 8.4.3 (Reeb's theorem)
If M is a compact m-dimensional manifold, which admits a Morse function f with only two critical points, then M is homeomorphic to the m-dimensional sphere.

PROOF Since M is compact, f must attain its minimum and maximum values: say $f(p_1) = c_1$ is the minimum and $f(p_2) = c_2$ is the maximum. Thus p_1 and p_2 are the critical points of f. See Figure 8.4.6. Morse's lemma implies the existence of local coordinates $(y_1, \ldots, y_m)$ near c_1 so that
$$f(q) = c_1 + y_1^2 + \ldots + y_m^2$$
for q near p_1. Therefore, the 'lower cap'
$$D_1 = \{q \mid f(q) \leq a\}$$
is diffeomorphic to an m-dimensional disk, provided that a is chosen close enough to c_1. Similarly, the 'upper cap'
$$D_2 = \{q \mid f(q) \geq b\}$$

8.4. THE TOPOLOGY OF LEVEL SETS

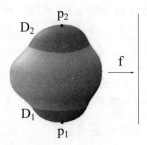

FIGURE 8.4.6
Manifold homeomorphic to a sphere.

is diffeomorphic to an m-dimensional disk, provided that b is chosen close enough to c_2.

Now, applying Theorem 8.4.1 (i) and (ii), it follows that M is homotopy equivalent to $M^b \cup D_2$, M^b is homotopy equivalent to M^a (the homotopy equivalence being obtained by descending along the gradient flow trajectories) and $M^a = D_1$. Hence M is homotopy equivalent to the topological space obtained by gluing the boundary of D_2, one-to-one, to the boundary of D_1. Thus M is homotopy equivalent to an m-dimensional sphere. ∎

REMARK 8.4.4 (i) The theorem is true even if the critical points are degenerate, but the proof is more technical. See Hirsch (1976).

(ii) It is not true, in general, that M must be diffeomorphic to the m-dimensional sphere (endowed with the standard smooth structure). Milnor found that the 7-dimensional sphere has 28 non-diffeomorphic smooth structures. The intersection between the 9-dimensional unit sphere S^9, defined by the equation $|z_1|^2 + \cdots + |z_5|^2 = 1$ in $\mathbb{C}^5 = \mathbb{R}^{10}$, and each of the 8-dimensional manifolds $z_1^{6k-1} + z_2^3 + z_3^2 + z_4^2 + z_5^2 = 0$, $k = 1, \ldots, 28$, is a 7-dimensional manifold homeomorphic to S^7. The 28 smooth structures induced on S^7 are mutually non-diffeomorphic. See Milnor (1956). ∎

8.5 Manifolds represented as CW complexes

The notion of a CW complex is a natural extension of the notion of a smooth manifold to a purely topological setting. While a manifold can be viewed as the result of 'gluing' together open disks of the same dimension, the gluing being done through diffeomorphisms on the common parts, CW complexes can be viewed as the results of gluing together closed disks of various dimensions, the gluing being done along the boundaries through continuous maps.

DEFINITION 8.5.1
A CW complex is a topological space constructed inductively as follows:

(i) Start with a collection X^0 of points (viewed as 0-dimensional disks), endowed with the discrete topology. This is called the 0-skeleton.

(ii) The set X^n is obtained by attaching n-dimensional closed disks D_α^n to X^{n-1} via continuous maps $\phi_\alpha : \partial D_\alpha^n \to X^{n-1}$; that is, X^n is defined as a quotient space of the disjoint union

$$\left(X^{n-1} \cup \bigcup_\alpha D_\alpha^n \right)$$

by the equivalence relation $x \sim \phi_\alpha(x)$ for $x \in \partial D_\alpha^n$. The set X^n is called the n-skeleton.

(iii) The union $X^0 \cup X^1 \cup \ldots$ is endowed with the following topology: a set $A \subseteq X$ is open provided $A \cap X^n$ is open for each n.

Note that when an n-dimensional closed disk is attached to a skeleton, the interior of the attached disk, with the topology induced by the CW complex, remains homeomorphic to the interior of an n-dimensional Euclidean disk.

Example 8.5.2
(i) In Figure 8.5.1, a CW complex is obtained by attaching three 2-dimensional disks to a point. The resulting topological space is not a manifold.

(ii) In the example discussed in Section 8.1, the sub-level sets M^a are homotopy equivalent to CW complexes. If $a < b$, the CW complex

8.5. MANIFOLDS REPRESENTED AS CW COMPLEXES

FIGURE 8.5.1
A CW complex consisting of a bouquet of three spheres.

associated to M^b is obtained from the CW complex associated to M^a by attaching a number of disks corresponding to the critical points contained in $f^{-1}[a,b]$, of dimensions equal to the Morse indices of those critical points. □

We will now see that every smooth manifold can be represented as a CW complex. This is useful in determining some of the topological properties of the manifold, as for example in the computation of the homology groups. For simplicity, we will prove this result only in the compact case.

THEOREM 8.5.3
Let M be a compact smooth manifold and $f : M \to \mathbb{R}$ be a Morse function on M with n_i critical points of index i, for $i = 1, \ldots, m$. Then M is homotopy equivalent to a CW complex with n_i disks of dimension i, for $i = 1, \ldots, m$.

PROOF Since M is compact, there are only finitely many critical points, and thus finitely many critical values $c_1 < c_2 < \cdots < c_k$. We start with the lowest one. If there is exactly one critical point corresponding to c_1, Theorem 8.4.1 (ii) implies that M^a is homotopy equivalent to a one l_1-dimensional disk (where l_1 is the index of the critical point), for each $c_1 < a < c_2$. If more than one critical point corresponds to c_1, Theorem 8.4.1 (iii) implies that M^a is homotopy equivalent to a disjoint union of disks of dimensions corresponding to the indices of the critical points for each $c_1 < a < c_2$.

Then we proceed to the next critical interval, attaching one or several disks of dimensions corresponding to the indices of the critical points, as in Theorem 8.4.1. We continue the procedure until we exhaust all of the critical values.

In the end, we only have to sort out the attached disks according to their dimensions. ∎

REMARK 8.5.4 (i) The theorem remains true for non-compact manifolds, with or without boundary. See Hirsch (1976).

(ii) We have to be aware of the limitations of this theorem. For the same manifold, different Morse functions may lead to different CW-representations. This ambiguity is because the only exact information information we obtain is about the number and the dimension of the disks; we obtain no information about how the disks are attached. We noticed at the beginning of Section 8.4 that the nature of the attachments determines the topology of the resulting space. In order to restore the full topological information, one has to consider attaching 'handles'. A handle is a product of two disks $D \times D'$, with the disk D designated as the 'axis' of the 'handle'. Handles are attached along the portion of the boundary $\partial D \times D'$ at the 'ends' of the 'axis'. Any smooth compact connected manifold is diffeomorphic to a finite sum of handles, attached in a way prescribed by the gradient flow. See Franks (1982) or Fomenko (1987). ∎

Example 8.5.5

Using Theorem 8.5.3 we can obtain a reasonable understanding of the topology of the complex projective plane $\mathbb{CP}^n$. Define $f: \mathbb{CP}^n \to \mathbb{R}$ by

$$f([z_0, \ldots, z_n]) = \sum_{i=0}^{n} a_i \left(\frac{|z_i|^2}{\sum_{i=0}^{n} |z_i|^2} \right),$$

where $a_0, \ldots, a_n$ are distinct positive numbers. Then f is a Morse function with precisely one critical point of index $2k$ for $k = 0, 1, \ldots, n$, and no critical points of odd index. We conclude that $\mathbb{CP}^n$ has the homotopy type of a CW complex of the form $D^0 \cup D^2 \cup \ldots \cup D^{2n}$. ☐

8.6 Morse inequalities

In this section we will see that the topology of a manifold M imposes constraints on the number of possible critical points of a Morse function on M. For a Morse function f on a compact m-dimensional manifold M, let us introduce the following polynomials

$$M_f(t) = \sum_{p \text{ critical point of } f} t^{\lambda(p)},$$

called the Morse polynomial of f, and

$$P_f(t) = \sum_{k=0}^{m} \beta_k \cdot t^k,$$

called the Poincaré polynomial of M. Here $\lambda(p)$ is the Morse index of the critical point p, and the numbers $\beta_k = \text{rank } H_k(M)$, $k = 0, \ldots, m$, are the Betti numbers of M.

The Morse polynomial can be written in the increasing order of the powers of t by combining the terms corresponding to critical points with the same index

$$M_f(t) = \sum_{k=0}^{m} n_k t^k, \tag{8.6.1}$$

where $n_k = n_k(f)$ denotes the number of critical points of index k.

THEOREM 8.6.1 *(Morse inequalities)*
Let f be a Morse function on the compact manifold M. Then there exists a polynomial $Q_f(t) = q_0 + q_1 t + \cdots + q_m t^m$ with non-negative coefficients such that

$$M_f(t) = P_f(t) + (1+t)Q_f(t). \tag{8.6.2}$$

Consequently, we have $M_f(t) \geq P_f(t)$, for all $t \geq 0$.

PROOF We follow Bott (1982). We study the changes in the Morse polynomial and in the Poincaré polynomial for f on the sub-level set M_f^a as the value of a increases. Let

$$M_f^a(t) = \sum_{f(p) \leq a} t^{\lambda(p)},$$

$$P_f^a(t) = \sum_{k=0}^{m} \beta_k^a \cdot t^k,$$

where $\beta_k^a = \text{rank}(M^a)$. We claim that for every a

$$M_f^a(t) = P_f^a(t) + (1+t)Q_f^a(t), \qquad (8.6.3)$$

for some polynomial Q^a with non-negative coefficients. Suppose that a is just slightly above the lowest critical value of f, and there are k critical points (of Morse indices equal to 0) corresponding to the lowest critical value. Then M^a is homotopy equivalent to k 0-dimensional disks, and so $H_0(M) \simeq \mathbb{R}^k$ and $H_i(M) \simeq 0$, for $i \geq 1$. Thus $M_f^a(t) = k$ and $P_f^a(t) = k$, and the claim is verified in this case.

Suppose now that (8.6.3) holds for M^a, $b > a$, and there is only one critical value of f in the range $a < f < b$. Moreover, for simplicity, we assume that there is only one critical point, of index l, corresponding to that value. By Theorem 8.4.1, the topology of M^b is the topology of M^a with an l-dimensional disk attached. The change in the Morse polynomial for the passage from level a to level b is $M_f^b(t) - M_f^a(t) = t^l$.

The change in the Poincaré polynomial depends on whether or not the $(l-1)$-dimensional sphere along which we attach D^l is trivial in the $(l-1)$-dimensional homology of M^a. We illustrate this in Figure 8.6.1 in the case when M^a is a solid torus and D^l is a 2-dimensional disk. If the boundary ∂D^l of the attached disk is the boundary of some simplicial l-chain c contained in M^a, then $c \cup D^l$ is a closed singular l-chain, which is not the boundary of any singular $(l+1)$-chain. This contributes a new generator to the l-th homology group, and so $P_f^b(t) - P_f^a(t) = t^l$. If the boundary ∂D^l of the attached disk is not the boundary of any simplicial l-chain c contained in M^a, it means that the simplicial $(l-1)$-chain ∂c becomes the boundary of a simplicial l-chain (representing D^l) contained in M^b. This suppresses an existing generator of the $(l-1)$-th homology group, and so $P_f^b(t) - P_f^a(t) = -t^{l-1}$.

The combined changes in the Morse polynomial and in the Poincaré polynomial imply

$$(t+1)Q_f^b(t) - (t+1)Q_f^a(t) = \begin{cases} 0 & \text{if } \partial D^l = \partial c \text{ for some } c \subseteq M^a, \\ (1+t)t^{l-1} & \text{if } \partial D^l \neq \partial c \text{ for any } c \subseteq M^a. \end{cases}$$

This shows that, when a passes through some critical value, the polynomial $Q_f^a(t)$ changes by a polynomial with non-negative coefficients.
∎

8.6. MORSE INEQUALITIES

FIGURE 8.6.1
Attachments of a disk to a solid torus.

REMARK 8.6.2 The inequality in Theorem 8.6.1 is sharp. Suppose that no consecutive powers of t occur in the Morse polynomial. Since $M_f(t) = P_f(t) + (1+t)Q_f(t)$, and since both $P_f(t)$ and $Q_f(t)$ have only non-negative coefficients, it follows $Q_f(t) = 0$ and $M_f(t) = P_f(t)$.

This is sometimes referred to as Morse's lacunary principle. This situation happens in the case of the Morse function on the complex projective space $\mathbb{CP}^n$ considered in Example 8.5.5. See Bott (1982). ∎

COROLLARY 8.6.3
Let f be a Morse function on the compact manifold M.

(i) The number of critical points n_k of Morse index k is at least as large as the k-th Betti number β_k, $n_k \geq \beta_k$.

(ii) The Euler characteristic equals the alternating sum of the number of critical points of index k with $k = 0, \ldots, m$,

$$\chi(M) = \sum_{k=0}^{m}(-1)^k \beta_k = \sum_{k=0}^{m}(-1)^k n_k.$$

PROOF (i) This follows by term by term comparison of the Morse polynomial $M_f(t)$, written as in (8.6.1), to the Poincaré polynomial.

(ii) This follows by setting $t = -1$ in (8.6.2). ∎

REMARK 8.6.4 There is a sharper form of the Morse inequalities: $\sum_{k=0}^{p}(-1)^{k+p}n_k \geq \sum_{k=0}^{p}(-1)^{k+p}\beta_k$, for $0 \leq p \leq m$. See Hirsch (1976). ∎

In Example 7.6.7 we showed that the intersection theory-based Euler characteristic of an orientable odd-dimensional manifold is zero. Now we will show that homology-based Euler characteristic is also zero, even if the odd-dimensional manifold is non-orientable. Note that we cannot directly apply Poincaré duality theorem to a non-orientable manifold.

PROPOSITION 8.6.5
The Euler characteristic of a compact odd-dimensional manifold is zero.

PROOF We pick a Morse function f on M. By Corollary 8.6.3 (ii),

$$\chi(M) = \sum_{k=0}^{m}(-1)^k \beta_k = \sum_{k=0}^{m}(-1)^k n_k(f).$$

Now notice that $-f$ is also a Morse function, and has exactly the same critical points. Since

$$\text{Hess}_p(-f) = \left(\frac{\partial^2(-f)}{\partial x_i \partial x_j}(p)\right) = (-1)^m \left(\frac{\partial^2 f}{\partial x_i \partial x_j}(p)\right) = (-1)^m \text{Hess}_p(f)$$

at each critical point p, it follows that the critical points of index k for f are critical points of index $m - k$ for $-f$. Hence $n_k(-f) = n_{m-k}(f)$. We apply 8.6.3 (ii) to $-f$ and obtain

$$\chi(M) = \sum_{k=0}^{m}(-1)^k \beta_k = \sum_{k=0}^{m}(-1)^k n_k(-f) = \sum_{k=0}^{m}(-1)^k n_{m-k}(f)$$
$$= \sum_{k=0}^{m}(-1)^m (-1)^{m-k} n_{m-k}(f) = (-1)^m \chi(M).$$

This equality implies that $\chi(M) = 0$ provided that m is odd. ∎

It is not a coincidence that we were able to prove Proposition 8.6.5 without using Poincaré duality. In fact, the Poincaré duality theorem can be proved by using Morse theory. See Fomenko (1987).

8.6. MORSE INEQUALITIES

THEOREM 8.6.6 (Poincaré-Hopf)
If X is a smooth vector field with isolated zeroes on a compact oriented manifold M, then the sum of the indices of the zeroes of X equals the Euler characteristic $\chi(M)$ — defined through homology.

PROOF We will use the Morse inequalities. Recall that

$$I(X_0, X) = \sum_{p \text{ zero of } X} I(X_0, X)_p = \sum_{p \text{ zero of } X} \text{Ind}_p(X)$$

is independent of the choice of a smooth vector field X on M transverse to the zero section X_0 (see Section 7.6). Choose a Morse function f on M and let $X = (1/2)(\text{grad} f)$. The non-degeneracy of the critical points of f is equivalent to the transversality of X to X_0.

Suppose that p is a critical point of f with Morse index equal to l. By the Morse lemma 8.2.4, there exists a coordinate system (x, y) near p, with $x = (y_1, \ldots, y_k)$ and $y = (y_{k+1}, \ldots, y_m)$, such that

$$f(x, y) = -|x|^2 + |y|^2 + f(p),$$
$$X(x, y) = (-x, y).$$

By (7.4.2), the index of X at p is $\text{Ind}_p(X) = (-1)^k$, hence

$$\sum_{p \text{ zero of } X} \text{Ind}_p(X) = \sum_{k=0}^{m} n_k (-1)^k.$$

From Corollary 8.6.3 (ii), we conclude

$$\sum_{p \text{ zero of } X} \text{Ind}_p(X) = \chi(M).$$

∎

In conclusion, on a compact oriented manifold, the index theory-based Euler characteristic, the triangulation-based Euler characteristic, and the homology-based Euler characteristic are all equal.

8.7 Exercises

8.7.1 Let M be a compact m-dimensional manifold embedded in $\mathbb{R}^{m+1}$, $u \in S^m$, and $f_u \colon M \to \mathbb{R}$ be given by $f_u(x) = u \cdot x$. Show that the set of all u for which f_u is a Morse function is an open and dense set in S^m.

8.7.2 Show that for almost every $x \in \mathbb{R}^4$, the equilibrium points of the gravitational potential from Example 8.2.10 are nondegenerate and finite in number.

Hint: Show that the function
$$x \in \mathbb{R}^3 \setminus \{x_1, \ldots, x_4\} \to (\|x - x_1\|, \ldots, \|x - x_4\|) \in \mathbb{R}^4$$
is an immersion.

8.7.3 Let $f \colon M \to \mathbb{R}$ be a smooth function on a compact manifold M, p be a critical point of f, and $(y_1, \ldots, y_m)$ be a coordinate system in a neighborhood V of p, as in Lemma 8.2.4. Assume that the Riemannian metric near p expressed in the local coordinates $(y_1, \ldots, y_m)$ is the Euclidean metric. Let
$$W^s_{loc}(p) = \{x \in M \mid \lim_{t \to \infty} \phi^t(x) = p\},$$
$$W^u_{loc}(p) = \{x \in M \mid \lim_{t \to -\infty} \phi^t(x) = p\}.$$
Show:

(i) $W^s_{loc}(p)$ and $W^u_{loc}(p)$ are connected smooth manifolds, of dimensions l and $m - l$, where l is the Morse index of p.

(ii) $W^s_{loc}(p)$ and $W^u_{loc}(p)$ intersect transversally at p.

8.7.4 Assume that f is a Morse function on M, M possesses a complete Riemannian metric for which $\mathrm{grad} f$ is bounded, and $a < b$ are real numbers. Show that if $f^{-1}[a, b]$ contains no critical point of f, then M^a is homotopy equivalent to M^b.

8.7.5 Use the Morse function $f \colon S^n \to \mathbb{R}$ given by $f(x_1, \ldots, x_{n+1}) = x_{n+1}$ to compute $\chi(S^n)$.

8.7.6 Let f and g be Morse functions on the compact manifolds M and N, respectively. Show that $h(p, q) = f(p) + g(q)$ defines a Morse function on $M \times N$. Derive the product formula for the Euler characteristic $\chi(M \times N) = \chi(M)\chi(N)$.

Chapter 9

Hyperbolic Systems

9.1 Introduction

A hyperbolic set for a diffeomorphism is an invariant set for which every point behaves like a hyperbolic fixed point. More precisely, the tangent space at each point of the invariant set splits into stable and unstable subspaces so that the derivative of the mapping takes the stable subspace at a point into the stable subspace at the image of that point, and the unstable subspace at a point into the unstable subspace at the image of that point. Moreover, the vectors in the stable subspace are uniformly contracted and the vectors in the unstable subspace are uniformly expanded by the derivative of the mapping.

To illustrate this type of behavior, consider the mapping $f : \mathbb{T}^2 \to \mathbb{T}^2$, induced by the linear transformation

$$A = \begin{pmatrix} 2 & 1 \\ 1 & 1 \end{pmatrix},$$

as in Example 7.4.5. The torus $\mathbb{T}^2$ is represented as the quotient space $\mathbb{R}^2/\mathbb{Z}^2$. The map f has only one fixed point in $\mathbb{T}^2$, which corresponds to $(0,0)$ in $\mathbb{R}^2$. The eigenvalues of A are

$$\lambda_1 = \frac{3 - \sqrt{5}}{2} \quad \text{and} \quad \lambda_2 = \frac{3 + \sqrt{5}}{2},$$

and the corresponding eigenvectors are

$$v_1 = \left(1, \frac{-\sqrt{5} - 1}{2}\right) \quad \text{and} \quad v_2 = \left(1, \frac{\sqrt{5} - 1}{2}\right).$$

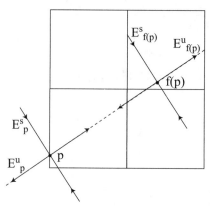

FIGURE 9.1.1
Hyperbolic automorphism of the torus.

Note that $|\lambda_1| < 1$, $|\lambda_2| > 1$. The fixed point of f is therefore hyperbolic; its stable and unstable subspaces are spanned by v_1 and v_2, respectively.

At every non-fixed point $p \in \mathbb{T}^2$, the tangent space at p can still be broken into stable and unstable subspaces

$$T_p\mathbb{T}^2 = E_p^s \oplus E_p^u,$$

where $E_p^s = \{tv_1 \mid t \in \mathbb{R}\}$ and $E_p^u = \{tv_2 \mid t \in \mathbb{R}\}$. Since $(df)_p = A$, we have $(df)_p(E_p^s) = E_{f(p)}^s$, and $(df)_p(E_p^u) = E_{f(p)}^u$. See Figure 9.1.1. Moreover, the magnitude of each vector in the stable subspace is contracted by a factor of $|\lambda_1|$, and the magnitude of each vector in the unstable subspace is expanded by a factor of $|\lambda_2|$. These conditions describe a hyperbolic structure on the whole torus.

The systematic study of hyperbolic systems was initiated in the 1960's by Steven Smale. Smale's program focused mainly on two problems: the structural stability, and the genericity of hyperbolic systems. Structural stability means that small perturbations of the system do not produce qualitative changes. It was shown that structural stability is tantamount to hyperbolicity. It was believed for a while that hyperbolic systems account for most dynamical systems. Although this turned out to be false, various extensions of the notion of hyperbolicity seem to be suitable for the majority of dynamical systems.

9.2 Hyperbolic sets

Let $f : M \to M$ be a C^1-diffeomorphism of a C^1-smooth manifold M, and $\langle \cdot, \cdot \rangle$ be a Riemannian metric on M. The Riemannian metric induces a norm $\|\cdot\|$ on each tangent space to M.

DEFINITION 9.2.1
A compact f-invariant set Λ is hyperbolic provided that:

(i) *at each point $p \in \Lambda$ the tangent space T_pM is the direct sum of two subspaces E_p^s and E_p^u;*

(ii) *at each point $p \in \Lambda$ we have $(df)_p(E_p^s) = E_{f(p)}^s$, and $(df)_p(E_p^u) = E_{f(p)}^u$;*

(iii) *there exist constants $C > 0$ and $0 < \lambda < 1$ such that, for each $p \in \Lambda$, $x \in E_p^s$, and $y \in E_p^u$, we have*

$$\|(df)_p^n(x)\| \leq C\lambda^n \|x\|, \tag{9.2.1}$$
$$\|(df)_p^{-n}(y)\| \leq C\lambda^n \|y\|, \tag{9.2.2}$$

for all $n \geq 0$.

The constant C in the growth conditions (9.2.1) and (9.2.2) determines the number of iterations necessary for the vectors in E_p^s to get contracted, and for the vectors in E_p^u to get expanded.

From the invariance of the stable and unstable subspaces under the derivative map and from the growth conditions, it can be shown that the stable and unstable subspaces E_p^s and E_p^u depend continuously on $p \in \Lambda$. The continuous dependence of the stable subspace means that for each $p \in \Lambda$ there exists a choice of an orthonormal basis $(v_q^1, \ldots, v_q^k)$ of E_q^s at each $q \in \Lambda$ in some neighborhood of p such that, whenever $q_n \to q$,

$$v_{q_n}^i \to v_q^i \quad \text{for all } i = 1, \ldots, k.$$

The continuous dependence of the unstable subspace is given by a similar condition. The stable and unstable subspaces at all points of the hyperbolic set form the (continuous) stable and unstable subbundles of the tangent bundle

$$E^s = \bigoplus_{p \in \Lambda} E_p^s, \quad E^u = \bigoplus_{p \in \Lambda} E_p^u.$$

These subbundles are invariant under df. Thus, the restriction $T_\Lambda M$ of the tangent bundle TM to Λ continuously splits as the direct sum of the stable and unstable subbundles E^s and E^u of the tangent bundle.

For $p \in \Lambda$ and $\epsilon > 0$, we can define the local stable and unstable manifolds of p by

$$W_\epsilon^s(p) = \{q \in \Lambda \cap B(p,\epsilon) \mid d(f^n(q), f^n(p)) \to 0\},$$
$$W_\epsilon^u(p) = \{q \in \Lambda \cap B(p,\epsilon) \mid d(f^{-n}(q), f^{-n}(p)) \to 0\},$$

where d is the distance on M induced by the Riemannian metric, and $B(p,\epsilon) = \{q \in M \mid d(p,q) < \epsilon\}$.

THEOREM 9.2.2 (Stable manifold theorem)
Assume that $f : M \to M$ is a C^k-diffeomorphism of a C^k-smooth manifold M, and Λ is a hyperbolic set for f. There exists $\epsilon > 0$ such that the local stable and unstable manifolds $W_\epsilon^s(p)$ and $W_\epsilon^u(p)$ are C^k-embedded submanifolds of M. Moreover, $T_p W_\epsilon^s(p) = E_p^s$ and $T_p W_\epsilon^u(p) = E_p^u$. The stable and unstable manifolds are locally unique.

For a proof, see Robinson (1998). The above theorem implies that there exist $\delta > 0$ and embeddings

$$\sigma_p^s : B(0,\delta) \cap (\mathbb{R}^s \times \{0\}) \to M, \quad \sigma_p^u : B(0,\delta) \cap (\{0\} \times \mathbb{R}^u) \to M,$$

where $s = \dim E^s$ and $u = \dim E^u$, such that the image of σ_p^s is $W_\epsilon^s(p)$ and the image of σ_p^u is $W_\epsilon^u(p)$. The local stable and unstable manifolds vary continuously with $p \in \Lambda$. This means that there exist embeddings σ_p^s and σ_u^p as above that vary continuously with $p \in \Lambda$, with respect to the uniform convergence topology.

The global stable and unstable manifolds are defined as follows:

$$W^s(p) = \bigcup_{n \geq 0} f^{-n} W_\epsilon^s(f^n(p)),$$
$$W^u(p) = \bigcup_{n \geq 0} f^n W_\epsilon^u(f^{-n}(p)).$$

It can be proved that the above definitions are independent of the particular choice of the local stable and unstable manifolds, and they are equivalent to the following topological description

$$W^s(p) = \{q \in M \mid \lim_{n \to \infty} d(f^n(q), f^n(p)) = 0\},$$
$$W^u(p) = \{q \in M \mid \lim_{n \to \infty} d(f^{-n}(q), f^{-n}(p)) = 0\}.$$

9.2. HYPERBOLIC SETS

Example 9.2.3 *(Smale's horseshoe)*
Consider a smooth planar mapping f that stretches the unit square U vertically, shrinks it horizontally, and folds as indicated in Figure 9.2.1 (right). The intersection $U \cap f^{-1}(U)$ consists of two disjoint rectangular boxes B_1 and B_2. Assume that f is affine on $B_1 \cup B_2$, and the stretching and shrinking of B_1 and B_2 are uniform. Let $\nu < 1/2$ be the contraction rate and $\mu > 2$ be the expansion rate. We extend this mapping to a diffeomorphism of $\mathbb{R}^2$ as follows. First we add two semi-circular caps to the top and bottom of the unit square, and prescribe that both caps get uniformly contracted and mapped inside the lower cap. Thus, the square with caps V is mapped into itself, and there is an attracting fixed point in the lower cap. Finally, we extend this mapping to the whole of $\mathbb{R}^2$ so that all points of the plane are taken inside V under sufficiently many forward iterations. One can eventually extend this mapping to S^2 by compactifying $\mathbb{R}^2$ with a point at infinity, and prescribing that this point is a repelling fixed point.

The images of B_1 and B_2 under f are vertical rectangles that stretch across the unit square, hence they also stretch across each of B_1 and B_2. Thus $f^{-1}(U) \cap U \cap f(U)$ consists of four rectangles of sides ν and $1/\mu$. Intersecting the negative and positive iterates up to the second order, we obtain 4^2 rectangles of sides ν^2 and $1/\mu^2$. Continuing this procedure, we end up with the maximal invariant set $\Lambda = \bigcap_{n \in \mathbb{Z}} f^{-n}(U)$ of U. It is easy to see that Λ is a Cantor set, that is, it is totally disconnected and without isolated points. See Figure 9.2.1 (left).

The set Λ is hyperbolic with respect to f. Indeed, for each point $x \in \Lambda$, there is a natural splitting $T_p\mathbb{R}^2 = E_p^s \oplus E_p^u$, with E_p^s being the horizontal line through p, and E_p^u the vertical line through p. We have $(df)_p(E_s^p) = E_{f(p)}^s$ and $(df)_p(E_p^u) = E_{f(p)}^u$, and

$$(df)_p = \begin{pmatrix} \pm\nu & 0 \\ 0 & \pm\mu \end{pmatrix}.$$

By choosing $C = 1$ and $\lambda = \max\{\nu, 1/\mu\}$, we obtain $\|(df^n)_p(x)\| < C\lambda^n\|x\|$ for each $x \in E_p^s$, and $\|(df^{-n})_p(y)\| < C\lambda^n\|y\|$ for each $y \in E_p^u$.

The local stable manifold of a point p in Λ consists of a horizontal line segment through that point, and the local unstable manifold consists of a vertical line segment through that point. The global stable manifold of each point p is an immersed curve in $\mathbb{R}^2$, containing horizontal line segments through a dense set of points in Λ. The global unstable manifold of each point p is an immersed curve containing vertical line segments through a dense set of points in Λ.

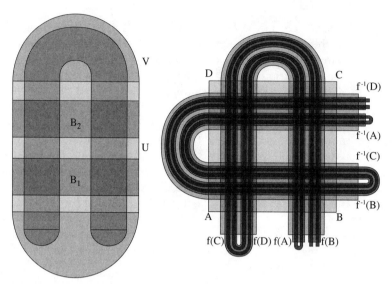

FIGURE 9.2.1
Smale's horseshoe.

We can easily check that the trajectories of any two nearby points will eventually spread apart. This means that the dynamics of Smale's horseshoe is chaotic. To show this, we describe the trajectory of each $p \in \Lambda$ by recording which of the boxes B_1, B_2 are successively visited by the iterates $f^i(p)$ of p. To each $p \in \Lambda$ we assign a string $\ldots a_{-2}a_{-1}a_0a_1a_2\ldots$ of symbols $a_i \in \{1, 2\}$, defined by $a_i = 1$ if $f^i(p) \in B_1$ and $a_i = 2$ if $f^i(p) \in B_2$. Each of the boxes B_1 and B_2 is uniformly contracted horizontally by the positive iterates of f, and uniformly contracted vertically by the negative iterates of f. This means that for every string $\ldots a_{-2}a_{-1}a_0a_1a_2\ldots$ of symbols $\{1, 2\}$, there exists a unique point p with $f^i(p) \in B_{a_i}$ for all i, where $\{p\} = \bigcap_{i \in \mathbb{Z}} f^{-i}(B_{a_i})$. Therefore, there is a one-to-one and onto correspondence between the points of Λ and the strings of symbols $\{1, 2\}$. In conclusion, for each pair of nearby points $p \neq q$, the corresponding strings will be different, which implies that $f^i(p)$ and $f^i(q)$ will land in different boxes, away one from the other, for some large (in absolute value) $i \in \mathbb{Z}$. □

Smale's horseshoe is an archetype of chaotic behavior. In some sense, almost every diffeomorphism of a compact manifold exhibits a Smale horseshoe. We describe the following typical situation. Let $f : M \to M$

9.2. HYPERBOLIC SETS

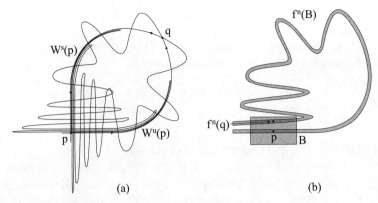

FIGURE 9.2.2
Homoclinic tangle.

be a diffeomorphism and p be a hyperbolic fixed point of f. Assume that the stable and unstable manifolds $W^s(p)$ and $W^u(p)$ have a transverse intersection at a point $q \neq p$. The point q is called a homoclinic point. Since q lies on both invariant manifolds, $f^n(q) \in W^s(p) \cap W^u(p)$ for all $n \geq 0$; furthermore, $W^s(p)$ is transverse to $W^u(p)$ at each point of the orbit $f^n(q)$ of q. In other words, if there is one transverse intersection point, then there are infinitely many such points. Moreover, we have $f^n(q) \to p$ as $n \to \pm\infty$. Therefore, the unstable manifold will fold in successive layers that approach p in forward time, and the stable manifold will also fold in successive layers that approach p in backwards time. See Figure 9.2.2 (a). Assume, for simplicity, that f is a diffeomorphism of the plane. Let B be a rectangular box centered at p. This box will contain some iterate $f^n(q)$ of q, with $n \geq 0$. If n is chosen large enough and the size of B is chosen carefully, the image $f^n(B)$ of B will stretch out along the unstable manifold of p and intersect B in the manner shown in Figure 9.2.2 (b). Geometrically, this represents a horseshoe for f^n, hence there exists a compact set Λ invariant under f^n, with Λ homeomorphic to a Cantor set. The size of B and the order of the iterate n of f can be chosen so that f^n satisfies the uniform contraction and expansion conditions on Λ. Under these circumstances, Λ is a hyperbolic invariant set for f^n. This type of argument leads to a general theorem, stated below. We recommend Katok and Hasselblatt (1995) for a proof.

THEOREM 9.2.4 (Birkhoff-Smale homoclinic orbit theorem)
Let $f : M \to M$ be a C^1-diffeomorphism and q a transversal homoclinic point for the hyperbolic fixed point p of f. There exists an integer n such that f^n contains a hyperbolic compact invariant set Λ homeomorphic to a Cantor set.

The situation described by the above theorem is typical.

THEOREM 9.2.5 (Kupka-Smale theorem)
Let M be a compact manifold and $1 \leq k \leq \infty$. Let $\mathcal{K}$ be the set of all diffeomorphisms of M for which all fixed points and all periodic orbits are hyperbolic, and the stable manifold of every fixed point (periodic orbit) intersects transversally the unstable manifold of every fixed point (periodic orbit). Then the set $\mathcal{K}$ is residual in the set of all C^k-diffeomorphisms of M.

See Robinson (1998) for a proof.
There exist maps for which the entire manifold is a hyperbolic set.

DEFINITION 9.2.6
A C^k-diffeomorphism of a compact manifold M is said to be Anosov if M is a hyperbolic set for f.

For an Anosov diffeomorphism, the local stable and local unstable manifolds are defined at each point $p \in M$. Although these manifolds are C^k-smooth, their assembly may not be smooth. That is, there may not exist embeddings σ_p^s and σ_p^u, with images $W_\epsilon^s(p)$ and $W_\epsilon^u(p)$, respectively, which depend smoothly on $p \in M$.

Example 9.2.7 (Hyperbolic automorphism of the torus)
Consider the hyperbolic automorphism of the torus discussed in Section 9.1. We found the eigenvalues of A

$$\lambda_1 = \frac{3 - \sqrt{5}}{2} \quad \text{and} \quad \lambda_2 = \frac{3 + \sqrt{5}}{2},$$

and the corresponding eigenvectors

$$v_1 = \left(1, \frac{-\sqrt{5} - 1}{2}\right) \quad \text{and} \quad v_2 = \left(1, \frac{\sqrt{5} - 1}{2}\right).$$

9.2. HYPERBOLIC SETS

FIGURE 9.2.3
The unstable manifold of a point for a hyperbolic automorphism of the torus.

We want to check that M is a hyperbolic set for f. Let $E_p^s = \{tv_1 \mid t \in \mathbb{R}\}$ and $E_p^u = \{tv_2 \mid t \in \mathbb{R}\}$ at each $p \in M$, and $0 < \lambda = |\lambda_1| = |1/\lambda_2| < 1$. Pick $C > 1$. In Section 9.1 we showed that $(df)_p(E_p^s) = E_{f(p)}^s$, and $(df)_p(E_p^u) = E_{f(p)}^u$. If $x = tv_1 \in E_p^s$, then $\|(df^n)_p(x)\| = \|\lambda_1^n(tv_1)\| < C\lambda^n\|x\|$. Similarly $\|(df^{-n})_p(y)\| = \|\lambda_2^{-n}(y)\| < C\lambda^n\|y\|$, for each y in E_p^u. Thus f is an Anosov diffeomorphism.

The global stable and unstable manifolds $W^s(p)$ and $W^u(p)$ of a point $p \in \mathbb{T}^2$ are the images of the lines $p + E_p^s$, and, respectively, $p + E_p^u$, through the canonical projection $\pi : \mathbb{R}^2 \to \mathbb{T}^2$. Since these lines have irrational slopes, the global stable and unstable manifolds wind densely around the torus (see Figure 9.2.3). □

Now we define a hyperbolic structure for flows. We consider a smooth flow ϕ^t on a compact manifold M. A set $\Lambda \subseteq M$ is said to be invariant under the flow if $\phi^t(\Lambda) = \Lambda$ for all $t \in \mathbb{R}$.

DEFINITION 9.2.8
A compact ϕ^t-invariant set Λ is hyperbolic provided that:

(i) at each point $p \in \Lambda$ the tangent space T_pM is the direct sum of three subspaces E_p^s, E_p^u, and E_p^0, where E_p^0 is tangent to the flow line through p, and E_p^u and E_p^s vary continuously with $p \in \Lambda$;

(ii) for each point $p \in \Lambda$ we have $(d\phi^t)_p(E_p^s) = E_{\phi^t(p)}^s$, $(d\phi^t)_p(E_p^u) = E_{\phi^t(p)}^u$, and $(d\phi^t)_p(E_p^0) = E_{\phi^t(p)}^0$;

(iii) there exist constants $C > 0$ and $0 < \lambda < 1$ such that, for each

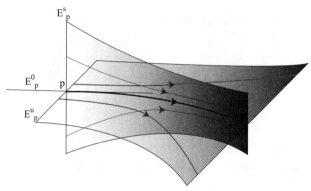

FIGURE 9.2.4
The splitting of the tangent space into stable, unstable and orbit subspaces along a hyperbolic orbit.

$p \in \Lambda$, $x \in E_p^s$ and $y \in E_p^u$, we have

$$\|(d\phi^t)_p(x)\| \leq C\lambda^t \|x\|, \qquad (9.2.3)$$

$$\|(d\phi^{-t})_p(y)\| \leq C\lambda^t \|y\|, \qquad (9.2.4)$$

for all $t \geq 0$.

The continuity of the stable and unstable subbundles E^s and E^u does not follow from the growth conditions as in the discrete case, and hence is required in the definition. On the other hand, the orbit subbundle E^0 is obviously continuous. A hyperbolic periodic orbit is a simple example of a hyperbolic set for a flow (see Figure 9.2.4).

DEFINITION 9.2.9
A flow ϕ^t on a compact manifold M is called Anosov if the set M is hyperbolic for the flow.

A quintessential example of an Anosov flow is the geodesic flow on a compact manifold of negative sectional curvature, which will be studied in Section 9.4.

REMARK 9.2.10 The situation described in the homoclinic orbit theorem 9.2.4 arises, for example, in the study of the three-body prob-

9.2. HYPERBOLIC SETS

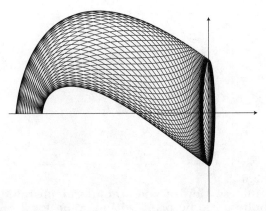

FIGURE 9.2.5
The local unstable manifold of some hyperbolic periodic orbit, projected on the configuration space (reproduced with permission of Josep Masdemont).

lem. As previously discussed in Section 2.1, the motion of two bodies under gravitation (the two-body problem) has explicit solutions. The analogous three-body problem does not have closed-form solutions. This was proved by Poincaré in 1890. Moreover, the three-body problem exhibits chaotic behavior, in the sense that there are solutions that depend with exquisite sensitivity on the initial conditions.

A simple model of the three-body problem in which this type of behavior can be studied analytically is the planar circular restricted three-body problem. In this model, two heavy bodies move along circular orbits about their common center of mass, and a third light body moves in the plane of the circles, subject to the gravitational attraction of the two other bodies. The motion of the heavy bodies is not affected by the motion of the infinitesimal body. One way to establish the existence of chaotic dynamics in this case is by showing that, within some constant energy manifold of the phase space, there is a hyperbolic periodic orbit whose stable and unstable manifolds intersect transversally (see Figures 9.2.5 and 9.2.6). Based on this, it is possible to choose an appropriate local cross section and show that a certain iterate of the Poincaré map possesses a horseshoe. This yields the existence of chaotic motions in the original system.

A modern treatment of the situation described above can be found in Llibre, Martinez and Simó (1985), and a geometrical-numerical argument in Canalias, Gidea and Masdemont (2004). ∎

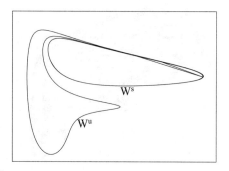

FIGURE 9.2.6
Transverse intersection of the stable and unstable manifolds of a hyperbolic periodic orbit within some local cross section (reproduced with permission of Josep Masdemont).

9.3 Hyperbolicity criteria

In order to verify the hyperbolicity of an invariant set, one needs to identify the stable and unstable subspaces of the tangent space at each point. However, one usually does not have an a priori knowledge of these subspaces. A practical way to overcome this difficulty is through families of cones that approximate the stable and unstable subspaces.

Such cones can be described through quadratic forms. Let $K(\xi,\xi)$ be a non-degenerate quadratic form on $\mathbb{R}^m$. With respect to a basis of $\mathbb{R}^m$, such a form can be represented as $K(\xi,\xi) = \xi^t A\xi$ for all $\xi \in \mathbb{R}^m$, where A is an $m \times m$ symmetric matrix. A quadratic form can always be reduced, through Jordan's procedure, to a diagonal form

$$K(\xi,\xi) = \lambda_1 \xi_1^2 + \cdots + \lambda_m \xi_m^2,$$

where $(\xi_1,\ldots,\xi_m)$ are the components of ξ with respect to some appropriate basis, and the real numbers $\lambda_1,\ldots,\lambda_m$ are the eigenvalues of A. The non-degeneracy of the quadratic form means that $K(\xi,\xi) = 0$ if and only if $\xi = 0$. In general, a quadratic form is indefinite, meaning that it has both positive and negative eigenvalues. The set

$$C^+ = \{\xi \in \mathbb{R}^m \mid K(\xi,\xi) \geq 0\}$$

is called the positive cone associated with K, and the set

$$C^- = \{\xi \in \mathbb{R}^m \mid K(\xi,\xi) \leq 0\}$$

9.3. HYPERBOLICITY CRITERIA

is called the negative cone associated with K. The closure of the complement of a positive (negative) cone is a negative (positive) cone.

A cone field is a family of (positive or negative) cones $C_p \subseteq T_p M$, varying continuously with $p \in M$. Continuity here means that the quadratic form $K_p : T_p M \to \mathbb{R}$ which defines C_p is continuous in p. A detailed discussion of cone families can be found in Katok and Burns (1994).

In the 'real world' cone fields are usually chosen as follows. One identifies a continuous splitting $T_p M = S(p) \oplus U(p)$ of the tangent space at each point of Λ which is not necessarily invariant under f, but appears to be 'almost invariant': the derivative $(df)_p$ maps $S(p)$ into a vector subspace of $T_{f(p)} M$ 'close' to $S(f(p))$, and maps $U(p)$ into a vector subspace of $T_{f(p)} M$ 'close' to $U(f(p))$. At each p one defines a positive cone and a negative cone corresponding to a quadratic form of the type

$$K((u,v),(u,v)) = \gamma^2 \|u\|^2 - \|v\|^2,$$

for some $\gamma > 0$. Then a family of positive cones is defined by

$$C_p^+ = \{(u,v) \in S_p \oplus (U_p \oplus E_0^p) \mid \|v\| \leq \gamma \|u\|\},$$

and a family of negative cones is defined by

$$C_p^- = \{(u,v) \in (S_p \oplus E_0^p) \oplus U_p \mid \|v\| \geq \gamma \|u\|\}.$$

Example 9.3.1
Positive and negative cones can have various shapes depending on the dimensions of $S(p)$ and $U(p)$, and do not always look like the traditional cones from Greek geometry. For example, in $\mathbb{R}^3$, if $\dim E_p^s = 2$ and $\dim E_p^u = 1$, the set

$$C_p^+ = \{(u,v) \in E_p^s \oplus E_p^u \mid v_1^2 + v_2^2 \leq \gamma^2 u^2\}$$

is a traditional cone, but the set

$$C_p^- = \{(u,v) \in E_p^s \oplus E_p^u \mid v_1^2 + v_2^2 \geq \gamma^2 u^2\}$$

is the closure of the complement of a traditional cone. See Figure 9.3.1. Cone families of this type are used in the proof of the Stable Manifold Theorem 9.2.2 (see Robinson (1998)). □

Now we discuss how to use cone families in order to prove the hyperbolicity of an invariant set.

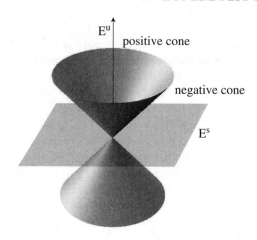

FIGURE 9.3.1
Positive and negative cones.

PROPOSITION 9.3.2

Assume that $\Lambda \subseteq M$ is a compact f-invariant set, and at each $p \in \Lambda$ there exist

- a positive cone field $C_p^+ = \{\xi \in T_pM \mid K_p(\xi,\xi) \geq 0\}$, and
- a negative cone field $C_p^- = \{\xi \in T_pM \mid K_p(\xi,\xi) \leq 0\}$,

that are complementary in the following sense: at each point p there exist complementary subspaces V_p^+ and V_p^- such that $V_p^+ \subset C_p^+$ and $V_p^- \subset C_p^-$. The dimensions of these spaces are required to be independent of p. If

(i) $(df)(C_p^+) \subseteq \text{int}\left(C_{f(p)}^+\right)$ and $(df^{-1})(C_p^-) \subseteq \text{int}\left(C_{f^{-1}(p)}^-\right)$,

(ii) $\|(df)(\xi)\| \geq \mu\|\xi\|$ if $\xi \in C_p^+$, and
$\|(df^{-1})(\xi)\| \geq \lambda^{-1}\|\xi\|$ if $\xi \in C_{f(p)}^-$,

for some $0 < \lambda < 1 < \mu$, then Λ is f-hyperbolic. See Figure 9.3.2.

Below we outline the proof of Proposition 9.3.2, and refer to Katok and Hasselblatt (1995) for a complete proof.

The main step in the proof is to show that

$$E_p^+ = \bigcap_{k=0}^{\infty}(df^k)(C_{f^{-k}(p)}^+) \quad \text{and} \quad E_p^- = \bigcap_{k=0}^{\infty}(df^{-k})(C_{f^k(p)}^-)$$

9.3. HYPERBOLICITY CRITERIA

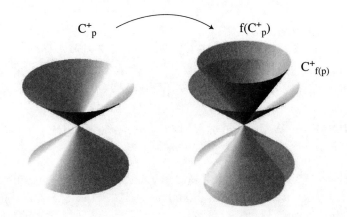

FIGURE 9.3.2
An invariant positive cone family.

are subspaces of T_pM with complementary dimensions. Since $E_p^+ \subset C_p^+$, $E_p^- \subset C_p^-$ and $C_p^+ \cap C_p^- = \{0\}$, it then follows that $T_pM = E_p^+ \oplus E_p^-$. This splitting is hyperbolic because df uniformly expands vectors in C^+ and uniformly contracts vectors in C^-.

We now outline the main step. Since the cone fields C^+ and C^- are complementary, we can choose subbundles V^+ and V^- of TM with constant and complementary dimensions (d_+ and d_-, respectively) such that $V_p^+ \subset C_p^+$ and $V_p^- \subset C_p^-$ for each p. We want to show that E_p^+ and E_p^- are subspaces with dimensions d_+ and d_-, respectively. Since the two cases are similar, we consider only E_p^+. Each of the cones $f^k(C_{f^{-k}(p)}^+)$ contains a d_+-dimensional subspace, namely $V_k^+ = (df^k)(V_{f^{-k}(p)}^+)$. Since these cones are closed and $C_p^+ \supset (df)(C_{f^{-1}(p)}^+) \supset (df^2)(C_{f^{-2}(p)}^+) \supset \cdots$, it is easy to show that E_p^+ contains a subspace of dimension d_+.

It remains to show that E_p^+ is no larger than this subspace. For $\alpha > 0$, let $C_k^+(\alpha)$ be the cone in T_pM consisting of vectors that can be written as $v + w$ with $v \in V_p^+$, $w \in V_p^-$, and $\|w\| \leq \alpha\|v\|$. It is enough to find a sequence $\alpha_k \to 0$ such that $(df^k)(C_{f^{-k}(p)}^+) \subset C_k^+(\alpha_k)$. In fact we can choose

$$\alpha_k = \left(\frac{\lambda}{\mu}\right)^k \frac{\mu + m}{\mu - \lambda},$$

where $m = \sup_{x \in \Lambda} \|df\|$, as we now show.

LEMMA 9.3.3
Suppose $\eta \in C_q^-$ and $\xi + \eta \in C_q^+$ for some $\xi \in T_q M$. Then
$$\|\eta\| \leq \frac{\mu + m}{\mu - \lambda} \|\xi\|.$$

PROOF We have $\mu \|\eta\| - \mu \|\xi\| \leq \mu \|\xi + \eta\| \leq \|df(\xi + \eta)\| \leq \lambda \|\eta\| + m\|\xi\|$. Hence $(\mu - \lambda)\|\eta\| \leq (\mu + m)\|\xi\|$. ∎

Since V_k^+ and V_p^- span $T_p M$, any vector $u \in T_p M$ can be written as $u = v + w$ with $v \in V_k^+$ and $w \in V_p^-$. Let $\xi = (df^{-k})(v)$ and $\eta = (df^{-k})(w)$. Then
$$\|v\| \geq \mu^k \|\xi\| \quad \text{and} \quad \|\eta\| \geq \lambda^{-k}\|w\|.$$
If $u \in (df^k)(C_{f^{-k}(p)}^+)$, then the vectors ξ and η will satisfy the hypotheses of Lemma 9.3.3 with $q = f^{-k}(p)$. Combining the conclusion of the lemma with the previous inequalities gives us $\|w\| \leq \alpha_k \|v\|$, so $u \in C_k^+(\alpha_k)$ as desired.

COROLLARY 9.3.4
If condition (ii) in Proposition 9.3.2 is replaced by the condition

(ii') $\|(df^n)(\xi)\| \geq \epsilon \mu^n \|\xi\|$ if $\xi \in C_p^+$, and
$\|(df^{-n})(\xi)\| \geq \epsilon \lambda^{-n}\|\xi\|$ if $\xi \in C_{f(p)}^-$,

for some $\epsilon > 0$ and some $n \geq 1$, then it also follows that Λ is hyperbolic.

PROOF We apply Proposition 9.3.2 to a suitable power of f. ∎

In the case of flows, we have the following similar criterion:

PROPOSITION 9.3.5
A compact ϕ^t-invariant set $\Lambda \subseteq M$ is hyperbolic provided that there exist $\epsilon > 0$ and $0 < \lambda < 1 < \mu$ such that for every $p \in \Lambda$ there exist

- a positive cone field $C_p^+ = \{(u, v) \in S_p \oplus (U_p \oplus E_0^p) \mid \|v\| \leq \gamma \|u\|\}$ with $\gamma > 0$,

- a negative cone field $C_p^- = \{(u, v) \in (S_p \oplus E_0^p) \oplus U_p \mid \|u\| \leq \gamma \|v\|\}$ with $\gamma > 0$, satisfying the following conditions

(i) $(d\phi^t)(C_p^+) \subseteq \text{int}\left(C_{\phi^t(p)}^+\right)$ and $(d\phi^{-t})(C_p^-) \subseteq \text{int}\left(C_{\phi^{-t}(p)}^-\right)$,

(ii) $\|(d\phi^t)(\xi)\| \geq \epsilon \mu^t \|\xi\|$ for all $\xi \in C_p^+$ and $t \geq 0$, and
$\|(d\phi^{-t})(\xi)\| \geq \epsilon \lambda^{-t} \|\xi\|$ for all $\xi \in C_{\phi^t(p)}^-$ and $t \geq 0$.

9.4 Geodesic flows on compact Riemannian manifolds with negative sectional curvature

The geodesic flow is one of the central examples in dynamical systems, especially in the study of hyperbolic systems. It is not too much of an exaggeration to view a considerable part of the history of hyperbolic dynamics as an attempt to prove the randomness of the behavior of the geodesic flow and of related objects (see Remark 9.4.2).

THEOREM 9.4.1
The geodesic flow $\phi : \mathbb{R} \times T^1 M \to T^1 M$ on the unit tangent bundle of a compact manifold M with negative sectional curvature is Anosov.

PROOF We first derive some consequences of the fact that the sectional curvature is negative. Let γ be any unit geodesic, and u be any vector tangent to M at $\gamma(t)$ independent of $\gamma'(t)$. The sectional curvature determined by the plane Π spanned by u and γ' is

$$K_\Pi = \frac{\langle R(u,\gamma')\gamma', u\rangle}{\|u\|^2 \|\gamma'\|^2 - \langle u, \gamma'\rangle^2}.$$

The fact that the sectional curvature is negative, implies, due to the compactness of the manifold, that there exists $b \neq 0$ such that

$$\langle R(u,\gamma')\gamma', u\rangle \leq -b^2,$$

for all γ and u orthogonal to $\gamma'(t)$ with $\|u\| = 1$. From the fact that the bilinear form
$$(u,v) \to \langle R(u,\gamma')\gamma', v\rangle$$
is symmetric, it follows that

$$|\langle R(u,\gamma')\gamma', v\rangle| \leq b^2 \|u\| \|v\|,$$

for all pairs of vectors u, v tangent to M at $\gamma(t)$, and so

$$|\langle R(u,\gamma')\gamma', u\rangle| \leq b^2 \langle u, u\rangle. \tag{9.4.1}$$

We would like to use Proposition 9.3.5 in order to verify that $T^1 M$ is a hyperbolic set for ϕ^t. We need to find a decomposition of $T_\theta(T^1 M) = S_\theta \oplus U_\theta \oplus E_\theta^0$, at each $\theta = (x, v) \in T^1 M$. Using the isomorphism i_θ defined in Section 5.9, we have a splitting $T_\theta(T^1 M) \simeq T_\theta^\perp(T^1 M) \oplus (\mathbb{R}\theta)$, where $T_\theta^\perp(T^1 M) \simeq v^\perp \oplus v^\perp$ under i_θ. Here $v^\perp$ denotes the orthogonal complement in $T_\theta(T^1 M)$ of the linear space spanned by v. The space $T_\theta^\perp(T^1 M)$ is invariant under the derivative of the geodesic flow. We define a norm on $T_x M \oplus T_x M$ by

$$\|(J, J')\| = \langle J, J\rangle + \langle J', J'\rangle.$$

Proposition 5.9.2 says that the differential of the geodesic flow can be written, modulo this identification, as

$$\left(d\phi^t\right)_\theta (\xi) = (J_\xi, DJ_\xi/dt),$$

where $\xi = (\xi_h, \xi_v) \in T_x M \oplus T_x M$ and J_ξ is the Jacobi field along the geodesic $\pi \circ \phi^t(x)$ of initial conditions ξ_h and ξ_v.

Now we should notice that the flow ϕ^t leaves invariant the orthogonal complement to the orbit. This fact can be expressed in terms of Jacobi fields along geodesics. Let $\gamma(t)$ be a geodesic obtained as the projection of a geodesic flow line from TM to M, through the canonical projection $\pi : TM \to M$. Suppose that $J(0)$ and $J'(0)$ are two tangent vectors perpendicular to the geodesic, that is $\langle J(0), \gamma'(0)\rangle = 0$ and $\langle J'(0), \gamma'(0)\rangle = 0$. From Proposition 5.5.6, there exist a unique Jacobi field $Y(t)$ along γ with initial conditions $J(0)$ and $J'(0)$. By Gauss' Lemma 4.5.3, the vector fields $J(t)$ and $DJ/dt(t)$ remain perpendicular to the geodesic for all t.

Define the fields of positive and negative cones

$$C^+_{(x,v)} = \{(w_h, w_v) \,|\, \langle w_h, w_v\rangle \geq 0\},$$
$$C^-_{(x,v)} = \{(w_h, w_v) \,|\, \langle w_h, w_v\rangle \leq 0\}.$$

The quadratic form here is the Riemannian metric $\langle \cdot, \cdot\rangle$ on M. In terms of Jacobi fields along a unit speed geodesic $\gamma(t)$ these cones are given by

$$C^+_{(\gamma(t),\gamma'(t))} = \{J(t) \,|\, \langle J, DJ/dt\rangle \geq 0\},$$
$$C^-_{(\gamma(t),\gamma'(t))} = \{J(t) \,|\, \langle J, DJ/dt\rangle \leq 0\},$$

9.4. GEODESIC FLOWS

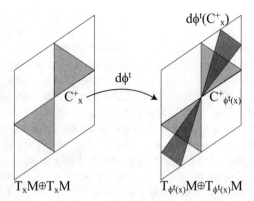

FIGURE 9.4.1
Invariance of cones under the derivative of the geodesic flow.

where $J(t)$ is a Jacobi field along $\gamma(t)$ that is orthogonal to $\gamma'(t)$.

We claim that the positive cones are positively invariant. Using the Jacobi equation, we have

$$\frac{d}{dt}\langle J, DJ/dt\rangle = \langle DJ/dt, DJ/dt\rangle + \langle J, D^2/Dt^2 J\rangle$$
$$= \langle DJ/dt, DJ/dt\rangle - \langle J, R(J,\gamma')\gamma'\rangle$$
$$\geq \langle DJ/dt, DJ/dt\rangle + b^2\langle J, J\rangle \geq 0,$$

with equality if and only if $J = 0 = DJ/dt$. This shows that $(d\phi^t)(C_x^+) \subseteq \text{int}\left(C_{\phi^t(x)}^+\right)$. See Figure 9.4.1.

Now we claim that the positive cones expand exponentially, in the sense specified by Proposition 9.3.5 (ii). We have

$$\frac{d}{dt}\langle J, DJ/dt\rangle = \langle DJ/dt, DJ/dt\rangle + b^2\langle J, J\rangle$$
$$\geq 2b\langle J, DJ/dt\rangle.$$

This implies that the quantity $\langle J, DJ/dt\rangle$ grows at an exponential rate with t, for any nontrivial $J \in C_x^+$ with $\langle J(0), DJ/dt(0)\rangle > 0$. Since $\|\langle J, DJ/dt\rangle\| \geq 2\langle J, DJ/dt\rangle$ it follows that $\|\langle J, DJ/dt\rangle\|$ grows at an exponential rate for any nontrivial $\xi \in C_x^+$. This says that the cone field C^+ satisfies the conditions of Proposition 9.3.5.

A similar argument shows that the cone field C^- also satisfies the conditions of Proposition 9.3.5. This proposition implies the hyperbolicity of the geodesic flow. ∎

REMARK 9.4.2 At the beginning of the section we used the term 'randomness' in a loose sense. One way to describe randomness in a dynamical system is through the notion of ergodicity. A dynamical system is ergodic if it preserves a probability measure and each measurable invariant set is either of zero or of full measure. A physical model for ergodicity is the natural mixing of two gases in a chamber, after they were initially separated by a wall. In the long run, the gasses mix thoroughly, and there will be no region in the chamber in which one of the gasses will refuse to mix. Ergodicity also means that the space averages of any measurable parameter of the system equal the time averages for almost all orbits. See Petersen (1983) for an introduction to this subject. For the geodesic flow, there is a natural measure on the unit tangent bundle, called the Liouville measure, which is invariant under the flow. The ergodicity of the geodesic flow was established in the 1930's for certain surfaces of curvature $K = -1$ by Hedlund, and for surfaces of curvature $K < 0$ by E. Hopf. The hyperbolicity of the geodesic flow for surfaces with $K < 0$ was proved by Anosov in the 1960's, and the so-called non-uniform hyperbolicity for surfaces with $K \leq 0$ was proved by Pesin in the 1970's-1980's. The persistence of ergodicity under small perturbations (stable ergodicity) for the time-1 map of the geodesic flow on surfaces with $K < 0$ was established by Grayson, Pugh and Shub in the 1990's. For the state of the art in this subject, the reader should consult Burns, Pugh, Shub and Wilkinson (2001) and Pugh and Shub (2004). ∎

9.5 Exercises

9.5.1 Show that a fixed point p of a flow $\phi : \mathbb{R} \times M \to M$ is hyperbolic if and only if p is hyperbolic as a fixed point for every map ϕ^t, with $t \neq 0$.

9.5.2 Assume that Λ is a hyperbolic set for $f : M \to M$, and the constants C, λ are as in Definition 9.2.1. Show that for any $0 < \lambda < \lambda_*$, there exists a norm $\| \cdot \|_*$ on $T_p M$, depending continuously on $p \in \Lambda$, such that, for any $p \in \Lambda$, $x \in E_p^s$ and $y \in E_p^u$, we have

$$\|df_p^n(x)\|_* < \lambda_*^n \|x\|_*,$$
$$\|df_p^{-n}(y)\|_* < \lambda_*^n \|y\|_*,$$

for all $n \geq 0$.

9.5. EXERCISES

9.5.3 Assume that Λ is a hyperbolic set for f. Show that the global stable manifolds $W^s(p)$ and $W^s(q)$ corresponding to two distinct points $p, q \in \Lambda$ are either disjoint or coincide. Similar statement holds for the unstable manifolds.

9.5.4 Consider the set Σ_n of all bi-infinite sequence $(a_i)_{i \in \mathbb{Z}}$ of symbols from $\{1, \ldots, n\}$. Endow $\{1, \ldots, n\}$ with the discrete topology and Σ_n with the product topology. Define the homeomorphism $\sigma : \Sigma_n \to \Sigma_n$ that shifts each sequence one place to the left, i.e. $\sigma((a_i)_i) = (b_i)_i$ where $b_i = a_{i+1}$ for all i. Show:

(i) the number of periodic orbits of σ of period p in n^p;

(ii) the periodic point of σ are dense in Σ_n;

(iii) there exists a sequence $a = (a_i)_{i \in \mathbb{Z}}$ in Σ_n whose orbit $(\sigma^n(a))_{n \in \mathbb{Z}}$ is dense in Σ_n.

9.5.5 In the Example 9.2.3, consider the map $h : \Lambda \to \Sigma_2$ that assigns to each $p \in \Lambda$ the sequence $h(p) = (a_i)_{i \in \mathbb{Z}}$ defined by $a_i = 1$ if $f^i(p) \in B_1$ and $a_i = 2$ if $f^i(p) \in B_2$. Show that h is a homeomorphism that takes orbits of f into orbits of σ, that is, $h \circ f = \sigma \circ h$. Deduce that the periodic orbits in Λ are dense, and there exists a dense orbit in Λ.

9.5.6 Consider the full shift (Σ_n, σ) and a $n \times n$ (transition) matrix $A = (\alpha_{ij})_{i,j=1,\ldots,n}$ of entries 0 and 1. Consider the subspace Σ_n^A of sequences $(a_k)_{k \in \mathbb{Z}}$ in which a symbol j can follow a symbol i if and only if $\alpha_{ij} = 1$. Let σ_A be the restriction of σ to Σ_A. The dynamical system (Σ_n^A, σ_A) is called a partial shift of finite type. The matrix A is said to be irreducible provided that for every $i, j \in \{1, \ldots, n\}$, there exists $n \geq 0$ such that the (i,j)-entry of A^n is non-zero. The matrix A is said to be eventually positive provided that there exists $n \geq 0$ such that all entries of A^n are non-zero. Show:

(i) the number of periodic points of period p equals $\operatorname{trace}(A^p)$;

(ii) if A is an irreducible matrix, then the periodic points of σ_A are dense;

(iii) if the matrix A is eventually positive, then there exists a sequence $a = (a_k)_{k \in \mathbb{Z}}$ whose orbit $\{\sigma_A^n(a)\}_{n \in \mathbb{Z}}$ is dense in Σ_A.

9.5.7 Consider the 'broken' horseshoe shown in Figure 9.5.1. Assume that the image of B_1 under f stretches across both B_1 and B_2, and the

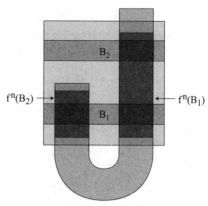

FIGURE 9.5.1
Broken horseshoe.

image of B_2 under f stretches only across B_1. Assume that f is affine on B_1 and on B_2, the expansion on B_2 is uniform and equal to $\mu_2 > 2$, the expansion rate on B_1 is uniform and equal to $\mu_1 > \mu_2 > 2$, and the contraction rate on $B_1 \cup B_2$ is uniform and equal to $\nu < 1/2$. Let $\Lambda = \bigcap_{n \in \mathbb{Z}} f^{-n}(B_0 \cup B_1)$.

Show:

(i) The set Λ is hyperbolic with respect to f.

(ii) Let
$$A = \begin{pmatrix} 1 & 1 \\ 1 & 0 \end{pmatrix},$$
and consider the map $h : \Lambda \to \Sigma_A$ that assigns to each $p \in \Lambda$ the sequence $h(p) = (a_i)_{i \in \mathbb{Z}}$ defined by $a_i = 1$ if $f^i(p) \in B_1$ and $a_i = 2$ if $f^i(p) \in B_2$. Show that h is a homeomorphism that takes orbits of f into orbits of σ, that is, $h \circ f = \sigma \circ h$.

(iii) Deduce that the periodic orbits in Λ are dense, and there exists a dense orbit in Λ.

Note: Horseshoe-type dynamical systems can be described in purely topological terms, by specifying how various boxes cross each other under iteration. When the image of a box meets another box, it has to cross it 'all the way through'. This can be rigorously described topologically using homotopy, homology (cohomology), or Brouwer degree. See Gidea (1999), Zgliczyński and Gidea (2004), and the references listed therein.

References

Abraham, R. and Marsden, J. (1985), *Foundations of Mechanics*, Addison-Wesley, Reading, MA.

Abraham, R., Marsden, J.E. and Ratiu, T. (1988), *Manifolds, Tensors, Analysis and Applications*, Springer-Verlag, New York.

Anderson, T. (2004), Geometrization of 3-manifolds via the Ricci flow, *Notices of the American Mathematical Society* **51**, 184–193.

Arnold, V.I. (1978), *Mathematical Methods of Classical Mechanics*, Springer Verlag, New York.

Bishop, R. and Crittenden, R. (1964), *Geometry of Manifolds*, Academic Press, New York.

Bott, R. (1982), Lectures on Morse theory, old and new, *Bulletin of the American Mathematical Society* **7**, 331–358.

Brin, M. and Stuck, G. (2002), *Introduction to Dynamical Systems*, Cambridge University Press, Cambridge, U.K.

Bröcker, T. and Jänich, K. (1982), *Introduction to Differential Topology*, Cambridge University Press, Cambridge, U.K.

Brown, R.F. (1971), *The Lefschetz Fixed Point Theorem*, Scott, Foresman and Company, Glenview, IL.

Burns, K., Pugh, C., Shub, M. and Wilkinson, A. (2001), Recent results about stable ergodicity, in *Smooth Ergodic Theory and Its Applications*, American Mathemtical Society, Providence, RI, 327–366.

Canalias, E., Gidea, M. and Masdemont, J. (2004), Numerical evidence for obstructions against integrability in a planar circular restricted three-body problem, preprint.

do Carmo, M.P. (1976) *Differential Geometry of Curves and Surfaces*, Prentice Hall, Upper Saddle River, NJ.

Coddington and Levinson (1955), *Theory of Ordinary Differential Equations*, McGraw-Hill, New York.

Conley, C. (1978), *Isolated Invariant Sets and the Morse Index*, CBMS Regional Conf. Series in Math. **38**, American Mathematical Society, Providence, RI.

Crampin, M. and Pirani, F.A.E. (1986) *Applicable Differential Geometry*, London Math. Soc. Lect. Notes Series 59, Cambridge University Press, Cambridge, U.K.

Dold, A. (1980), *Lectures on Algebraic Topology*, Springer-Verlag, Berlin.

Fathi, A. (1997), Partitions of unity for countable covers, *American Mathematical Monthly* **104**, 720–723.

Fomenko, A.T. (1987), *Differential Geometry and Topology*, Consultants Bureau, New York.

Fonseca, I. and Gangbo, W. (1995), *Degree Theory in Analysis and Applications*, Oxford University Press, New York.

Franks, J.M. (1982), *Homology and Dynamical Systems*, CBMS Regional Conference Series in Mathematics **49**, American Mathematical Society, Providence, RI.

Gidea, M. (1999), The Conley index and countable decompositions of invariant sets, in *Conley Index Theory*, edited by K. Mischaikow, M. Mrozek, and P. Zgliczyński, Banach Center Publications, Warsaw.

Greenberg, M.J. (1994), *Euclidean and Non-Euclidean Geometries: Development and History*, W.H. Freeman, New York.

Gromov, M. (1981), *Structures métriques pour les variétés riemanniennes*, edited by J. Lafontaine and P. Pansu, CEDIC, Paris.

Gottlieb, D.H.(1996), All the way with Gauss-Bonnet and the sociology of mathematics, *American Mathematical Monthly* **103**, 457–469.

Guillemin, V. and Pollack, A. (1974), *Differential Topology*, Prentice-Hall, Englewood Cliffs, NJ.

Hamilton, R. (1982), Three-manifolds of positive Ricci curvature, *Jour-*

nal of Differential Geom. **17**, 255-306.

Hatcher, A. (2002), *Algebraic Topology*, Cambridge University Press, Cambridge, U.K.

Henle, M. (1994), *A Combinatorial Introduction to Topology*, Dover Publications, Toronto, ON.

Hirsch, M. (1976), *Differential Topology*, Springer-Verlag, New York.

Hirsch, M. and Smale, S. (1974), *Differential Equations, Dynamical Systems, and Linear Algebra*, Academic Press, New York.

Katok, A. and Burns, K. (1994), Infinitesimal Lyapunov functions, invariant cone families and stochastic properties of smooth dynamical systems, *Ergod. Th. & Dynam. Sys.* **14**, 757–785.

Katok, A. and Hasselblatt, B. (1995), *Introduction to the Modern Theory of Dynamical Systems*, Cambridge University Press, Cambridge, U.K.

Klingenberg, W. (1995), *Riemannian Geometry*, de Gruyter, Berlin.

Lang, S. (1972), *Differentiable Manifolds*, Addison-Wesley, Reading, MA.

Lee, J.M. (1997), *Riemannian Manifolds: An Introduction to Curvature*, Springer-Verlag, New York.

Lee, J.M. (2000), *Introduction to Topological Manifolds*, Springer-Verlag, New York.

Llibre, J., R. Martinez and C. Simo (1985), Transversality of the invariant manifolds associated to the Lyapunov family of perodic orbits near L2 in the restricted three-body problem, *Journal of Diff. Eq.* **58**, 104–156.

Milnor, J. (1965), *Topology from the Differentiable Viewpoint*, The University Press of Virginia, Charlotesville, VA.

Milnor, J. (1978), Analytic Proofs of the "Hairy Ball Theorem" and the Brouwer Fixed Point Theorem, *American Mathematical Monthly* **84**, 521-524.

Milnor, J. (1963), *Morse Theory*, Annals of Math. Studies, Princeton University Press, Princeton, NJ.

Milnor, J. (1956), On manifolds homeomorphic to the 7-sphere, *Annals of Mathematics* **64**, 399–405.

Morgan, F. (1997), *Riemannian Geometry: A Beginners Guide*, A.K. Peters, Natick, MA.

Moser, J. (1965), On the volume elements of a manifold, *Transactions of the American Mathematical Society* **120**, 286–294.

Munkres, J.R. (2000), *Topology*, Prentice Hall, Upper Saddle River, NJ.

Nash, J. (1956), The imbedding problem for Riemannian manifold, *Annals of Mathematics* **53**, 20–63.

Olver, P.J. (1993), *Applications of Lie Groups to Differential Equations*, Graduate Texts in Mathematics, Vol. 107, Springer-Verlag, New York.

O'Neill, B. (1983), *Semi-Riemannian Geometry with Applications to General Relativity*, Academic Press, New York.

Oprea, J. (2004), *Differential Geometry and Its Applications*, Prentice Hall, Upper Saddle River, NJ.

Palais, R. (1963), Morse theory on Hilbert manifolds, *Topology* **2**, 299–340.

Perelman G. (2002), The entropy formula for the Ricci flow and its geometric applications, preprint, math.DG/0211159.

Perelman G. (2003a), Ricci flow with surgery on three-manifolds, preprint, math.DG/0303109.

Perelman G. (2003b), Finite extinction time for the solutions to the Ricci flow on certain three-manifolds, preprint, math.DG/0307245.

Petersen, K. (1983), *Ergodic Theory*, Cambridge University Press, Cambridge, U.K.

Pollard, H. (1976), *Celestial Mechanics*, Mathematical Association of America.

Pugh, C. and Shub, M. (2004), Stable ergodicity, *Bulletin of the American Mathematical Society* **41**, 1–41.

Robinson, C. (1998), *Dynamical Systems: Stability, Symbolic Dynamics, and Chaos*, CRC Press, Boca Raton, FL.

Rudin, W. (1966), *Real and Complex Variables*, McGraw-Hill, New York.

Schutz, B.F. (1985), *A First Course in General Relativity*, Cambridge University Press, Cambridge, U.K.

Spivak, M. (1965), *Calculus on Manifolds. A Modern Approach to Classical Theorems of Advanced Calculus*, W. A. Benjamin, Inc., New York.

Spivak, M. (1999), *A Comprehensive Introduction to Differential Geometry*, Vols. 1-5, Publish or Perish, Houston, TX.

Sternberg, S. (1958), On the structure of local diffeomorphisms of Euclidean n-space II, *Amer. J. Math.* **80**, 623–631.

Sternberg, S. (1964), *Lectures on Differentiable Geometry*, Prentice-Hall, Englewood Cliffs, NJ.

Szlenk,W. (1984), *An Introduction to the Theory of Smooth Dynamical Systems*, John Wiley and Sons, Chichester.

Warner, F.W. (1971), *Foundations of Differentiable Manifolds and Lie Groups*, Scott, Foresman and Company, Glenview, IL.

Whitney, H. (1944), The self-intersections of a smooth n-manifold in $2n$-space, *Annals of Mathematics* **45**, 220–246.

Yoseloff, M. (1974), Topological proofs of some combinatorial theorems, *Journal of Math. Anal. Appl.* **46**, 95-111.

Zgliczyński, P. and Gidea, M. (2004), Covering relations for multidimensional dynamical system, *Journal of Diff. Eq.* **202**, 32–58.

Index

(ϵ, τ)-chain, 340
(ϵ, τ)-pseudo-orbit, 340
α-limit
    for a flow, 107
    for a map, 107
ϵ-chain, 96
ϵ-pseudo-orbit, 96
ω-limit
    for a flow, 107
    for a map, 107
n-body problem, 72

affine connection, 131
    compatible with the metric, 136
    symmetric, 139
attracting fixed point, 101

Betti number, 258
Bianchi identity, 181
    differential, 222
Birkhoff-Smale homoclinic orbit theorem, 364
Borsuk-Ulam theorem, 324
Brouwer fixed point theorem, 59, 78, 270, 305
bundle
    cotangent, 228
    fiber, 230
    normal, 230
    section in a, 230
    tangent, 24
    trivial, 228
    vector, 227

central force problem, 72
chain recurrence, 96
chain recurrent set, 96, 340
chaos, 103, 367
chart, 10
Christoffel symbols, 131
Clairaut equation, 170
codimension, 16
complex projective space, 67
cone field, 369
configuration space, 25
conjugate points, 211
Conley's fundamental theorem of dynamical systems, 96, 340
connection
    affine, 131
    Levi-Civita, 141
    Riemannian, 141
constant
    Lipschitz, 55
continuous
    map, 7

coordinate change, 10
covariant derivative, 133
critical point, 45
    nondegenerate, 330
critical value, 45
curvature
    extrinsic, 171
    Gaussian, 172, 188, 192
    geodesic, 189
    intrinsic, 171
    mean, 172
    principal, 172
    sectional, 195
    tensor, 176
curve
    regular, 143
CW complex, 348

D'Alembert principle, 170
de Rham cohomology group, 258
de Rham theorem, 271
deformation retraction, 340
degree, Brouwer, 286
diffeomorphism, 11, 18
    Anosov, 364
    orientation preserving, 35
    orientation reversing, 35
differential form, 239
    closed, 243
    exact, 243
distance, 6
    Riemannian, 162
dynamical system
    continuous time, 82
    discrete time, 94

elliptic geometry, 146
embedding, 38
    theorem, 42
Euler characteristic, 306
Euler-Lagrange equation, 221

existence and uniqueness of solutions of differential equations on manifolds, 82
existence and uniqueness of solutions of ordinary differential equations, 80

fiber bundle, 230
first fundamental form
    for a surface, 110
    for an immersed manifold, 114
fixed point
    attracting, 101
    hyperbolic, 97
    isolated, 293
    non-degenerate, 296
    repelling, 101
    saddle, 101
flow, 82
    geodesic, 218
    Ricci, 200
flow box coordinates, 85
Frobenius' theorem, 93
fundamental theorem of algebra, 278, 289
fundamental theorem of Riemannian geometry, 141

Gauss equation, 189
Gauss' lemma, 153
Gauss' Theorema Egregium, 173, 191
Gauss-Bonnet theorem, 318, 321
geodesic, 143
    ball, 152
    frame, 169
    segment, 143
    sphere, 152
geometrization conjecture, 125
global stable manifold, 102

INDEX

global unstable manifold, 102
gradient vector field, 329
group action, 121

Hénon attractor, 103
hairy ball theorem, 77, 290
Hartman-Grobman theorem
    for flows, 106
    for maps, 100
Hessian, 329
holonomy, of a simple closed curve, 317
homeomorphism, 9
homoclinic point, 363
homotopy, 63
homotopy equivalence, 340
Hopf's theorem of turning tangents, 316
Hopf-Rinow theorem, 164
hyperbolic fixed point
    for a flow, 104
    for a map, 97
hyperbolic periodic orbit
    for a flow, 104
    for a map, 103
hyperbolic plane, 123
hyperbolic set
    for a flow, 365
    for a map, 359

immersion, 38
    theorem, 39
index
    of a fixed point, 294
    of a semi-Riemannian metric, 120
    of a vector field, 300
inner product, 112
invariant set
    of a flow, 107, 365
    of a map, 95

Jacobi field, 203
Jacobi identity, 90
Jacobi matrix, 31

Kepler problem, 71
Kepler's first law, 73
Kepler's second law, 73
Kepler's third law, 73
Koszul formula, 141
Kronecker symbol, 122
Kupka-Smale theorem, 364

Lefschetz fixed point theorem, 303
Lefschetz number, 303
Levi-Civita connection, 141
Lie
    algebra, 93
    bracket, 89
    derivative, 87
    group, 93
local coordinate system, 10
    compatible, 10
local coordinates, 9
local parametrization, 9
local stable manifold, 102
local unstable manifold, 102

Möbius strip, 33
manifold
    contractible, 244
    orientable, 34
    Riemannian, 112
    simply connected, 125
    smooth, 9
    with boundary, 48
map
    antipodal, 35
    continuous, 4, 7
    derivative of, 30
    exponential, 149
    Gauss, 32

Lipschitz, 55
Poincaré, 95
proper, 42
transverse, 60
Maupertuis' least action principle, 222
metric, 6
Lorentz, 120
Minkowski, 120
Riemannian, 112
Sasaki, 220
Schwarzschild, 120
semi-Riemannian, 120
metric space, 6
Minkowski metric, 120
Morse function, 335
Morse index, 330
Morse inequalities, 351
Morse's lemma, 331
Moser's theorem, 275
multiplicity of conjugacy, 211

Nash's embedding theorem, 119
Newton's first law of motion, 143
Newton's second law of motion, 25, 220
no-retraction theorem, 58, 79, 256, 262
non-degenerate fixed point, 296
nondegenerate critical point, 330
nonwandering point, 108
normal coordinates, 222

orientable
    surface, 33
oriented intersection number, 291
osculating circle, 171
osculating plane, 171

parallel transport, 135
partition of unity, 11

Perron-Frobenius theorem, 324
phase portrait, 105
phase space, 26
Poincaré conjecture, 125
Poincaré disk, 124
Poincaré duality, 258, 274, 354
Poincaré half-plane, 123
Poincaré lemma, 244
Poincaré map, 95
Poincaré-Hopf theorem, 310, 311, 355
projective space
    complex, 67
    real, 14
pseudosphere, 194

real projective space, 14
Reeb's theorem, 346
regular point, 45
regular value, 45
repelling fixed point, 101
retraction, 58
Ricci curvature tensor, 199
Riemannian connection, 141
    naturality of, 168
Riemannian distance, 162
Riemannian manifold, 112
    geodesically complete, 164
Riemannian metric, 112

saddle point, 101
Sard's theorem, 54
    for manifolds with boundary, 57
Sasaki
    metric, 220
Schwarzschild metric, 120
second fundamental form, 187
semi-Riemannian metric, 120
set
    Cantor, 70

INDEX

closed, 4, 7
dense, 7
fractal, 57
full measure, 53
measure zero, 53
nowhere, 7
open, 4, 5
residual, 53
simplex, 263
    singular, 264
simply connected manifold, 125
singular chain, 264
solid angle form, 249
space form, 198
Sperner's lemma, 269
stable manifold theorem, 102
Stokes' theorem, 251
    for singular chains, 266
sub-bundle, horizontal, 216
sub-bundle, vertical, 216
submanifold, 15
    transverse, 62
submersion, 38
    theorem, 43
surface, 9
    crosscap, 15
    embedded, 16
    of revolution, 169
    regular, 2

tangent
    bundle, 24
    space, 24
    vector, 20
tensor, 179, 233
    alternating, 238
tensor bundle, 234
tensor field, 234
three-body problem, 367
topological space, 5
topology, 5

discrete, 5
Hausdorff, 7
torus, 12, 16
transversality theorem, 65
triangulation, 267
tubular neighborhood, 231
two-body problem, 71

Uniformization theorem, 122, 193

variation through geodesics, 202
vector bundle, 227
vector bundles
    equivalent, 229
vector field
    parallel, 134
    smooth, 74
volume
    form, 241
        canonical, on a Riemannian manifold, 249
    of a manifold, 249

wandering point, 108
Weingarten equation, 187
Whitney's embedding theorem, 44
winding number, 279